Springer Theses

Recognizing Outstanding Ph.D. Research

Aims and Scope

The series "Springer Theses" brings together a selection of the very best Ph.D. theses from around the world and across the physical sciences. Nominated and endorsed by two recognized specialists, each published volume has been selected for its scientific excellence and the high impact of its contents for the pertinent field of research. For greater accessibility to non-specialists, the published versions include an extended introduction, as well as a foreword by the student's supervisor explaining the special relevance of the work for the field. As a whole, the series will provide a valuable resource both for newcomers to the research fields described, and for other scientists seeking detailed background information on special questions. Finally, it provides an accredited documentation of the valuable contributions made by today's younger generation of scientists.

Theses may be nominated for publication in this series by heads of department at internationally leading universities or institutes and should fulfill all of the following criteria

- They must be written in good English.
- The topic should fall within the confines of Chemistry, Physics, Earth Sciences, Engineering and related interdisciplinary fields such as Materials, Nanoscience, Chemical Engineering, Complex Systems and Biophysics.
- The work reported in the thesis must represent a significant scientific advance.
- If the thesis includes previously published material, permission to reproduce this must be gained from the respective copyright holder (a maximum 30% of the thesis should be a verbatim reproduction from the author's previous publications).
- They must have been examined and passed during the 12 months prior to nomination.
- Each thesis should include a foreword by the supervisor outlining the significance of its content.
- The theses should have a clearly defined structure including an introduction accessible to new PhD students and scientists not expert in the relevant field.

Indexed by zbMATH.

More information about this series at https://link.springer.com/bookseries/8790

Georg Maximilian Stockinger

Fracturing in Deep Boreholes

Stress, Structural and Lithology-controlled Fracture Initiation and Propagation in Deep Geothermal Boreholes in the Upper Jurassic Carbonate Rocks of the North Alpine Foreland Basin

Doctoral Thesis accepted by
Technical University of Munich, Munich, Germany

Springer

Author
Dr. Georg Maximilian Stockinger
Chair of Engineering Geology
Technical University of Munich
Munich, Germany

Supervisor
Prof. Dr. Kurosch Thuro
Chair of Engineering Geology
Technical University of Munich
Munich, Germany

ISSN 2190-5053 ISSN 2190-5061 (electronic)
Springer Theses
ISBN 978-3-030-94568-8 ISBN 978-3-030-94569-5 (eBook)
https://doi.org/10.1007/978-3-030-94569-5

This Springer imprint is published by the registered company Springer Nature Switzerland AG
The registered company address is: Gewerbestrasse 11, 6330 Cham, Switzerland

Supervisor's Foreword

The expansion of the baseload capable, climate-friendly, although not regenerative, but practically inexhaustible energy source "geothermal energy" represents an important pillar of the energy supply of the future. If geothermal energy production could be extended accordingly, Germany could generate 100% of its energy in a climate-neutral manner by 2050.

This thesis presented by Georg Stockinger complements a series of Ph.D. thesis at the Chair of Engineering Geology and adjacent chairs at the Technical University of Munich, which cope with various geological topics on deep geothermal energy in Bavaria. The topic of a geomechanical and numerical characterization of the Upper Jurassic reservoir initially emerged from the joint project "Dolomitkluft," which was funded by the German Federal Ministry of Economic Affairs and Energy from 2016 to 2018. The vast results from this research project were quickly adopted and implemented as part of the joint research project "Geothermal Alliance Bavaria," which is funded by the Bavarian State Ministry of Education and Culture, Science and the Arts.

Georg Stockinger's Ph.D. thesis focused on the geomechanical characterization of deep hydro- and petrothermal reservoirs in carbonate rocks—limestones and dolostones—of the Upper Jurassic in the North Alpine Foreland Basin. The work has specifically addressed fracturing, particularly fracture initiation and propagation in the vicinity of geothermal wells and its controlling factors such as the principal in situ stress conditions, existing discontinuities intersecting the well and host rock, and the geomechanical rock properties. He has thus addressed the most important aspects for the extraction of geothermal energy at its point of origin—namely the (deep) borehole.

This research has delved deeply into nondestructive and geotechnical laboratory testing, as well as adapting and using classical, geological petrographic analysis methods for reservoir characterization. New ground was broken in the area of modeling, as Irazu's FEMDEM code is not yet widely used and has not yet been applied to reservoir geomechanics. Thus, in this study an incredibly wide range of methods has been used. Since this research topic is also set at the interface between general geology on the one hand and engineering geology, respectively, rock mechanics and reservoir technology, on the other hand, the extraordinarily high complexity of the topic had to be faced—among other things—in different scales.

Therefore, with his dissertation, Georg has made an extremely valuable contribution to the field of fracture initiation and fracture propagation around deep geothermal boreholes in carbonate rocks at a very high technical level combining field-based, laboratory, as well as numerical approaches. His work demonstrates that the research on reservoir geomechanics that he has presented has enormous potential and can be further developed. He shows many approaches where and how one could go deeper into the analysis and evaluation methodology and thus offers starting points for further research work—especially for the generation of the following deep geothermal boreholes.

By now, parts of Georg's work, results and expertise have already been requested and taken under consideration for the research project "ZoKrates," which is the successor of "Dolomitkluft" and carries out the very first hydraulic stimulation tests in the Northern Alpine Foreland Basin. The "Geothermie-Allianz Bayern," which is just successfully starting its second stage, is also adopting the numerical models to represent the fracture processes in the reservoir even more realistically, including sensitivity analyses of the geotechnical parameters and advanced 3D models.

Munich, Germany
November 2021

Prof. Dr. Kurosch Thuro

Abstract

Geothermal energy from deep (> 4000 m) fluid-bearing rocks is becoming increasingly important in meeting a fraction of the demand for renewable and sustainable energy, for both domestic heating and power generation. Particularly in the North Alpine Foreland Basin (NAFB), SE Germany, southward-descending Upper Jurassic carbonates (formerly referred to as Malm) act as a reservoir for geothermal fluids encountered by several geothermal wells. Although shallow carbonates up to 3000 m have been intensely studied stratigraphically and mechanically, comparable knowledge is lacking for rocks at greater depth. This knowledge gap concerning mechanical rock properties and structural elements within the rock mass is paralleled by insufficient and inconclusive investigation of the differential stress field. Such knowledge, however, is crucial as rock mechanical properties, discontinuities, and stresses are all essential to evaluating borehole stability and assessing fracture initiation and propagation when it comes to considering the application of enhanced geothermal system (EGS). This dissertation sets out to characterize the mechanical behavior of reservoir rocks in the geothermal well Geretsried GEN-1 and its sidetrack GEN-1ST-A1, which reach depths of 4840 and 4740 m TVD and lengths of 5980 m and 5700 m MD, respectively.

For an initial geomechanical characterization, eleven analog rocks matching the lithological properties of the reservoir rocks of GEN-1 were sampled in outcrops of the Franconia Alb, the Swabian Alb, and the Helvetian Facies in Vorarlberg, Austria. Alongside the analog samples, 20 m of in situ rock cores, recovered from 4600 m to 4715 m TVD, was subjected to nondestructive testing to compare rock mechanical parameters beyond lithology. The nondestructive tests comprise parameters from ultrasonic testing: P-wave (v_P) and S-wave (v_S) velocity as well as the resonance frequency. The dynamic Young's modulus, dynamic Poisson's ratio, the v_P/v_S ratio, and the acoustic impedance proved sufficient to roughly identify equivalent rocks. However, to ensure accurate equivalency, heterogeneities, tectonic history, and accessory minerals must also be considered. Four analog rocks were identified in comparison with the in situ rock cores that best represent the reservoir in Geretsried. Three limestone samples were assessed: (i) Solnhofener Plattenkalk (SPK), the most homogeneous sample with evenly distributed fine pores, and the strongest candidate;

(ii) Dietfurter Kalk, which is of intermediate strength with minor inhomogeneities but limited porosity; (iii) the heterogenic and porous Bankkalk (BK), the weakest candidate; and (iv) one dolostone sample, Pfraundorfer Dolomit (PFD), representing the dolomitic facies. All analog samples were tested destructively for their uniaxial compressive and tensile strength as well as their static elasticity parameters.

Furthermore, nondestructive testing on the drill cores provided evidence that time-dependent effects do not lead to progressive failure in carbonates over a longer period as brittle failure occurs rapidly. Nevertheless, it was possible to register scale-related and stress-related effects. Scale-related effects, tied to properties of preexisting discontinuities, allowed recovery of specimens that represented credible rock properties. Stress-related effects, such as core disking, allowed determination of the stress regime and delimitation of the orientation and magnitudes of the principal stresses.

This dissertation introduces a novel and simple approach that rotates in situ stresses onto drill trajectories, primarily developed to apply differential in situ stress fields for 2D numerical simulations. Shear stresses revealed by this method suggest that coring oblique to the principal stresses is the most likely cause of core jamming. For the Upper Jurassic carbonates of the NAFB, a NS drilling direction is best for coring.

A combination of the stress rotation method with saddle-shaped disks from core disking confirms a strike-slip (SS) stress regime with S_{Hmax} striking NS and a ratio of $S_{Hmax} : S_v : S_{hmin} = 1.7 : 1.0 : 0.7$ or with a probably even higher differential ratio. By contrast, the orientation of borehole breakouts suggests a normal faulting (NF) stress regime. However, numerical models show that these breakouts are likely to be bound to the bedding of weak lithologies, which might have led to previous misinterpretations of the stress regime.

Asymmetrical breakouts, mapped in the boreholes, agree with the fracture traces in the image logs and on the rock cores. Although the rock cores lacked orientation, tectonic indicators allowed to back-rotation of the true fracture network. Thus, five joint sets emerge in the reservoir: two sub-vertical, NNE-SSW striking, which agree with SS faults; two EW striking, southward and northward dipping joint sets, according with a paleo NF regime; and lastly, the bedding, which dips shallowly to the south.

The numerical finite-discrete element models developed for this study (FEMDEM) simulate fracturing around an excavated borehole. These models use the static rock properties from the analog rocks BK, SPK, and PFD by reenacting in situ reservoir conditions in 4600 m TVD for two different scenarios with twelve models each. Discrete fracture networks (DFNs) derived from the fracture traces of the rock cores supplement the models. The models show that the permeability due to natural fracturing decreases rapidly around the borehole, and a good hydraulic connectivity of the borehole to the rock mass develops only in a 3–7 cm radius. However, severe borehole instabilities only occur in purely weak rock (BK). Realistically, an interstratification of various rock strengths plus implemented DFNs diminishes this effect and limits the extent of the loosened rock to a lesser degree—thus keeping the borehole stable.

A comparison of asymmetrical breakouts with the numerical models implies that the NF faults dipping southward and the strike-slip faults are mechanically active. Consequently, for a profitable well productivity at this depth and latitude, the drill trajectory should point in an azimuthal direction of 111°–124° and 291°–304°, utilizing the dilatation potential of strike-slip faults. At the same time, the presence of southward dipping normal faults should be avoided since they alter the stress field and nullify the dilatation. Hence, a lack of dilatation combined with a low fracture permeability might have caused the poor fluid flow rates in Geretsried.

For future research, extensive triaxial compression tests on rock cores from GEN-1ST-A1 and 3D FEMDEM models should elucidate the dilatation tendency and the hazardous potential of slip to minimize potentially geological hazards and economic risks and maximize the fluid flow, i.e., the efficiency of geothermal wells in the NAFB.

Parts of this thesis have been published in the following journal articles:

STOCKINGER, G., THURO, K., MOECK, I. & STRAUBINGER, R. (in print): The rock mass as the governing factor for successfully developing deep geothermal systems in Southern Germany.—Geomechanics & Tunneling.

STOCKINGER, G., KÄSLING, H., MENSCHIK, F. & THURO, K. (2019): 3D rotation applied to in situ stress fields for 2D numerical modelling, borehole stability and drill core recovery in deep geothermal wells.—In: FONTOURA, S.D., ROCCA, R.J. & MENDOZA, J.P. (Eds.): Rock Mechanics for Natural Resources and Infrastructure Development—Full papers, 3144-3151, London (CRC Press).

STOCKINGER, G., BOHNSACK, D., MOECK, I., KÄSLING, H. & THURO, K. (2019): Möglichkeiten und Grenzen der Erhebung geomechanischer Parameter an tiefen Bohrkernen.—In: Fachsektionstage Geotechnik 2019 mit Forum für junge Ingenieurgeologen, Würzburg, Deutschland, 28.-30. Oktober 2019.

STOCKINGER, G., MRAZ, E., MENSCHIK, F. & THURO, K. (2018): Geomechanical Model for a Higher Certainty in Finding Fluid Bearing Regions in Non-porous Carbonate Reservoirs.—In: SHAKOOR, A. & CATO, K. (Eds.): IAEG/AEG Annual Meeting Proceedings, San Francisco, California, 2018, 193-198, Cham (Springer International Publishing).

STOCKINGER, G., MENSCHIK, F. & THURO, K. (2017): Geomechanische Charakterisierung tiefer geothermischer Aquifere mit Feld- und Labormethoden.—In: Fachsektionstage Geotechnik 2017 mit Forum für junge Ingenieurgeologen, Würzburg (Deutsche Gesellschaft für Geotechnik, DGGT)), 06.-08. September 2017.

BACKERS, T., KAHNT, R. & STOCKINGER, G. (in print): Structural dominated geothermal reservoir reaction during proppant emplacement in Geretsried.—Geomechanics & Tunneling.

THURO, K., ZOSSEDER, K., BOHNSACK, D., HEINE, F., KONRAD, F., MRAZ, E. & STOCKINGER, G. (2019): Dolomitkluft—Erschließung, Test und Analyse des ersten kluftdominierten Dolomitaquifers im tiefen Malm des Molassebeckens zur Erhöhung der Erfolgsaussichten. Teilprojekt B: Geomechanische und hydrogeologische Parametrisierung und Modellierung.—413 p., München (Technische Universität München—Ingenieurfakultät Bau Geo Umwelt—Lehrstuhl für Ingenieurgeologie;).

MOECK, I., STRAUBINGER, R., THURO, K., STOCKINGER, G. & DUSSEL, M. (2019): The deep well GEN-1ST-A1 (Geretsried, Bavaria) as case study to explore a geothermal reservoir in the Southernmost German Molasse Basin.—Celle Drilling 2019, Celle (GeoEnergy Celle e.V.).

Acknowledgements

While finalizing and revising this thesis, my thoughts wander over the last four and a half years, and retrospectively, this period feels like a wild roller-coaster ride. My time conducting experiments in the field, at the drill site, in the laboratory, in the office evaluating data, rotating stresses, and worrying about correlations, numerical models created over and over again as the mesh did not do as I wanted, or the pursuit of credible input parameters, came with several ups and downs, looping and twists, temporarily losing orientation, or turnarounds forcing me to reconsider my thoughts. Many intelligent ideas turned out to be dead ends, foolish concepts developed into smart solutions, and impossibilities suddenly became possible approaches.

Now, almost at the end of this exciting ride, I want to share my happiness and express my gratitude to the people who guided and accompanied me through this stage of my life and without whom all of this would not have been possible:

My supervisor Prof. Dr. Kurosch Thuro, Chair of Engineering Geology at the Technical University of Munich (TUM), and Lead of the BMWi project "Dolomitkluft," at which I started my postgraduate career and to whom I initially owe the pleasure of starting this roller-coaster ride.

> Kurosch—Thank you so much for all the trust, support, and freedom you provided me with over the last few years. Beginning with the completion of my Bachelor's thesis, to my Master's thesis, the initiation of my stay abroad in Canada, which broadened my worldview, and finally for the fruitful discussions and improvements to my dissertation.

My co-supervisors, Prof. Dr. Tobias Backers, Chair of Engineering Geology and Rock Mechanics at the Ruhr-Universität Bochum, and Prof. Dr. Michael Drews, Assistant Professorship of Geothermal Technologies at the TUM, for their time and patience working through this dissertation and their evaluation.

> Tobias—Thank you for your efforts and time revising this thesis, for discussions and beers at congresses—hopefully, more are to come in the future.

> Michael—Thank you for committing as a co-supervisor for my thesis and the time spent in Zoom discussing stress regimes, structural indicators and disclosing the effects of pore pressure in the Upper Jurassic carbonates of the NAFB.

The project partners of Dolomitkluft, who contributed a huge amount of data and findings, which partly provided the foundation of this dissertation and supported its development: Dr. Robert Straubinger, CEO of Enex Power Germany GmbH; Andreas Gahr, CEO; Bettina Siebenlist of Enex Geothermieprojekt Geretsried Nord GmbH & Co. KG; Prof. Dr. Inga Moeck, Dr. Michael Dussel, and Vladimir Shipilin from the Leibniz Institute for Applied Geophysics; Dr. René Kahnt of G.E.O.S. Ingenieurgesellschaft mbH; and Dr. Markus Wolfgramm from Geothermie Neubrandenburg GmbH (GTN). In particular, special thanks go to project partners who provided metadata beyond published work:

Robert, Andreas, and Bettina—Thank you for providing me with the facilities to establish a provisional laboratory for the on-site examinations at Gut Breitenbach and a couch to spend the nightshifts on during coring. Also, I very much appreciate the uncomplicated clearance of data for GEN 1.

Inga, Michael, and Vladimir—Thank you for the wonderful collaboration at the well site and the data, including the HMI logs, the 360° core images, and the fault geometries you provided to support my dissertation.

My mentor Dr. Florian Menschik, Laboratory Dr. Ettl and Dr. Schuh, Munich, for the continuous professional support, personal guidance, and the friendship over the last few years.

Florian, together with Tina—thank you for all your support, discussions, and scientific input, which helped me successfully finish this dissertation.

My former and present colleagues and friends at the Chair of Engineering Geology and associated chairs and professorships at the TUM, who contributed strategies to reconcile interdisciplinary fields of geology with rock mechanics. Prof. Dr. Michael Krautblatter for his advice and improvements to my disputation and for chairing my doctoral defense. Laboratory Head Dr. Heiko Käsling, Dr. Florian Duschl, my office roomies at TUM, my home–office companion during the times of COVID, Amelie Schroth, as well as all of the academic, technical staff, team assistants, and students who contributed so much. Explicitly, I would like to thank Karl Hughes, TUM language center, for the joint proofreading of the most essential parts of my thesis.

Michael—thank you for your precious time being the defense's chairman and for commonly working through my disputation presentation, giving it a final finishing touch.

Heiko—thank you for providing me with your knowledge of rock mechanics, for fortifying my further understandings, and for offering ideas and resources to implement new approaches—as adventurous as they might have sounded.

Florian—thank you for the as yet not long but still fruitful discussions.

Karl—thank you for conveying my wild strain of thoughts into comprehensive English.

Omid Mahabadi, Ph.D., Andrea Lisjak, Ph.D., and Liqiang He, Ph.D., from the company Geomechanica, Inc., who provided the finite-discrete element modeling code Irazu. Beyond providing the code, any problems concerning the software, setting up models, or running the models were resolved instantly.

Omid and Andrea—thank you for your swift reactions and the abrupt assistances, also with minor and insignificant problems. Although I know that I am far from mastering your software, I have never lost the motivation to keep going and improve.

Bernd Ruehlicke, President of Eriksfiord, Inc., who provided the software Vinland, to evaluate HMI logs.

Bernd—thank you for the uncomplicated and free-of-charge supply of Vinland. The intuitive handling of your program contributed vast results to my thesis.

Last but not least, the most valuable thing in life: my family and friends, who are sometimes indistinguishable. When the ride gets too bumpy, I can always rely on the support, serenity, expertise, and experience of my friends from Munich, my hometown Passau, or from my university peeps.

Timi—your skills in coding lifted mine to another level—thank you.

Katha—despite your stressful job, you still find time refining my dissertation—thank you.

Serjoscha—thanks for your finalizing remarks on my abstract.

Finally, the seatbelt that keeps you safe and in place even if the ride flips over:

Thank you, Mama, and Opa, for your unconditional and loving support over all the years. Regardless of the situation, you never doubted me, whether over my chosen path nor over what I am capable of.

A huge thanks to all the unmentioned people, who were part of my life during this stage, before and after—your support, friendship, and kindness are of no less importance. Finalizing this dissertation marks the start of a new chapter and the start of a new ride. I am eager to see what the future will bring, and I am looking forward to it!

Contents

Symbols | Notations | Abbreviations

This dissertation uses units of the SI system. Occurring unconventionalities are distinctly referred to.

asl	above sea level (height above sea level, German reference height)
bsl	below sea level (depth below sea level, German reference height)
DFN	Discrete Fracture Network
EGS	Enhanced/Engineered Geothermal System
FEMDEM/FDEM	Finite-Discrete Element Method
FPZ | EDZ	Fracture Process Zone | Excavated Damaged Zone
GeotIS	Geothermisches Informationssystem/Geothermal Information System
LVDT	Linear Variable Differential Transformer (analog sensor for the measurement of strain)
Ma	Megaannum (Million years ago)
MD	Measured Depth (total length of an inclined, non-vertical well)
mD	millidarcy (permeability, $1*10^{-3}*9{,}86923*10^{-13}$ m^2)
NAFB	North Alpine Foreland Basin
TVD	Total Vertical Depth (vertical depth of a well, measured from surface)

Stress Terms

σ_1, σ_2, σ_3	Maximum, intermediate, and minimum stress
S_v, or σ_v	Vertical stress (the three principal stresses)
S_{Hmax}, SH, or σ_H	Maximum horizontal stress (components of any local)
S_{hmin}, Sh, or σ_h	Minimum horizontal stress (3D stress tensor)
p_w	Mud weight, pressure/water column in the borehole

p_f	Formation pressure, pore pressure
σ'	Effective stress
σ_r	Radial stress
$\sigma_{\theta,max}$	Maximum tangential (hoop) stress at the borehole wall
$\sigma_{\theta,min}$	Minimum tangential (hoop) stress at the borehole wall
σ_{xx}, σ_{yy}, σ_{zz}	Principal stresses in a Cartesian coordinate system
τ_{xy}, τ_{yx}, τ_{xz}, τ_{zx}, τ_{yz}, τ_{zy}	Shear stresses in a Cartesian coordinate system
σ_n, τ_s	Normal stress and shear strength in the Mohr–Coulomb criterion, combined with
μ, φ	The friction coefficient (μ) (–), and its inverse tangents, the angle of inner friction (φ) (°)

Abbreviations Defined by This Dissertation

BBOs	Borehole Breakouts (symmetrical, opposing breakouts in a borehole)
UPs	Unilateral Patterns (asymmetrical breakouts/undefined areal pattern in a borehole)

Rock Mechanical Parameters

E_{dyn} | E_{stat} | V_{stat}	Dynamic | static | average Young's modulus (GPa)
ν_{dyn} | ν_{stat}	Dynamic | static Poisson's ratio (–)
v_P | v_S	P-wave | S-wave velocity (m/s or km/s)
v_{Pa} | v_{Pr}	P-wave velocity in axial | in radial direction
σ_u	Uniaxial compressive strength (MPa)
σ_t	Indirect tensile strength (MPa)
AI	Acoustic impedance in general (km/s * g/cm^3)
Z_P	Acoustic impedance of the P-wave
Z_S	Acoustic impedance of the S-wave
τ_s	Shear strength (MPa)

List of Figures

List of Tables

Chapter 1
Introduction

Climate change, the finite availability of and dependency on fossil fuels, and the relentless search for disposal sites for radioactive waste are pushing renewable energies back into public awareness and political debates. As noted by [1], the German government has declared that, by 2050, renewable and sustainable energy sources should cover 95% of district heating. Among these sources, geothermal energy occupies a unique place, as its availability is independent of external influences. It is therefore suitable not only for heating but also for base-load electrical energy [2], which disqualifies most other renewable energy sources, whose energy cannot be stored and retrieved in an efficient way to suit the fluctuating power consumption within the power grid.

The [3] defines geothermal energy as energy stored below the surface of the solid earth in the form of heat. This energy can be extracted from shallow geothermal systems, defined to a maximum of 400 m depth, for both domestic cooling and heating. Deep geothermal systems, deeper than 400 m, produce energy suitable for spa applications, district heating or even the production of electricity and are subdivided into hydrothermal and petrothermal systems. Both of these are distinguished by the presence of fluids [2]. In a hydrothermal system, fluids are naturally available and may migrate freely through pore and karst spaces or fracture openings, all of which contribute to the permeability of the system. Petrothermal systems, by contrast, lack of the presence of fluids. The rock itself is dense, and the fluids, which must be supplied externally, can only conduct through fractures. In petrothermal systems, natural (fracture-)permeability is typically increased by stimulating existing fractures or even creating new ones by pressurizing the borehole. These systems are so-called Enhanced- or Engineered Geothermal Systems (EGS).

Two main factors limit the use of this energy source. One is the temperature, which classifies the system as either low, moderate (medium), or high enthalpy type. In general, high enthalpy systems, which are mostly convection-driven, are found in magmatic or extensional domain regions, where maximum temperatures of >300 °C are possible [4]. Intermediate and low enthalpy regions, as they occur in Germany, do not exceed 180 °C, and are predominantly conduction-driven with

G. M. Stockinger, *Fracturing in Deep Boreholes*, Springer Theses,
https://doi.org/10.1007/978-3-030-94569-5_1

an average geothermal gradient of 32 °C/km [2]. The second limiting factor is fluid flow, which can be specified by an approach to evaluating rocks based on their porosity/permeability ratio. The reservoir is accordingly classified as a hydrothermal or petrothermal system [4]. The density of the fluids and the specific heat capacity must also be taken into consideration, but as these are largely constant [5, 8], they will not further be discussed. Applied to Germany, considering the best combination of these factors, three particularly suitable regions have been identified for geothermal use: the North German Basin, the Upper Rhine Graben, and the North Alpine Foreland Basin, also referred to as the South German or Bavarian Molasse Basin [2].

Currently, there are 37 well-known active projects in these three major geothermal regions (with 3 in planning), which produce a total of 1500 GWh/a thermal energy and 175 GWh/a electrical energy [65, 66, 6]. Based on extrapolations from the drilling performance of petroleum wells, Klaus et al. [7] calculates conservatively that 2100 geothermal power plants and thus 4200, 5 km deep boreholes could be established by 2050 and could contribute to a total energy production of 50 TWh/a (net amount). Hence, the development of deep geothermal wells is not even close to being exhausted and offers huge potential for future projects. The South German Molasse Basin accommodates 22 projects and offers sufficient space for more well sites ([5, 8]: 3).

1.1 Deep Geothermal Energy in the North Alpine Foreland Basin

In the South German Molasse Basin, geothermal wells have profited greatly from the experience gathered from hydrocarbon wells, which was oriented to the exploitation of fossil fuels in the Tertiary strata. The underlying Upper Jurassic Carbonates were soon identified as favorable for geothermal exploitation. Hence, the first geothermal well was established from the former hydrocarbon well at Erding in 1983 [2], followed by the first, exclusive geothermal well in 2006 in Unterhaching [9], and was succeeded by many similar projects. Until 2012, the exploration strategy was aimed at massive dolomitized facies, where a satisfactory porosity/permeability fulfills the required fluid flow rate for hydrothermal use ([10], Fig. 1.1, green circle). However, this approach failed in 2013 with the Geretsried GEN-1 well and caused a huge setback to the geothermal industry. Since 2017, the geothermal industry has been gradually recovering through new successes, with drillings, e.g., at Holzkirchen, targeting a combination of facies, structural elements (pores and fractures), and karstification [11]. Further successes led to the realization of a 3D seismic exploration campaign in Munich, which facilitated the detection of these structures [12] and the subsequent development of successful wells in the Schäftlarnstraße ([13–15], Fig. 1.1, blue circle). The same approach was applied to the second GEN-1ST-A1 well in Geretsried by targeting a conjugated fault system with a sidetrack drilled

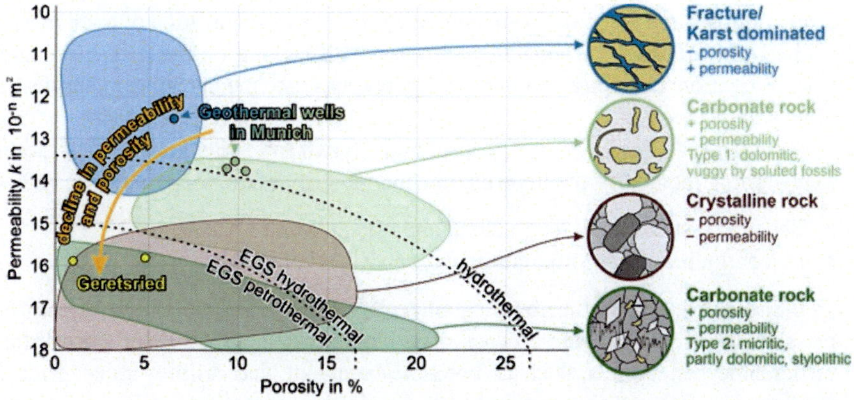

Fig. 1.1 Rock porosity and reservoir permeability of fracture/karst dominated (blue) or porosity dominated carbonate rocks (green) compared to crystalline rocks (red) with their applicability of hydrothermal, or hydro- or petrothermal Enhanced Geothermal Systems (EGS) with the locations of geothermal wells ins Munich and the geothermal wells of Geretsried (yellow dots), modified after [16]

from the existing well. However, GEN-1ST-A1, along with remote geothermal wells Weilheim and Icking on the same latitude could not achieve sufficient flow rates [67, 68, 69]. Investigations by [16] on the permeability and porosity of GEN-1ST-A1 showed that both Geretsried wells must be assigned to a petrothermal EGS (Fig. 1.1, yellow circles). The orange arrow in Fig. 1.1 indicates the decline in permeability and porosity from wells in Munich to those in Geretsried.

Hence, on a longitudinal distance of only 25 km, the method to realize a geothermal system changes from hydrothermal to a petrothermal EGS as it is typical in crystalline rocks. [5, 8] shows that with increasing depth compaction and diagenesis become the governing elements for permeability development in carbonate rock, whereas facies lose its significance. Faults and fractures in these carbonates, nevertheless, become the driving features for constituting flow zones in this geology, which is comparable to EGS sites in igneous rocks like Soultz-sous-forêts or Rittershofen [16]. Considering these facts—and keeping in mind that for the utilization of geothermal energy in the Bavarian Molasse Basin a vast potential is available—it is inevitable that geothermal wells must be developed within fractured rocks or, if necessary, as artificially stimulated EGS.

1.2 Motivation and Research Goals

The development of wells in fractured rocks or the artificial stimulation of existing discontinuities requires a high degree of mechanical knowledge about the reservoir rock. The redistribution of stresses induced by both methods substantially affects the rock mass and may lead to borehole instability, as well as the activation of present discontinuities and the initiation of new fractures, or the propagation of both. In rocks of the Upper Jurassic carbonates, research is still lacking on the rock mechanical parametrization, characterization of rock properties, and the plausible quantification of its parameters.

For deep-seated cavities, [17] considered the two following groups of properties as governing factors for the deformation and integrity of the rock mass:

- The in situ stress field, its gradient, and orientation
- Discontinuities, their orientation, frequency, and condition

Indisputably, these contribute significantly to the characterization of the rock mass. However, state-of-the-art rock mass classifications [18–20] and observations of borehole instabilities, such as borehole breakouts and drilling induced tensile fractures [21–24], are also considered as equally important:

- The material parameters of the encountered rock

- Elastic parameters: Young's modulus and Poisson's ratio
- Strength parameters: uniaxial compressive strength and tensile strength.

Besides classic analytical approaches that use rock mechanical input parameters to derive elastic or plastic deformations from empirical relations, advances in numerical modeling have introduced new paths to working with these parameters. These new simulation techniques precisely model fracture initiation and propagation and provide new insights into the mechanical reservoir behavior.

The geothermal wells of Geretsried offer a unique opportunity to determine these parameters on drill cores, recovered from a depth of 4600 to 4700 m TVD from the sidetrack well. Although this thesis refers to these wells exclusively, it presents a novel holistic approach to the mechanical characterization and classification of the aquifer in the Upper Jurassic carbonates of the Molasse Basin at depths over 4000 m. These new cores allow the comparison of rock properties with samples from near-surface outcrops with equivalent lithology and age—so-called analog samples. Although researchers (e.g., [25–27]) previously conducted lithological comparisons of drill cuttings with analog samples, to date, a valid geotechnical comparison was not possible due to the lack of in situ cores. This thesis closes this research gap and complements the understanding of the mechanical behavior of the aquifer, taking into account the impact of in situ stresses, the effect of mechanically effective discontinuities, along with the magnitude and range of likely occurring rock mechanical parameters.

Finite element—discrete element (FEMDEM) models combine the information and create mechanical scenarios that account for a better reservoir understanding and hence, engineering. This allows, e.g., the strategic management of mud weight or drawdown pressure in order to prevent the formation of an extended excavation damaged zone (EDZ). It further allows the mapping out of specific lithologies, which are suitable for EGS or evaluation of fracture opening and fracture slip within existing or newly emerging fractures. The simulations in this thesis, explicitly, do not model reservoir stimulation.

1.3 Scope of this Thesis and Link to the "Dolomitkluft" Research Report

This thesis divides the rock-mechanical and geomechanical characterization of the carbonate reservoir of the geothermal wells at Geretsried into four major parts (Fig. 1.2). Emerging from the "Dolomitkluft" research report (BMWi project 0324004 B; [28]), publications with a contribution (Fig. 1.2, grey), or a main authorship (Fig. 1.2, green) contribute with exclusive contents (Fig. 1.2, red) to these four parts. Chapter 3 presents the project "Dolomitkluft" with its partners and its published data.

Part 1 deals with the rock properties from analog rocks and in situ rock cores. Mraz et al. [29] sampled the analog rocks based on the lithological description of drill cuttings from the first well in Geretsried (GEN-1) and supplied the lithological and stratigraphical description and classification. Results published by Stockinger et al. [30], and Mraz et al. [29] were synthesized by Thuro et al. [28] with the corresponding rock mechanical and petrophysical parameters for analog rocks and in situ rock cores. This thesis combines all previous data, statistically evaluates these parameters, and correlates them with properties of the in situ rock cores. Steiger and Uhlig [31], appended in Thuro et al. [28], were commissioned with the lithological classification of the in situ cuttings and rock cores and thereby presented a reinterpretation of the geological environment.

Part 2, exclusively in this thesis, deals with stress- and structural-related features that define the rock mass properties, including discontinuities, borehole breakouts, asymmetric breakouts, and drilling-induced tensile fractures in image logs. Furthermore, computed tomographic images, back rotated fracture networks, stress indicators, and quantifiers measured on rock cores characterize the rock mass.

Part 3 deals with the in situ stresses and their impact on the borehole. Stockinger et al. [32] published a method for 3D stress rotation, which is applied to a 3D in situ stress field in order to adapt it to a 2D cross-section as input parameters for modeling, but also for the evaluation of borehole stability and the formulation of a theory of why coring in great depths with an oblique angle to one of the principal stresses is likely to fail.

Finally, the properties from Parts 1 to 3 serve as input for Part 4—the FEMDEM models. The models published by Thuro et al. [28] present only the dynamic properties—derived from ultrasonic tests—of the in situ rock cores and do not include a discrete fracture network (DFN). In this thesis, the models consider the static properties of the analog rocks and the discontinuities found in the rock cores. Further, the models incorporate multiple lithologies.

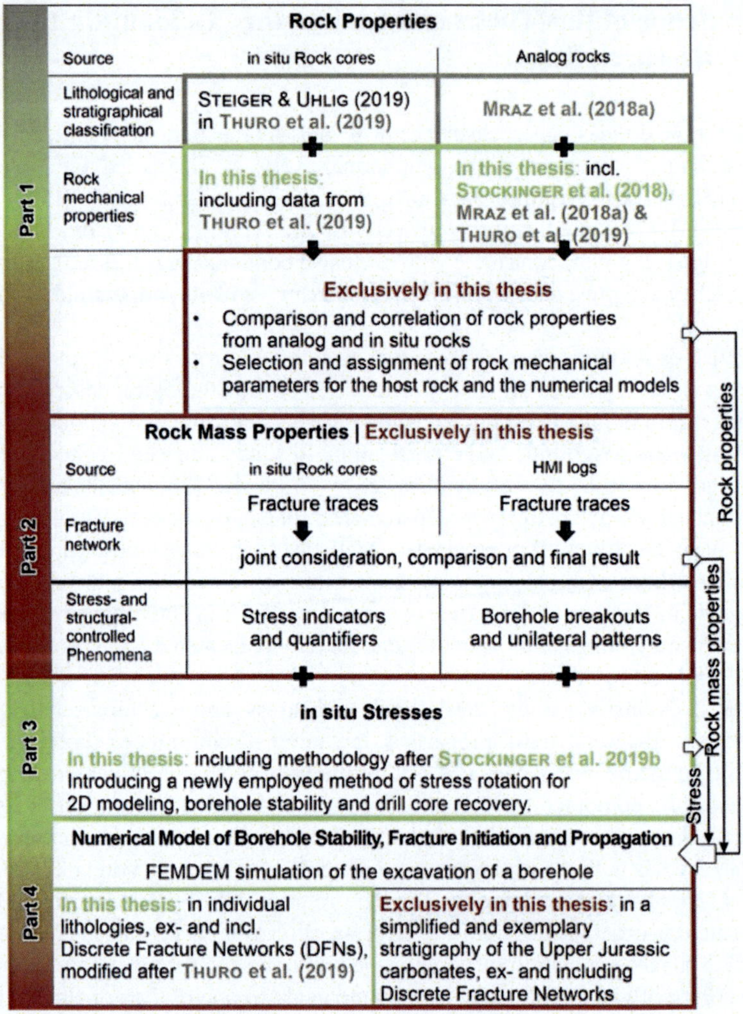

Fig. 1.2 Scope of the thesis; (Part 1) Determining and comparing rock properties from analog rocks and in situ rock cores; (Part 2) Determining stress- and structural-related rock mass properties from in situ rock cores and image logs; (Part 3) Evaluating in situ stresses and their impact on the boreholes' trajectories; (Part 4) Numerical model setup; Citations in green are publications as the main author, in grey as co-author, and exclusive content, as yet unpublished data, in this thesis is marked with red headings/frames

1.4 Geology of the Study Area

The mechanical characterization of the deep carbonate host rock from the Upper Jurassic requires an understanding of the geological situation. Although lithology, facies, grain size, or fossil content are essential for chronological placement and the evaluation of porosity and permeability, this information is insufficient for mechanical characterization. Further knowledge about deposition, consolidation, recent and former overburden, the amount and direction of the stresses, and occurring rock mass discontinuities (e.g., bedding, fracture networks, faults or joints) is required as mentioned in the previous Chapters. The hypothesis and Interpretations in this work are in strong connection with the modern understanding of the geological genesis. All absolute ages in the following Chapters refer to the latest International Chronostratigraphic Chart (ICS), published by Cohen et al. [33].

1.4.1 Evolution of the North Alpine Foreland Basin

The North Alpine Molasse Basin is a wedge-shaped, asymmetric orogenic foreland basin that extends from the Lake Geneva (Switzerland) in the west to Lower Austria in the east [34]. From the west to the east its composition differs strongly. Therefore, this Chapter focuses on the development and state of the central part of the North Alpine Foreland Basin, also known as the Bavarian Molasse Basin, comprising Munich and its greater area to the south as schematically shown in Fig. 1.3.

Its geological units date back to Permo-Carboniferous times (~300 Ma). Pre-Mesozoic, Variscan crystalline rocks, mainly gneisses, and granites, shape the basement. This basement outcrops to the east in the Bohemian Massif and the west in the Black Forest [34]. Sedimentation between the Permo-Carboniferous period and the

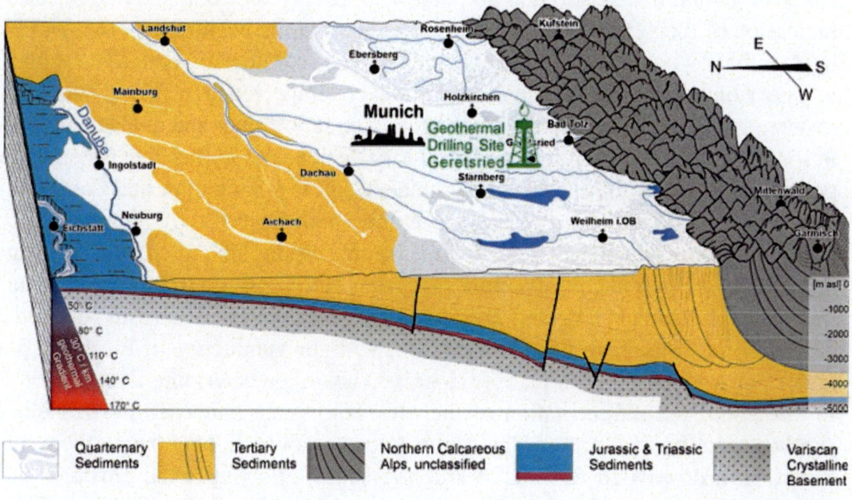

Fig. 1.3 Generalized 3D block image of the simplified geological units within and to the border of the Bavarian Molasse Basin. The green drill rig marks the location of the geothermal well Geretsried. The 2D cross section shows the geological situation with depth, complemented with the geothermal gradient of 30 °C/km (modified after Wellnhofer in [35] from Meyer & Schmidt-Kaler (2002: 8f))

Early Jurassic is only of minor importance. Firstly, Permian sediments of Rotliegend only occurred in SW-NE striking, isolated troughs, which presumably did not exist in the greater area and south of Munich. Secondly, with the beginning of the Mesozoic era, around 250 Ma, sediments from the Triassic Period were deposited in the north and the west of the modern Molasse Basin but its depositional area still never reached Munich or its south.

Same applies for sediments from the Early Jurassic (201.3 Ma), also known as "Lias", where neglectable heights of sediments are documented in the north of the modern North Alpine Foreland Basin. Erosion of the crystalline basement continued until the beginning of the Middle Jurassic (Dogger), 174.1 Ma, as the sea transgressed south and southeastward and covered the whole area of the modern Molasse Basin, depositing layers with a thickness of around 100 to 0 m (N–S) from the northern Molasse Basin to its south. Progressing uplift of the Rhenish-Bohemian Massif subsequently led to a separation of the North German Basin from the Franconian Platform and to the integration of South Germany into the Tethys, connecting with the Helvetic shelf (Fig. 1.8)—which before was separated by the Vindilician Land.

In contrast to deposits from the Triassic to the Middle Jurassic, the thickness of Late Jurassic (Malm) sediments, mainly carbonates and subordinately marls, increase towards the southeast and can reach up to 600 m [34]. Although the common name "Malm" is still widely used and present in daily practice, the scientifically correct form is Weißjura-Group [36]. Alternatively, this work refers to as Upper Jurassic carbonates. The facies and lithology of the Late Jurassic, beginning at 165.1 Ma, are elucidated comprehensively in Sect. 1.4.2 since these carbonates are the main target of geothermal production and object to this works examinations. A eustatic regression of the Tethys sea towards the southeast at the beginning of the (145 Ma) led to a shallowing of the Tethys shelf and reduced sedimentation rate. The exposure of Upper Jurassic carbonates, in areas without coverage, led to increased erosion and fostered profound karstification of the mighty strata of lime- and dolostones, which in some regions lasted for more than 100 Ma. Residues of Purbeckian (intermediate period of Jurassic and Cretaceous), Turonian, Conician–Santonian, and Campanian–all Cretaceous—sediments only appear in the South-East of the Molasse Basin. In most of the Molasse Basin, these sediments are missing due to erosion, yet, small occurrences help to identify the transition from Tertiary strata to Mesozoic strata.

The North Alpine Foreland Basin—in its actual form as a basin—started its development in the Late Eocene (∼38 Ma) with the subduction of European plate under the Adriatic-African plate [37]—also known as the Alpine Orogeneses. Up to this point, the sedimentation of the Mesozoic strata happened horizontally or sub-horizontally. The thrust of the Alpine Orogen, however, bent the northern lithosphere in a downward flexure towards the south, allowing fresh erosional sediments from the Bohemian Massif in the east and from the Alps in the south [38] to fill this newly created and continuously growing basin. The Mesozoic sediments, abutting the lithosphere, subsided with it. Inevitably, following the principle of a seesaw, a forebulge formed behind the inflection point of the down-flexure, lifting the Swabian/Franconian Alb by several hundred meters of altitude (Fig. 1.4). Gener-

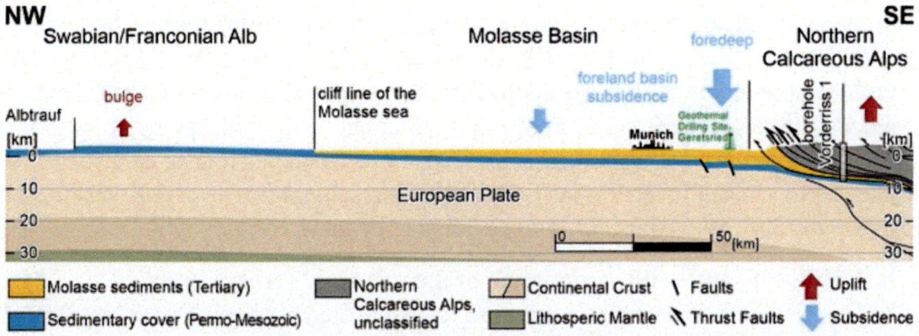

Fig. 1.4 Northern part of a NW–SE cross-section through the Eastern Alps modified after Meschede and Warr ([39]: 225, Fig. 13.16); With a N/NW thrust of the Northern Calcareous Alps, the Permo-Mesozoic Sediments (blue), covered by Molasse sediments, subside in a foredeep, while the Swabian/Franconian Alb in the NW uplifts in a bulge. The green drill rig marks the geothermal well Geretsried

ally, the Jurassic carbonate platform dips a few degrees to the southeast [39]. Lemcke [38] proposes—independently from tectonic faults—a southward inclination of the basis of the Tertiary strata of 1.7° north, and 3.8° south of Munich. Lemcke ([40]: 84) further adds that "*the inclination of the bottom of the Tertiary strata increases towards the Alpine thrust and reaches a presumed depth of 6600 m bsl in the borehole Vorderriss 1* " (see Fig. 1.4), resulting in an even steeper inclination.

Figure 1.5 displays recent data from the German geothermal information system (GeotIS, [6]), where a 2D cross-section is extracted from Dachau (north of Munich) to approximately 40 km south of Munich, showing the top of Weißjura-Group. The course of the Weißjura-Group, supported by data from deep drillings (Fig. 1.5, orange labels), agrees with Lemckes [40] hypothesis that the inclination continuously increases and reaches 8.1° at the intersection with the geothermal well Geretsried (Fig. 1.5, green drill rig).

Concurrently, during the Eocene and Oligocene, the subsidence and bending of the lithosphere and its attached overlay was "*accompanied by the development of a complex set of tensional faults often parallel to the basin as evident in the Molasse Basin of Germany and Upper Austria*" ([41]: 413). Both synthetic (dipping to the

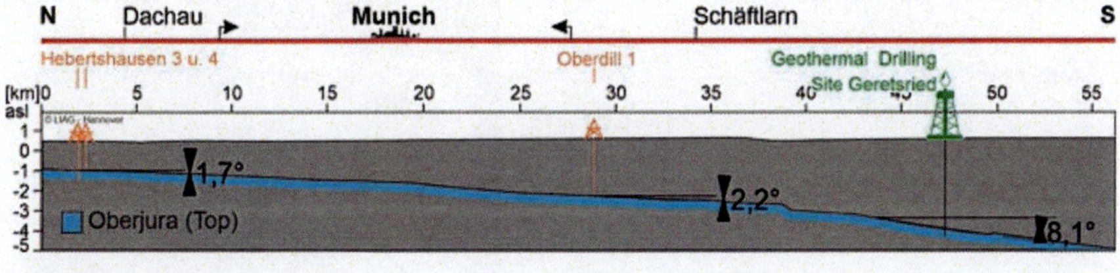

Fig. 1.5 NS 2D-cross-section from Dachau, intersecting Munich towards the well of Geretsried with the course of the top surface of the Weißjura-Group, inclining by 1.7° in Hebertshausen and 2.2° in Oberdill towards the south. At the geothermal well Geretsried, the inclination increases to 8.1° (modified after GeotIS, [5])

South) and, more common, antithetic (dipping to the North) normal faults incline with 50° to 70° and are vertically displaced by 150–200 m ([40, 42]: 63). These faults embarked on the crystalline basement, traversed the Mesozoic sediments and reached up to the sediments from the Aquitanian [40]. With the beginning of the Miocene, compressional forces countered these tensional forces, exceeded those, and led to a basin inversion [41]. These compressional forces mainly distorted the tertiary sediments. After [40]: 66ff) evidence for these compressional stresses are:

- Upwarping and partly overturning of the youngest Tertiary strata
- Basin wide narrowing/compression, causing the creation of an Alpine-parallel major through
- Shallow buckling and thrusting in the area of the Chiemsee, Bad Aibling and East of Munich
- Horizontal compaction of the Upper Freshwater Molasse, seen in seismic wave velocities
- Excessive fluid pressure (see Fig. 1.7)
- Stress-induced failure in underground coal mines in the Folded Molasse, accompanied by reverse faulting of several centimeters.

Asides from the compressional component, Unger [43], and Kraemer [44] also find indications for an NNE-SSW strike-slip faulting in the Mesozoic and Cenozoic strata in the Peißenberg area, close to the Alpine thrust belt. This kind of shearing causes only small vertical displacements but may show noticeable horizontal displacements combined with "pull-apart" or "pop-up" structural elements ([45]: 56). According to the indenter model after [46], the alpine thrust fault penetrates the foreland (the Molasse sediments) in Fig. 1.6. While in the west, the alpine units experience an uplift due to the indention, eastwards the alpine unis extrude and spread. The extrusion causes a wedge-shaped system of faults that penetrate the northern foreland creating a set of sinistral shear faults (Fig. 1.6).

Frequent alterations in sea regressions and transgressions define the depositional regime of the Tertiary strata and its stratigraphical and lithological features [47].

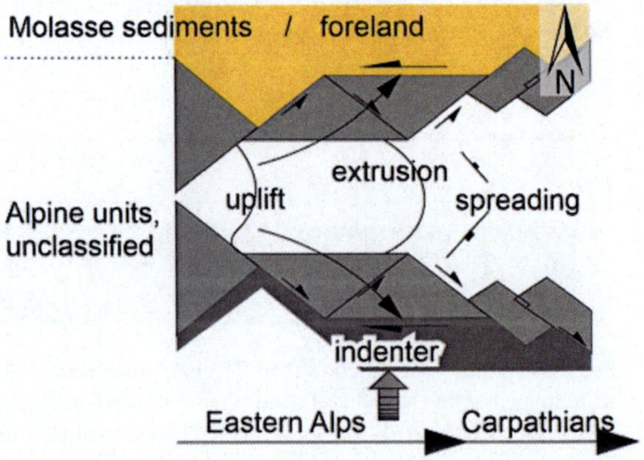

Fig. 1.6 Indenter model modified after Ratschbacher et al. ([46]: 267, Fig. 1.8), as a mechanical explanation for occurring sinistral shear faults at the northern margin of the Alpine orogen

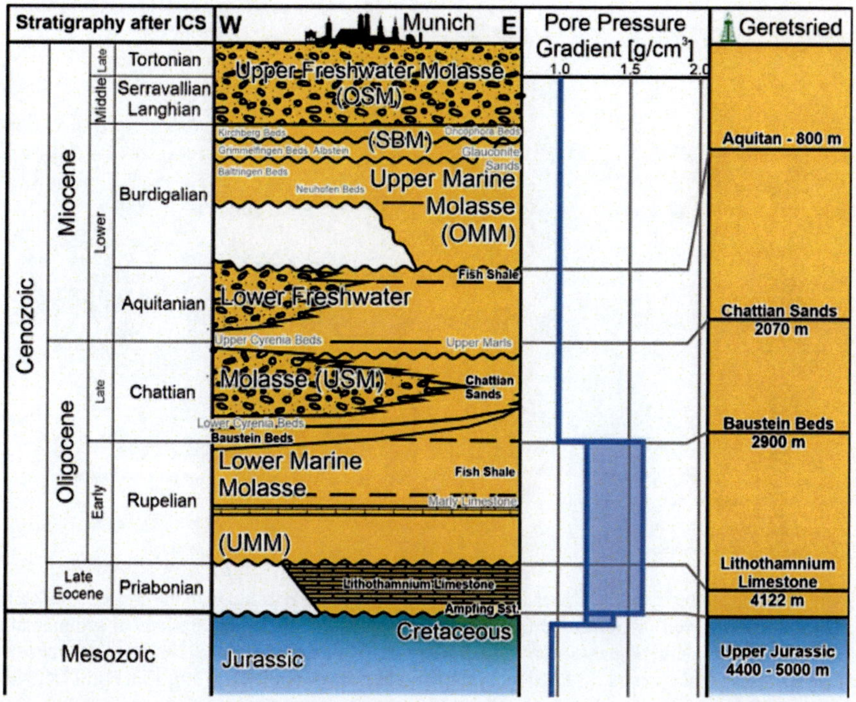

Fig. 1.7 Stratigraphic overview of the North Alpine Foreland Basin (or Bavarian Molasse Basin), composed from the chronology after [33], the stratigraphy after ([50]: 48, Fig. 1.2), the pore pressure gradient after [49] and the stratigraphy of the geothermal wells of Geretsried after [51]

At the end of the Eocene (33.9 Ma), the first transgression reached its climax with the sedimentation of the Ampfinger Schichten and the Lithothamnium Limestone. Two phases of regression followed during the Oligocene (33.9 to 23.03 Ma). Marly limestones, marls, and shales prevailed in the Lower Marin Molasse (Rupelian). The most prominent representative is the Fish Shale. The second regression followed in the Chattian and led to even more clastic sediments like the Baustein Beds or Chattian Sands, deposited in the Lower Freshwater Molasse. The Upper Marine Molasse and the Upper Freshwater Molasse complete the Tertiary Strata. In total, the Tertiary sediments nowadays comprise a thickness of up to 5000 m at the margin to the Alpine thrust front [34]. Lemcke [48] adds that the Tertiary strata used to be even thicker, with additional overburden of 1900 m in the western part and 100 to 200 m in the central part of the Molasse Basin, which already eroded. Lemcke [48] emphasizes that these values are rather conservative estimations and may even be underestimated. Fluids within these sediments are over-pressured, and hence, drilling operations must consider mud weight, casing design, reduction of non-productive time, and safety and cost-efficiency [49].

Figure 1.7 shows the generalized stratigraphy of the Cenozoic sediments covering the Mesozoic layers, as evaluated in the previous paragraphs after [50], with the

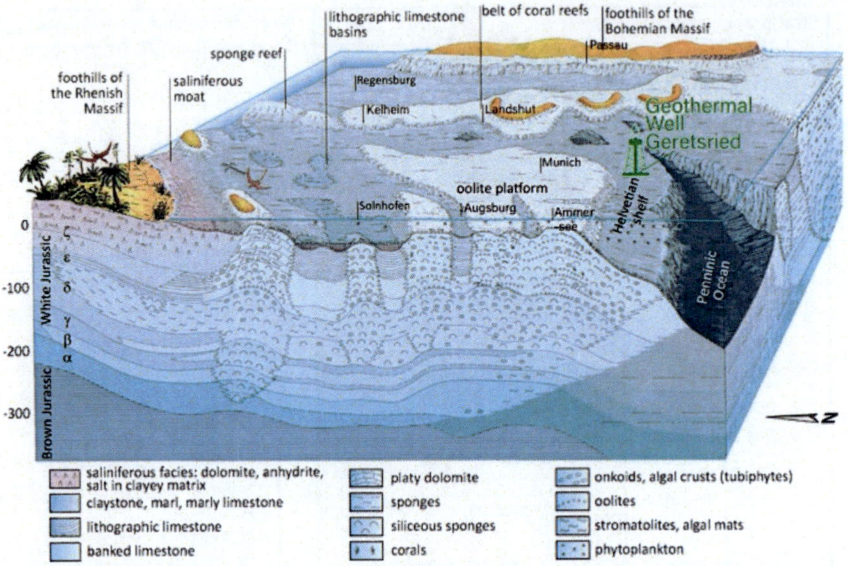

Fig. 1.8 Stratigraphy of the Upper Jurassic of the Franconian Alb and with the area of Munich and the geothermal well of Geretsried (green rig) in the South, with different kinds of sedimentation environments and lithologies (according to the legend). Noticeably, lagoonal lithographic limestones and reef limestones prevail ([39, 59] modified from Meschede and Warr ([39]: 173, Fig. 11.62) after

possible occurring excess pressure, which lies in a range of 1.25 to 2 g/cm^3 after [49] and the stratigraphy of the geothermal wells of Geretsried after [51].

For this thesis the most important particular features of the Mesozoic geothermal host rock and the Molasse Basin can be summarized as:

- Up to 5000 m Tertiary overburden (certainly higher in former times)
- The strata is inclined to the south (8° in Geretsried)
- Fault systems striking E–W, within the Basement, Mesozoic, and lower Tertiary sediments
- NNE-SSW striking faults indicate a strike-slip regime
- Fluids are under excessive hydrostatic pressure in the Cenozoic sediments, and under negative pressure in the Mesozoic strata.

1.4.2 Geology of the Upper Jurassic Formation

Due to its high permeability the Weißjura-Group is the target formation for geothermal wells. A combination from porosity, karstification and fractured habitus makes these carbonate rocks the perfect source for geothermal fluids. In reality, it is unlikely to sharply distinguish these types of aquifers since porosity and karst is linked and so is an increased karstification with advancing faulting. Section 1.4.1

Table 1.1 Classification and subdivision of the Upper Jurassic Epoch and its Stages, all ICS references are in accordance with [33], the Upper Jurassic term after [36], and the former stages after [52]

Series after ICS	Upper Jurassic in Southern Germany after [36]	Stage/Age after ICS	Stage after [52]	Absolute Age after ICS in Ma
Upper Jurassic	Weißjura-Group (Middle Oxfordian to Lower Tithonian)	Tithonian	Zeta 1–6 (ζ)	145.0–152.1
		Kimmeridgian	Epsilon (ε)	152.1–157.3
			Delta 1–4 (δ)	
			Gamma (γ)	
		Oxfordian	Beta (β)	157.3–163.5
			Alpha (α)	

elucidates the external tectonic forces leading to faults and joints within the Weißjura-Group. Porosity and karstification, in contrast, is closely connected with the rock itself, its sedimentation history and chemistry.

Table 1.1 gives a review of the recent classification of the Upper Jurassic Series after ICS [33] compared to the former subdivision of [52] and their corresponding absolute ages.

1.4.2.1 Genesis of the Weißjura-Group

In her thesis, [5, 8] gives a concise description of five facies realms, introduced by Koch [53], which subdivide the Weißjura-Group of the NAFB in the Franconian facies, the Swabian facies, the Argovian facies, the Rauracien facies and the Helvetian Facies. The Franconian and the Swabian facies daylight in the Franconian and Swabian Alb and can be found in drillings within the Molasse Basin [10]. While the Swabian facies is dominated by marly and poor dolomitic sediments, the Franconian facies—also known as "*Frankendolomit*" after [54]—is characterized by its dominating dolomite content [10, 36]. Both are characterized by white to light grey carbonates, which are either deposited in a massive (reef) facies or a bedded facies [55]. The Southern Franconian facies, after [36], undergoes a sedimentation cycle of a steady increase and growth of reefs in the first 140 m (Middle Oxfordian to mid Upper Kimmeridgian, α to δ, compare to Fig. 1.8). Organisms like (siliceous) sponges, corals, ooids, onkoids, algal crusts and stromatolites built up the massive facies, or reef facies [56]. A sudden change in reef-growth as a consequence of the Torleite-Transgression allowed the formation of basins, which were coevally filled with debris and bioclasts of the surrounding reefs (mid Upper Kimmeridgian to Lower Tithonian, ε to ζ [36]. This resulted in the bedded facies with banked, micritic (lithographic) and marly limestones ([56], Fig. 1.8). Targets of geothermal wells that aim for a porous aquifer, are preferably massive, high permeable dolostones from the massive facies [57, 58]

Furthermore, the Helvetian facies in the Bavarian Molasse Basin is revealed by drillings in the Lake Constance region autochthonously [10] and also allochthonous in the Helvetic nappes of Vorarlberg and the Walensee region in Switzerland [5, 8, 60]. Mraz [5, 8] mentions that the occurrence of tintinnids on a shelf—here the Helvetian Shelf—proves this depositional environment (Fig. 1.8). Chronologically, the Helvetian facies is composed of three members: the marly Schilt-Fm., the calcareous Quinten-Fm., and with the transition to the Cretaceous, the Zementstein-Fm. Thuro et al. ([28]: 56). Schneider [61] found out that the Upper Jurassic carbonates are comparable with limestones of the Quinten-Fm. from the Kanisfluh. Mraz [5, 8] classifies the lithology of this Quinten-Fm as micritic limestones with pelagic microfauna or, in rare cases, as an arenitic Wackestone. With reference to Fig. 1.8, the location of the geothermal well of Geretsried locates directly vertically above the depositional area of the Helvetian Shelf. However, recent research carried out in the project "Dolomitkluft" rejects that theory.

1.4.2.2 Updates on the Geological Situation from Recent Research

In anticipation to Chap. 3, which presents the wells GEN-1 and its sidetrack GEN-1ST-A1, this Chapter will provide latest insights into geology, particularly the lithology and facies as derived from cores and cuttings from both boreholes. Accordingly, after [31], the lithostratigraphic sequence in both boreholes corelates well and can be chronologically ordered from late Kimmeridigian to Tithonian. Although, these dense, greyish/black, micritic limestones, as described by Steiger and Uhlig [31] and Mraz [5, 8], appear to have an identical appearance as the limestones from the Helvetian facies, their microfacies distinguished them. Wolfgramm [51] explicitly mentions the absence of calpionellids in the cores and cuttings, which are abundant in the Quinten-Fm. Mraz [5, 8] calls this a transitional facies in the southern part of the Molasse Basin, which can be matched to the Franconian facies in the area of Munich (also see [31]). The depositional environment is described as a basin—not a shelf as the Helvetian facies suggests. Components within the limestone from neighboring shallow water platforms support this argument. The macroscopic appearance issues from euxinic sedimentation conditions lead to a dark color and a bituminous habitus.

Table 1.2 lists the four formations which can be distinguished within this depositional environment and adds age, depth in MD and TVD and equivalent surface rock formation after [31].

The Solnhofen-Fm. and the "Treuchtlinger Marmor", as named by Steiger and Uhlig [31] as equivalent, will be described in the following.

Solnhofen Formation: Barthel [62] describes the Solnhofen Limestone lithologically as a light gray, dark bluish (particle size <5 μm) limestone where bedding thicknesses range between <1 mm and 300 mm. He further describes the bedding planes as perfectly smooth with exception of slumping faults, load casts, and other penecontemporaneous structures. A distinction between "Flinz", a pure micritic limestone (95–98% $CaCO_3$ and only minor proportions of clay and quartz) and "Fäulen" (can be translated as "rotten"), thinner bedded, fissile shaly limestones (77–87%

Table 1.2 Stratigraphic distinction of the Upper Jurassic Carbonates into Platform-Carbonates and Platform- and Basin Carbonates according to the depths of the geothermal wells of Geretsried, including the Purbeck facies, with their equivalent stratigraphic units in the Franconian Alb (after [31])

Rock formation	Stage/age after ICS	Top depth m MD		Top depth m TVD		Equivalent
		GEN-1	GEN-1ST-A1	GEN-1	GEN-1ST-A1	
Purbeck facies	Tithonian-Lower cretaceous	4470	4500	4318	4353	–
Platform-carbonates	Tithonian	4600	4670	4398	4461	–
Platform- and basin carbonates	Kimmeridgian-tithonian	4820	5000	4490	4589	Solnhofen-Fm (Lower Tithonian)
Platform-carbonates	Kimmeridgian	5090	5300	4595	4683	Treuchtlinger Marmor

$CaCO_3$, 10–20% clay and up to 3% quartz), characterizes the Solnhofen Formation [62]. Intermediates between flinz and fäule are called "Blätterflinz" or "zähe Fäule". Besides differing in volumetric mineral content, they mainly diverge in porosity and permeability and can be classified after [63] as follows:

Solnhofen Formation, terms:	"Flinz"	"Blätterflinz"	"zähe Fäule"	"Fäule"
Range of porosity (Vol. -%):	0.2–ca. 6	ca. 6–ca. 13	ca. 13–ca. 21	ca. 21–ca. 26
Permeability (mD):	From 0.012	➔		Up to 0.741

Worth mentioning is that the "*stacks of flinz split into individual sheets along bedding planes, which are defined by surfaces where there is a concentration of clay minerals*" ([64]: 38).

Treuchtlingen Formation: The Treuchtlichen Formation, to which the "Treuchtlinger Kalkstein" (coll. Treuchtlinger Marmor) belongs to, consist of thick-bedded spongue-micorobalit-biostroma which can reach up to 1.5 m, sometimes 2 m and can also transgress in a massive facies, which can also be dolomitized [36].

References

1. Günther J, Lehmann H, Lorenz U, Purr K (2017) Den Weg zu einem treibhausgasneutralen Deutschland ressourcen-schonend gestalten. 72p., Dessau-Roßlau (Umweltbundesamt)
2. Agemar T, Weber J, Schulz R (2014) Deep geothermal energy production in Germany. Energies 7(7): 4397–4416
3. VDI 4640 (2010) Thermal use of the underground. VDI-Gesellschaft Energie und Umwelt (GEU), Berlin
4. Moeck IS (2014) Catalog of geothermal play types based on geologic controls. Renew Sustain Energy Rev 37: 867–882
5. Mraz E, Wolfgramm M, Moeck I, Thuro K (2019) Detailed fluid inclusion and stable isotope analysis on deep carbonates from the north alpine foreland basin to constrain paleofluid evolution. Geofluids 23
6. Agemar T, Alten J, Ganz B, Kuder J, Kühne K, Schumacher S, Schulz R (2014) The geothermal information system for Germany - GeotIS. ZDGG 165: 129–144
7. Klaus T, Vollmer C, Werner K, Lehmann H, Müschen K (2010) Energieziel 2050: 100% Strom aus erneuerbaren Quellen. 194p., Dessau-Roßlau (Umweltbundesamt)
8. Mraz E (2019) Reservoir characterization to improve exploration concepts of the Upper Jurassic in the southern Bavarian Molasse Basin, 126p. Dissertation, Chair of Engineering Geology, Technical University of Munich, Munich
9. Wolfgramm M, Bartels J, Hoffmann F, Kittl G, Lenz G, Seibt P, Schulz R, Unger HJ (2007) Unterhaching geothermal well doublet: structural and hydrodynamic reservoir characteristic; Bavaria (Germany). In: Proceedings European Geothermal Congress 2007, Unterhaching
10. Böhm F, Koch R, Höferle R, Baasch R (2010) Der Malm in der Geothermiebohrung Pullach Th2—Faziesanalyse aus Spülproben (München, S-Deutschland). Geol Bl NO-Bayern 60:17–49
11. Böhm F, Dorsch K, Savvatis A, Beichel K, Steiner U (2017) Reservoircharakterisierung des Malms - neue Erkenntnisse aus dem Geothermieprojekt Holzkirchen. Der Geothermiekongress DGK 2017, München, Deutschland (Bundesverband Geothermie e.V.)

12. Rioseco EM, Ziesch J, Buness H, Thomas R (2017) Strukturanalyse und TH-Modellierung auf Basis der GRAME 3D Seismik in München – Projekt GeoParaMoL. – Der Geothermiekongress DGK 2017, München, Deutschland (Bundesverband Geothermie e.V.)
13. Dirner S (2018) Geothermie Schäftlarnstraße: Explorationsstrategie und erste Projektergebnisse. Der Geothermiekongress DGK 2018, Essen, Deutschland (Bundesverband Geothermie e.V.)
14. Meinecke M (2018) Geothermieprojekt Schäftlarnstraße in der Millionenmetropole München - Von der Idee zur Realisierung. Der Geothermie Kongress DGK 2018, Essen, Deutschland (Bundesverband Geothermie e.V.)
15. Meinecke M (2019) Das Geothermieprojekt Schäftlarnstraße in der Millionenmetropole München: Das Projekt, die Explorationsstrategie und Projektergebnisse. Der Geothermie Kongress DGK 2019, München, Deutschland (Bundesverband Geothermie e.V.)
16. Moeck IS, Dussel M, Weber J, Schintgen T, Wolfgramm M (2019) Geothermal play typing in Germany, case study Molasse Basin: a modern concept to categorise geothermal resources related to crustal permeability. Netherlands J Geosci 98: 10
17. Haimson BC (1996) Can inclined exploratory boreholes be used for in situ stress measurements? A review of three case histories. In: ISRM International Symposium - EUROCK 96, 403–409, Turin, Italy (International Society for Rock Mechanics and Rock Engineering), 1996/1/1/, ISRM
18. Bieniawski ZT (1989) Engineering rock mass classifications: a complete manual for engineers and geologists in mining, civil, and petroleum engineering, 272p. John Wiley & Sons
19. Barton N (1978) Recent experiences with the Q-system of tunnel support design. In: Reprinted from proceedings of the symposium on exploration for rock engineering, Johannesburg, pp 107–117
20. Hoek E (1994) Strength of rock and rock masses. ISRM News J 2(2): 4–16
21. Haimson BC, Herrick CG (1989) Borehole breakouts and in situ stress. In: Rowley JC (ed) Drilling Symposium 1989 [12th annual energy-sources technology conference and exhibition], vol 22. Houston (American Society of Mechanical Engineers, New York), pp 17–22
22. Haimson BC (1997) Borehole breakouts and core disking as tools for estimating in situ stress in deep holes. In: Sugawara K, Obara Y, Sato A (eds) Rock stress, Balkema, pp 35–42
23. Zoback MD, Moos D, Mastin L, Anderson RN (1985) Well bore breakouts and in situ stress. J Geophys Res: Solid Earth 90(B7): 5523–5530
24. Zoback MD, Barton CA, Brudy M, Castillo DA, Finkbeiner T, Grollimund BR, Moos DB, Peska P, Ward CD, Wiprut DJ (2003) Determination of stress orientation and magnitude in deep wells. Int J Rock Mech Mining Sci 40(7–8): 1049–1076
25. Beichel K, Koch R, Wolfgramm M (2014) Die Analyse von Spülproben zur Lokalisierung von Zuflusszonen in Geothermiebohrungen. Beispiel der Bohrungen Gt Unterhaching 1/1a und 2. (Süddeutschland, Molassebecken, Malm). Geol Bl NO-Bayern 64:43–65
26. Steiger T, Uhlig S, Moeck I (2015) Thin section based cutting analysis as a new approach in rock type determination while drilling deep geothermal wells. In: Proceedings—world geothermal congress 2015, 1–8, Melbourne, Australia
27. Wolfgramm M, Buse C, Mraz E, Thiem S (2017) Analyse von Bohrklein: methode sowie Cuttingatlas für Karbonate des Oberjura im Molassebecken. Der Geothermie Kongress DGK 2017, München, Deutschland (Bundesverband Geothermie e.V.)
28. Thuro K, Zosseder K, Bohnsack D, Heine F, Konrad F, Mraz E, Stockinger G (2019) Dolomitkluft - Erschließung, Test und Analyse des ersten kluftdominierten Dolomitaquifers im tiefen Malm des Molassebeckens zur Erhöhung der Erfolgsaussichten. Teilprojekt B: Geomechanische und hydro-geologische Parametrisierung und Modellierung, 413p. München (Technische Universität München - Ingenieurfakultät Bau Geo Umwelt - Lehrstuhl für Ingenieurgeologie)
29. Mraz E, Bohnsack D, Stockinger G, Käsling H, Zosseder K, Thuro K (2018) Die Bedeutung von Analogaufschlüssen des Oberjura für die Interpretation der Lithologie der geothermalen Tiefbohrung Geretsried. Jahresberichte und Mitteilungen des Oberrheinischen Geologischen Vereins 100: 517–548

30. Stockinger G, Mraz E, Menschik F, Thuro K (2018) Geomechanical model for a higher certainty in finding fluid bearing regions in non-porous carbonate reservoirs. In: Shakoor A, Cato K (eds) IAEG/AEG annual meeting proceedings, San Francisco, California. Springer International Publishing, Cham, pp 193–198
31. Steiger T, Uhlig S (2018) Bio- und Lithostratigraphie der Geothermie-Bohrung Geretsried GEN- 1 mit Sidetrack GEN-1ST-A1. (Anlage 1). In: Thuro K, Zosseder K, Bohnsack D, Heine F, Konrad F, Mraz E, Stockinger G (eds) Dolomitkluft - Erschließung, Test und Analyse des ersten kluftdominierten Dolomitaquifers im tiefen Malm des Molassebeckens zur Erhöhung der Erfolgsaussichten. Teilprojekt B: Geomechanische und hydro-geologische Parametrisierung und Modellierung, München (Technische Universität München - Ingenieurfakultät Bau Geo Umwelt - Lehrstuhl für Ingenieurgeologie)
32. Stockinger G, Käsling H, Menschik F, Thuro K (2019) 3D rotation applied to in situ stress fields for 2D numerical modelling, borehole stability and drill core recovery in deep geothermal wells. In: Fontoura SD, Rocca RJ, Mendoza JP (eds) Rock mechanics for natural resources and infrastructure development–full papers. CRC Press , London, pp 3144–3151
33. Cohen KM, Harper DAT, Gibbard PL (2020) ICS International Chronostratigraphic Chart 2020/03. p. (International Commission on Stratigraphy, IUGS)
34. Bachmann GH, Müller M, Weggen K (1987) Evolution of the Molasse Basin (Germany, Switzerland). Tectonophysics 137(1–4): 77–92
35. Plessen M-L (1983) Die Isar - ein Lebenslauf. 374p., München (Hugendubel)
36. Niebuhr B, Pürner T (2014) Plattenkalk und Frankendolomit - Lithostratigraphie der Weißjura-Gruppe der Frankenalb (außeralpiner Oberjura, Bayern). SDGG 83: 5–72
37. Frisch W (1981) Plate motions in the Alpine region and their correlation to the opening of the Atlantic Ocean. Geol Rundschau 70:402–411
38. Lemcke K (1973) Zur nachpermischen Geschichte des nördlichen Alpenvorlands. Geol Bavarica 69:5–48
39. Meschede M, Warr LN (2019) The geology of Germany: a process-oriented approach, 304p. Springer, Cham
40. Lemcke K (1988) Das Bayerische Alpenvorland vor der Eiszeit: Erdgeschichte, Bau, Bodenschätze. Geologie von Bayern I, 175p. Stuttgart (Schweizerbart)
41. Ziegler PA (1987) Late Cretaceous and Cenozoic intra-plate compressional deformations in the Alpine foreland—a geodynamic model. Tectonophysics 137(1–4): 389–420
42. Bachmann GH, Dohr G, Müller M (1982) Exploration in a classic thrust belt and its foreland: Bavarian Alps, Germany. AAPG Bull 66(12): 2529–2542
43. Unger HJ (1999) Die tektonischen Strukturen der bayerischen Ostmolasse. Documenta naturae, 124: 1–16
44. Kraemer C (2009) Identifikation horizontaler Blattverschiebungen im Untergrund des Molassetrogs - neue Interpretationsergebnisse auf Grundlage von 2D-Seismik. Der Geothermiekongress 2009, Bochum, Deutschland (Bundesverband Geothermie e.V.)
45. Schneider M, Birner J, Steiner U, Thomas L, Böhm F, Savvatis A, Baumann T, Mayrhofer C, Nießner R (2012) Wissenschaftliche und technische Grundlagen zur strukturgeologischen und hydrogeologischen Charakterisierung tiefer geothermisch genutzter Grundwasserleiter am Beispiel des süddeutschen Molassebeckens, 237p. Berlin (Freie Universität Berlin, Fachbereich Geowissenschaften, Institut für Geologische Wissenschaften)
46. Ratschbacher L, Frisch W, Linzer H-G, Merle O (1991) Lateral extrusion in the eastern Alps, Part 2: structural analysis. Tectonics 10(2): 257–271
47. Haq BU, Hardenbol J, Vail PR (1988) Mesozoic and Cenozoic chronostratigraphy and cycles of sea-level change. In: Wilgus CK, Hastings BS, Posamentier HW, Wagoner JV, Ross CA, Kendall CGCS (eds) Sea-level changes: an integrated approach, vol 42. SEPM Society for Sedimentary Geology Special Publication, pp 71–108
48. Lemcke K (1974) Vertikalbewegungen des vormesozoischen Sockels im nördlichen Alpenvorland vom Perm bis zur Gegenwart. Eclogae Geologicae Helvetiae 67: 121–133
49. Drews MC, Bauer W, Caracciolo L, Stollhofen H (2020a) Corrigendum to "Disequilibrium compaction overpressure in shales of the Bavarian Foreland Molasse Basin: results and

geographical distribution from velocity-based analyses [Mar. Pet. Geol. 92 (2018) 37–50]. Marine and Petroleum Geology 112: 104102
50. Kuhlemann J, Kempf O (2002) Post-eocene evolution of the north alpine foreland basin and its response to alpine tectonics. Sedimentary Geology 152: 45–78
51. Wolfgramm M, Thiem S, Zimmermann J, Budach I, Buse C, Kabus F (2018) Forschungsbericht GTN: Erschließung, Test und Analyse des ersten kluftdominierten Dolomitaquifers im tiefen Malm des Molassebeckens, 93p. Berlin (Geothermie Neubrandenburg GmbH)
52. Quenstedt FA (1858) Der Jura. 842p. Tübingen (Laupp)
53. Koch R (1997) Daten zu Fazies und Diagenese von Massenkalken und ihre Extrapolation nach Süden bis unter die Nördlichen Kalkalpen. Geol Bl NO-Bayern 47:117–150 (Erlangen)
54. Gümbel CW (1868) Geognostische Beschreibung des Königreichs Bayern. 2. Abtheilung. Geognostische Beschreibung des Ostbayerischen Grenzgebirges oder des Bayerischen und Oberpfälzer Waldgebirges, 968p. Kassel
55. Meyer RKF, Schmidt-Kaler H (1989) Paläogeographischer Atlas des süddeutschen Oberjura (Malm). Geol Jb Reihe A 115:3–77
56. Meyer RKF, Schmidt-Kaler H (1994) Treuchtlingen, Solnhofen, Mörnsheim, Dollnstein. - Wanderungen in der Erdgeschichte, 1, 96p, München
57. Wolfgramm M, Dussel M, Lüschen E, Schulz R, Thomas R, Koch R (2011) Zuflusszonen im Malm-Untersuchungen des geothermischen Hauptgrundwasserleiters im süddeutschen Molassebecken. BBR Fachmagazin für Brunnen- und Leitungsbau Sonderheft Geothermie 2012:75–82
58. Böhm F, Birner J, Steiner U, Koch R, Sobott R, Schneider M, Wang A (2011) Tafelbankiger Dolomit in der Kernbohrung Moosburg SC4: Ein Schlüssel zum Verständnis der Zuflussraten in Geothermiebohrungen des Malmaquifers (Östliches Molasse-Becken, Malm Süddeutschland). Z Geol Wiss 39:117–157
59. Röper M, Rothgänger M, Rothgänger K (2000) Die Plattenkalke von Schernfeld (Landkreis Eichstätt), 128p. Eichendorf (Eichendorf Verlag)
60. Koch R (2000) Die neue Interpretation der Massenkalke des Süddeutschen Malm und ihr Einfluss auf die Qualität von Kalksteinen für technische Anwendungen. Archaeopteryx 18: 43–65
61. Schneider J (1958) Der Jura in Erdölbohrungen des westlichen Molassetrogs. Z Dtsch Ges Geowiss 110:1–16
62. Barthel KW (1978) Solnhofen: ein Blick in die Erdgeschichte. 393p
63. Weiß EG (1987) Porositäten, Permeabilitäten und Verkarstungserscheinungen im mittlerwen und oberen Malm der südlichen Frankenalb, 211p. Dissertation, Erlangen-Nürnberg
64. Barthel KW, Swinburne NHM, Morris SC (1990) Solnhofen, 245p. Cambridge University Press, Cambridge
65. www-01: https://www.geotis.de/geotisapp/templates/heatsumstatistic.php, Accessed on: 03.03.2020. Accessed on: 03.03.2020
66. www-02: https://www.geotis.de/geotisapp/templates/powersumstatistic.php. Accessed on: 03.03.2020
67. www-03: https://www.tiefegeothermie.de/news/projekt-in-weilheim-gestoppt. Accessed on: 09.12.2020
68. www-04: https://www.tiefegeothermie.de/news/bohrplaetze-in-icking-werden-zurueckgebaut. Accessed on: 09.12.2020
69. www-05: https://www.tiefegeothermie.de/projekte/geretsried. Accessed on: 03.03.2020
70. Meyer RKF, & Schmidt-Kaler H (2002) Auf den Spuren der Eiszeit südlich von München, östlicher. p2, Aufl (ed.,) Vol. 8, Teil, München (F. Pfeil)

Chapter 2
Rock Mechanical Basics

In the last decades, enormous efforts in engineering sciences and geological sciences, such as structural geology and rock structure, have produced new rock mechanical methods and state-of-the-art applications for tunneling, deep geological repositories, or petroleum engineering. This Chapter presents the fundamental state-of-the-art concepts used in this thesis.

2.1 Deformation of Rocks

Rocks deform under the effect of stress. For rocks and rock masses, the special case is that the deformation happens first elastically, and after reaching a specific limit, the yield stress, plastically. For both terms, the following Chapter gives a basic understanding of the governing equations and transitions to the stresses taken into account for these deformations.

2.1.1 Elastic Deformations

Elasticity implies that deformations in a material are completely reversible [1]. Hooke's law expresses, here for isotropic material only, the linear relation of the strain tensor ε in dependency of the stress tensor σ and the stiffness of the material, here the Young's modulus E:

$$\varepsilon = \frac{\sigma}{E}$$

The elastic behavior is limited by the maximum cohesive strength of the inner-atomic bonds of the material. Once these forces are exceeded, these bonds fail either in tension or in shearing [2].

G. M. Stockinger, *Fracturing in Deep Boreholes*, Springer Theses,
https://doi.org/10.1007/978-3-030-94569-5_2

2.1.2 *Plastic Deformations*

The process during the failure in the bonding of atoms defines as yielding. The yield stress describes the point where, theoretically, the first atomic bonds fail. In terms of uniaxial compression, this would be the crack initiation. This process is irreversible, and the deformation becomes plastic. Fairhurst [2] describes two conditions for rupture to occur:

- Exceeding the inner-atomic cohesive strength
- Supplying sufficient energy to create two distinct surfaces

2.1.2.1 Failure Criterions

A rupture—according to Fairhurst [2]—is "*a continuous plane of separation traversing the solid*", which ultimately leads to the failure of the rock. This requisite for failure can be expressed by different criteria, only to name a few the Griffith Criterion [3, 4], the Mohr–Coulomb failure criterion, where the shear strength is plotted versus the normal stress [5–7] or the Hoek–Brown criterion where major principal stress is plotted versus minor principal stress [8, 9]. The Mohr–Coulomb criterion, often applied to determine borehole stability, can be written in multiple but equivalent forms, which appear throughout this thesis in several Sects. (e.g., 2.2.2, 2.3, 5.5.4.1, 7.4). However, a pure linearity of Mohr–Coulomb's theory, according to Jaeger et al. [10], does not match in two aspects:

- The tensile strength is overestimated
- With increasing σ_1, σ_3 increases less than linear [6]

When the two limitations are accounted for, the line becomes an envelope—a parabolic Mohr envelope. Figure 2.1 shows this parabolic envelope and its four major states: (I) uniaxial tension, where the major principal stress σ_1 is 0, and the minor principal stress σ_3 is negative; the failure is purely tensional. (II) confined tension, where a negative σ_3 exceeds the tension, although a positive stress σ_1 acts confining. Both kinds of failures are tensional and happen perpendicular to σ_1. (III) is the failure in uniaxial compression with $\sigma_3 = 0$ and $\sigma_1 > 0$. (iv) is the stress state

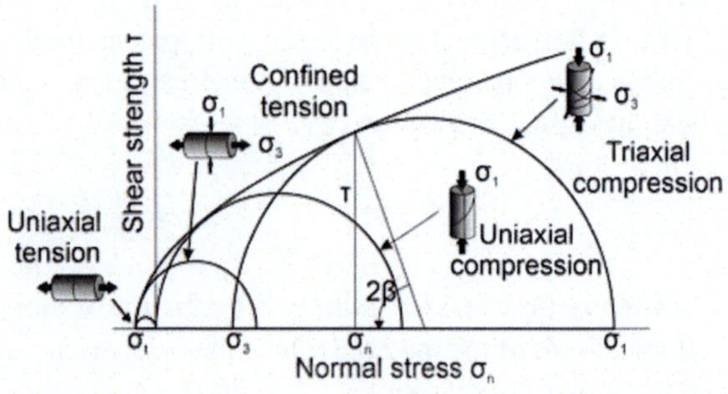

Fig. 2.1 Typical failure modes of intact rock plotted in terms of shear strength τ and normal stress σ_n with the major and minor principal stress in a Mohr circles and Mohr envelope diagram (modified after [11]: 288, Fig. 1.1)

in triaxial compression. States (III) and (iv) fail in shearing. Failure happens when the circle cuts the envelope.

2.1.2.2 Modes of Failure

After Scholz [12], three modes (Fig. 2.2) can describe the failures from Fig. 2.1. Mode I represents the tensile mode, where stresses act normal to the crack and lead to a tensile opening. Mode II and Mode III both represent shear failures. The displacement in Mode II acts perpendicular to the crack, and in Mode III parallel to the crack. In a three-dimensional space, all of these three can occur individually or combined, while in a two-dimensional space, Mode II and Mode III are indistinguishable.

2.1.3 Elastic and Plastic Deformation in Uniaxial Compression

In rocks, Martin and Chandler [13] describe 5 major stages in which the behavior of rocks under uniaxial compression can be classified. Stage I—"Crack Closure" - existing cracks in the rock specimen close under load (Fig. 2.3, σ_{cc}). Stage II describes the "Elastic Region". This section defines the slope region that defines the static elastic parameters of the rock:

- Young's modulus E_{stat} [GPa] from the slope of axial strain and axial load
- Poisson's ratio ν_{stat} "GPa" from the ratio of lateral and axial strain

This linear region is indistinctively linearly continued by Stage III, the "Crack Initiation" stage, where the yielding stress marks the breakage of the first atomic bonds and the beginning of the irreversible plastic deformation. Many authors (e.g., Jaeger et al. [10]: 83) and standards [86, 85] suggest the possibility of determining the Young's modulus in this linear region from either tangent or secant modulus. Mutschler [14], however, suggests determining the Young's modulus from a curve of unloading, which has an upper and lower limit within the linear region. This method allows neglecting the plastic deformations and accounts for more realistic

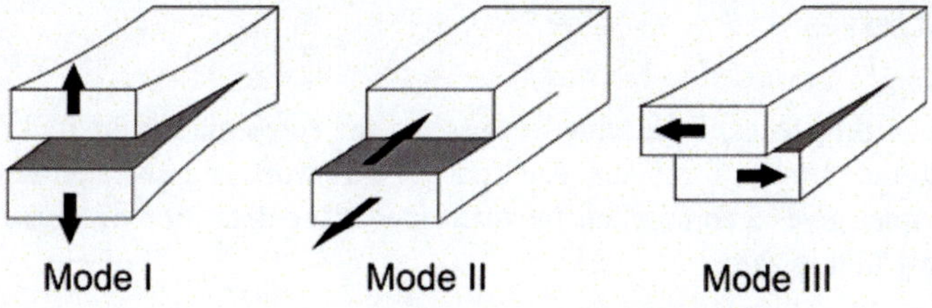

Fig. 2.2 The three modes of fracture propagation in rocks, modified after Scholz ([12]: 8, Fig. 1.5)

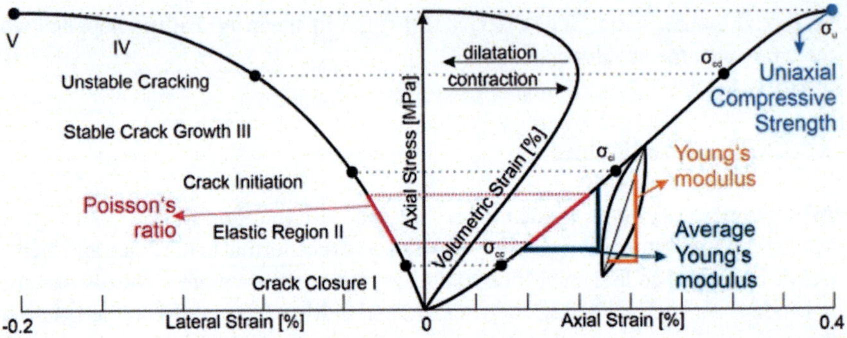

Fig. 2.3 Stress–strain curve for the axial strain and the lateral strain of a uniaxial compression strength test, with the parameters determined by this test (modified after Martin and Chandler [13]: 644, Fig. 1)

values of the elastic parameters. An complete unloading, as mentioned by Jaeger et al. [10], is not intended since closed cracks should not reopen, and the testing frame should contain its stiffness. This work follows the approach of Mutschler [14] to estimate the Young's modulus from the slope of reloading (Fig. 2.3, orange slope). The initial slope of elastic loading is further designated as the average Young's modulus (Fig. 2.3, green slope) after [86].

While the rock is still intact during the first three stages, further loading leads to an uncontrolled coalescence of cracks and initiates Stage IV "unstable crack growth" [15]. Martin and Chandler [13] claims that unstable crack growth is 70–85% of the short-term peak strength. Jaeger et al. [10] states that this may vary even stronger according to the rock's lithology. Neither the axial nor the lateral stress–strain curves directly indicate the point of unstable cracking. However, the volumetric strain, defined as the sum of axial strain and twice the lateral strain [16], changes its course from a contraction to dilatation and marks σ_{cd} (Fig. 2.3). Upon reaching this turning point, failure under consistent load is inevitable in the long-term view. If loading increases beyond "unstable crack growth", a short-term peak strength is achieved, known as the uniaxial compressive strength, referred to as σ_u in this work (Fig. 2.3, blue).

2.2 Stresses

From rock properties, to in situ stresses, to the reorientation of rock cores, and finally to the numerical models, each part of this work is associated with stresses. This Chapter gives a comprehensive overview of the definition of stresses and how they apply in this work.

2.2.1 *Stress acting on a Plane and Stresses acting at a Point*

Generally, there are two different ways to define stress. Firstly, stress is a force acting evenly on a defined area. Accordingly, this stress vector with a specific direction and magnitude can be differentiated in normal stress and shear stress [17]. Shear strength or the Uniaxial Compressive Strength (σ_u) is an example of this kind of stress.

Secondly, stress acting at a point can be defined by a three-dimensional stress tensor, introduced by Cauchy [18]. This stress tensor allows describing a stress state at a specific point in three dimensions without any knowledge of geometric reference [10, 18, 19]. Generally, this stress tensor is written in a stress matrix [20]:

$$\sigma = \begin{bmatrix} \sigma_{11} & \sigma_{12} & \sigma_{13} \\ \sigma_{21} & \sigma_{22} & \sigma_{23} \\ \sigma_{31} & \sigma_{21} & \sigma_{33} \end{bmatrix} = \begin{bmatrix} \sigma_{11} & \tau_{12} & \tau_{13} \\ \tau_{21} & \sigma_{22} & \tau_{23} \\ \tau_{31} & \tau_{21} & \sigma_{33} \end{bmatrix}$$

This matrix with nine components consists of the normal stresses σ_{11}, σ_{22}, and σ_{33} that act normal on the planes of an infinitely small cube [20] and six shear stresses that act parallelly to these planes and perpendicular to their associated normal stresses (Fig. 2.4). Figure 2.4 illustrates the shear forces countering each other, and due to the symmetry of the stress tensor, the following applies: $\sigma_{12} = \sigma_{21}$, $\sigma_{13} = \sigma_{31}$, and $\sigma_{23} = \sigma_{32}$.

Due to the conservation of momentum, as explained by Heidbach et al. [20], there is always a coordinate system where the shear stresses annulate each other, and only principal stresses apply. A special case—but a general assumption in structural geology—considers that in the earth crust, the vertical stress is a principal stress, which calculates from the rock's weight as a function of depth [20]. Three different stress regimes derive from this assumption.

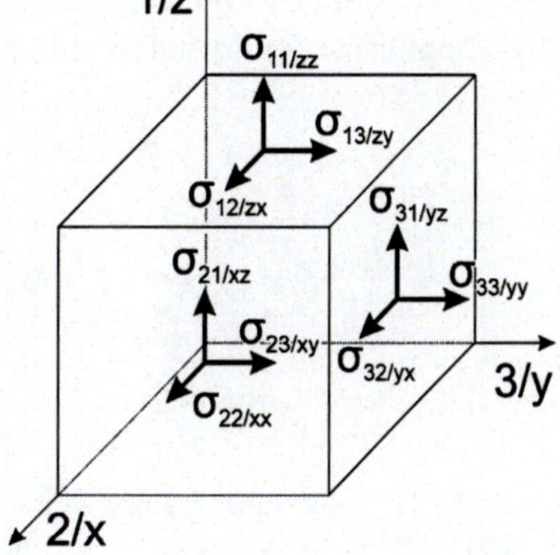

Fig. 2.4 Illustration of the Cauchy 3D stress tensors

2.2.2 Stress Regimes and the Mechanism of Faulting

Anderson [21, 22] distinguishes three types of faulting regimes occurring under different principal stress ratios.

1. In a normal faulting regime (NF), S_v is greater than both S_{Hmax} and S_{hmin} (Fig. 2.5a).
2. In a strike-slip faulting regime (SS), S_{Hmax} is larger than S_v, and S_{hmin} is the smallest principal stress;
3. The faulting has no vertical motion (Fig. 2.5b).
4. In a horizontal thrust or reverse faulting regime (TF), S_{Hmax} and S_{hmin} exceed S_v (Fig. 2.5c).

Transitional regimes can also exist when two principal stresses are equal. Accordingly, there can be a transitional NF/SS Regime when $S_v = S_{Hmax} > S_{hmin}$ or a transitional TF/SS Regime when $S_{Hmax} > S_{hmin} = S_v$ [23].

The prerequisite for faulting requires that the shear stresses exceed the shear strength of the rock and can, after the Mohr–Coulomb Criterion [7], be expressed as:

$$\tau = \sigma * tan\varphi + c$$

Byerlee [24] adapted this equation for rocks and specified two generalizations for different normal stresses, firstly $\sigma_n < 200$ MPa (2 kBars) and secondly 200 MPa $< \sigma_n < 2000$ MPa. These express as:

$$\tau = 0.85 * \sigma_n \text{ for } \sigma_n < 200 \text{ MPa.}$$
$$\tau = 0.5 + 0.6 * \sigma_n \quad \text{for } 200 \text{ MPa} < \sigma_n < 2000 \text{ MPa.}$$

From the equations above, the friction coefficients (μ) of 0.85 and 0.6 equal an angle of inner friction of 40° and 31°, respectively.

Zoback et al. [25] and Alber et al. [26] also find friction coefficients of 1 or even 1.1 applicable for crustal rocks and carbonate rocks.

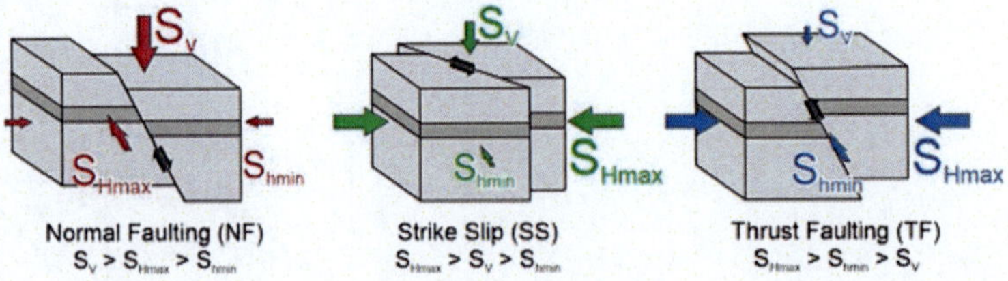

Fig. 2.5 Possible stress regimes in the Earth crust, defined by the orientation of S_{Hmax} and the magnitudes of S_v, S_{Hmax} and S_{hmin}; left: the vertical stress (S_v) as largest principal stress in a Normal Faulting (NF) regime, mid: S_V as the intermediate principal stress in a Strike-slip (SS) regime and to the right: S_V as the minor principal stress in a Thrust Faulting (TF) regime (after Heidbach et al. [20])

2.2.3 *Effective Stress*

In the presence of fluid or formation pressure, the absolute stress must be reduced in order to account for the effective stress. This effective stress σ' derives from the actual stress tensor σ minus the hydrostatic fluid pressure tensor p_f [27–29]. In simple terms, this equation writes as:

$$\sigma' = \sigma - \alpha p_f \ with \ \alpha = 1 - \frac{K}{K_s}$$

The factor α is the so-called Biot's coefficient, which depends on the bulk modulus of the rock K_s and the rock with pores K [30]. Hence, for rocks without any porosity, the absolute stress applies, while the stress for rocks with high porosity equals the differential pressure. Nur and Byerlee [30] emphasize that α is 1 in the presence of fluid pressure.

2.2.4 *Frictional Faulting Limitations*

Jaeger and Cook [31] initially introduced the concept, connecting Anderson's faulting theory, the Coulomb criterion, the effective stress, and Byerlees coefficient to determine the prevailing stress regime: the frictional faulting limit—graphically illustrated as a stress polygon by Zoback et al. [32] and Moos and Zoback [33].

The principal stresses are limited to both the principal effective stresses and the critical coefficient of friction,

$$\frac{\sigma_1'}{\sigma_3'} = \frac{\sigma_1 - p_f}{\sigma_3 - p_f} \le \left[(\mu^2 + 1)^{1/2} + \mu\right]^2$$

For the stress regimes NF, SS, or TF (see Sect. 2.2.2), σ_1 is $S_{Hmax,}$ S_{Hmax} or S_v and σ_3 is S_v, S_{hmin} or S_{hmin}, respectively.

2.3 Borehole Stability in Vertical and Deviated Wells

When a circular cavity, e.g., a tunnel, a shaft, or a borehole, is advanced and rock mass under the influence of high in situ stress is removed, this stress rearranges and, depending on the material parameters (in comparison to the previous Chapters), results in elastic strain or additionally in plastic deformation - in other terms fracturing. Pioneered by Kirsch [34], the stress distribution around a circular hole in an infinite plate is now a general assumption and was further developed for petroleum rock mechanics by Hiramatsu and Oka [35, 36], Haimson and Fairhurst [37, 38] and Peška and Zoback [39].

2.3.1 *Elastic and Plastic Deformation*

A typical first approach is an assumption that stresses around the borehole lead to elastic plane strain only. A second, more realistic approach is that stresses also lead to plastic deformation.

2.3.1.1 Elastic Strain

Three major far-field stresses are important for calculating the plane strain for a vertical borehole with an impermeable wall in an anisotropic stress field. Firstly, the radial stress σ_r acts perpendicular to the borehole. Secondly, the tangential stresses σ_θ (or hoop stress) acts tangential to the borehole. Finally, σ_z acts parallel to the borehole—and in vertical borehole equals the principal stress σ_v. Figure 2.6 illustrates these three far-field stresses, which are dependent on:

- σ_H (S_{Hmax}) and σ_h (S_{hmin}), perpendicular to σ_v (see Sect. 2.2.2)
- θ, which allows the fluctuation of stresses around the borehole if σ_H and σ_h are different
- R_w, the radius of the borehole and r the distance from the borehole
- p_w as mud weight and p_f as formation pressure
- ν_{fr} as Poisson's ratio of the rock

After Fjær et al. [40], the governing, derived after Kirsch [34], equations are:

$$\sigma_r = \frac{\sigma_H+\sigma_h}{2}\left(1-\frac{R_w^2}{r^2}\right)+\frac{\sigma_H-\sigma_h}{2}\left(1+3\frac{R_w^4}{r^4}-4\frac{R_w^2}{r^2}\right)cos2\theta+p_w\frac{R_w^2}{r^2}$$

$$\sigma_\theta = \frac{\sigma_H+\sigma_h}{2}\left(1+\frac{R_w^2}{r^2}\right)+\frac{\sigma_H-\sigma_h}{2}\left(1+3\frac{R_w^4}{r^4}\right)cos2\theta-p_w\frac{R_w^2}{r^2}$$

$$\sigma_z = \sigma_v - 2\nu_{fr}(\sigma_H-\sigma_h)\frac{R_w^2}{r^2}cos2\theta$$

$$\tau_{r\theta} = -\frac{\sigma_H-\sigma_h}{2}\left(1-3\frac{R_w^4}{r^4}+2\frac{R_w^2}{r^2}\right)\sin 2\theta$$

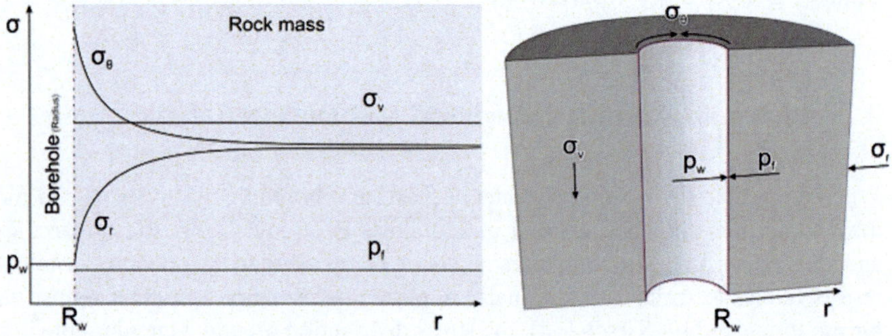

Fig. 2.6 Stresses around a borehole under a linear elastic consideration (merged and modified after Fjær et al. [40]: 137, Figs. 4.2 and 4.3)

At the borehole wall, R_w divided by r becomes one, and the previous equations simplify to:

$$\sigma_r = p_w \text{or with formation pressure} \sigma_r = \Delta p = p_w - p_f$$
$$\sigma_\theta = \sigma_H + \sigma_h - 2(\sigma_H - \sigma_h)cos2\theta - 2 * p_f - \Delta p$$
$$\sigma_z = \sigma_v - 2\nu_{fr}(\sigma_H - \sigma_h)cos2\theta \quad \text{and} \quad \sigma_z = \sigma_v \text{ for isostatic stresses}$$

The tangential stresses can be reduced to a minimum and a maximum value since the term cos 2θ defines a range between −1 and 1:

$$\sigma_{\theta,max} = 3\sigma_H - \sigma_h - 2 * p_f - \Delta p$$
$$\sigma_{\theta,min} = 3\sigma_h - \sigma_H - 2 * p_f - \Delta p$$

Stockinger et al. [41] developed a new solution for deviated boreholes, which uses a stress rotation in a cartesian coordinate system. Section 4.3 presents the single steps of this method in greater detail than yet published.

2.3.1.2 Plasticity

Pure elastic strain is an approximation that disregards the rock's yielding strain and, thus, the plastic behavior. However, the plastic approach gives a more accurate view of rock deformation (i.e., fracture extent) and stress development in the vicinity of the well [42]. Unlike the elastic approach, fracturing shifts the maximum tangential stresses from the borehole wall into the rock mass. With growing distance from the borehole wall, fracturing decreases, and the maximum tangential stress culminates until the fracture criterion is satisfied and subsequently drops. Although analytical solutions exist from Feder and Arwanitakis [43], these processes can be simulated much more easily with the hybrid finite-discrete element method employed in this thesis. Hence, this thesis relinquishes a deepening of this topic.

2.3.2 *Borehole Failure*

Derived from the equations in Sect. 2.3.1.1, the failure of boreholes can be described with the stresses around a borehole in an isotropic, elastic, homogenous and impermeable rock mass [19].

2.3.2.1 Tensile Failure in Boreholes

Hubbert and Willis [44] initially presented the theory of tensile failures, or so-called hydrofracs (HF). This type of failure represents a Mode I fracture. An opening happens when $\sigma_{\theta, \min}$ becomes smaller than the difference between fluid pressure in the well p_w and the formation pressure p_f, here Δp. The equation for the minimal tangential stress, derived from Sect. 2.3.1.1, can be rewritten to the condition $\Delta p > \sigma_\theta$, when existing fractures will open, or when new fractures emerge as the tensile strength σ_t is exceeded. This equation incorporates σ_t, and p_f in brackets since these terms only apply if no fracture preexists or fluid pressure is encountered. Brudy and Zoback [45] argue that for low porous and low permeable rocks, p_f may be neglected.

$$p_{w,max}^{frac} = 3\sigma_h - \sigma_H(+|\sigma_t|)(+p_f)$$

These tensile fractures, which are also known as Drilling Induced Tensile Fractures (DITFs), open in the direction of σ_H and pose a significant challenge because they lead to loss of drilling mud and circulation and thus reduce control over formation pressure and may result in gas kicks or blowouts [46, 47]. The term hydraulic fracturing applies to a method where fractures are created intentionally to either measure in situ stresses or to stimulate the reservoir. Section 2.4.2.3 expands on the orientation and extent of DITFs since they also indicate the strike and the gradient of the far-field *in situ* stresses. Approaches for porous and permeable rocks, e.g., by Schmitt and Zoback [48], will not be further discussed.

2.3.2.2 Compressive Failure in Boreholes

Zang and Stephansson [19] give a detailed overview of the research history on borehole breakouts, representing the contradiction to tensile failures. Although partly a failure in tension, most likely, a combination of Mode I and Mode II fractures leads to these instabilities [49]. The simplest failure criterion is mentioned by Zang and Stephansson [19], who describe the failure with the Uniaxial Compressive Strength σ_u:

$$p_{w,min}^{breakout} = 3\sigma_H - \sigma_h - \sigma_u(+p_f)$$

Martin and Christiansson [50] mention that these breakouts—in the form of spalling—already occur at 50% of σ_u. However, concerning the evolution of fractures in rocks in Sect. 2.4.2.3, this may vary significantly based on the crack initiation and unstable crack growth of individual rock types and should not be generalized. Considering the elastic stresses acting on a borehole, Zoback et al. [51] present a different method, based on the Coulomb fracture criterion, with the coefficient of friction μ, to calculate the maximum value of σ_u at which the borehole fails:

$$\sigma_u^{max} = \sqrt{1+\mu^2}\sqrt{\left(\frac{\sigma_\theta - \sigma_r}{2}\right)^2 + \tau_{r\theta}^2} - \mu\left(\frac{\sigma_\theta - \sigma_r}{2}\right)$$

2.3.2.3 Empirical Estimation of the Depth of Excavation Induced Damage

Several authors, e.g., Martin [11], Diedrichs [52], Hoek and Martin [53], suggest an empirical determination of the depth of an excavation damaged zone (EDZ). Particularly, in deep geological repositories for nuclear waste, the EDZ is of great interest since microfractures pose potential fluid pathways. Therefore, investigations were carried out mainly in granitoid rocks [54]. Former publications from Martin et al. [55] use a relationship to determine the depth of failure, calculated from the cavity center. Figure 2.7, altered by the terms used in this thesis, presents the modern empirical version after Hoek and Martin [11], which determines the depth of the excavation zone d_f directly.

The ratio of breakout depth d_f and the radius (a) of the circular cavity is a linear function of the maximum tangential stress $\sigma_{\theta,max}$ divided by the uniaxial compressive strength σ_u. From the empirical line, Hoek and Martin [11] derive a slope of 1.25 and a y-intercept of –0.51 (Fig. 2.7).

The x-intercept of the line indicates that failure already starts at a $\sigma_{\theta,max}$ of $0.4*\sigma_u$, which according to Diederichs [53] is the approximate crack initiation threshold (Sect. 2.1.3) of rocks, which is between $(0.3–0.5)*\sigma_u$.

2.4 Mechanical Data from the Weißjura-Group

The petrophysical parametrization of the host rocks in the geothermal industry is good practice, and key parameters such as porosity, permeability, thermal conductivity or

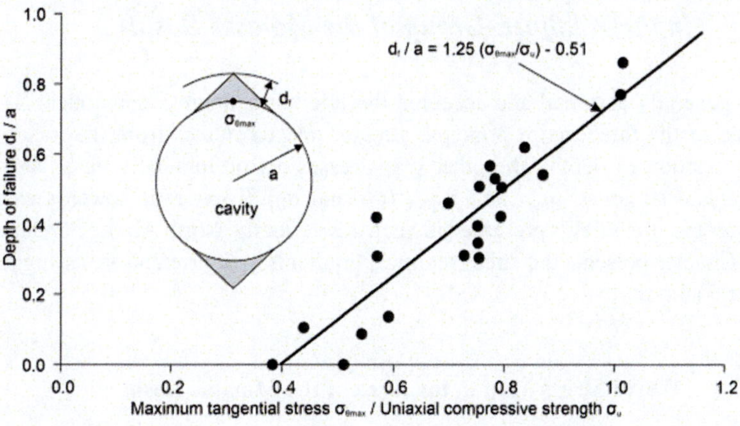

Fig. 2.7 Empirical estimation of depth of failure / breakout depth around circular cavities based on the ratio of the maximum tangential stress at the cavity wall with the uniaxial compressive strength (modified after Hoek and Martin [11]:297, Fig. 2.14)

diffusivity, and specific heat capacity play an important role in the success of a project. Still, rock mechanical data is barely collected, although Sect. 1.2 states that the importance of rock mechanical parameters grows.

2.4.1 Rock Mechanical Properties

Up to this point of this thesis, four major rock mechanical parameters are introduced that have a major influence on the rock's mechanical behavior: The Young's modulus and the Poisson's ratio, both elastic material parameters, the tensile strength, and the uniaxial compressive strength. Hitherto, the publications by Todera et al. [56], Homuth et al. [57] as well as Hedtmann and Alber [58] dealt with the measurement of those parameters for carbonates of the Weißjura-Group. They follow an approach to collect rocks from outcrops, earlier introduced as analog samples, that are stratigraphically equivalent to the host rock. However, all authors concluded that these values fluctuate strongly, mainly due to great deviations within the batch of samples such as fractures, porosity, depositional environment, history, and mineral structure. These factors are circumvented by the approach of Mraz et al. [59], who chose analog samples by their lithological composition from cuttings of the geothermal well GEN-1 and accordingly selected the outcrops, which are illustrated in Fig. 3.12, Sect. 4.1.1. Additionally, for the first time, the rock mechanical results on analog samples can be compared and validated with in situ rock specimens from a depth of over 4500 m TVD.

2.4.2 Stress Regimes and Stress Indicators in the Weißjura-Group of the Molasse Basin

It is generally assumed and accepted that the vertical stress component S_v (or σ_v) is one of the three major principal stresses and calculates from the rock's weight as a function of depth. The other two stresses act orthogonally on S_v and are the horizontal stresses S_{Hmax} and S_{hmin} (σ_H and σ_h). Their ratio towards each other defines the stress fields presented in Sect. 2.2.2. For the North Alpine Foreland Basin, this Chapter presents the stress regimes, gradients, and orientation, as suggested by several authors.

2.4.2.1 World Stress Map in the Area of the Molasse Basin

The World Stress Map (WSM) encompasses worldwide stress orientations and magnitudes of the maximum horizontal stress (S_{Hmax}). This data, collected from earthquake focal mechanisms, borehole breakouts, drilling-induced tensile fractures,

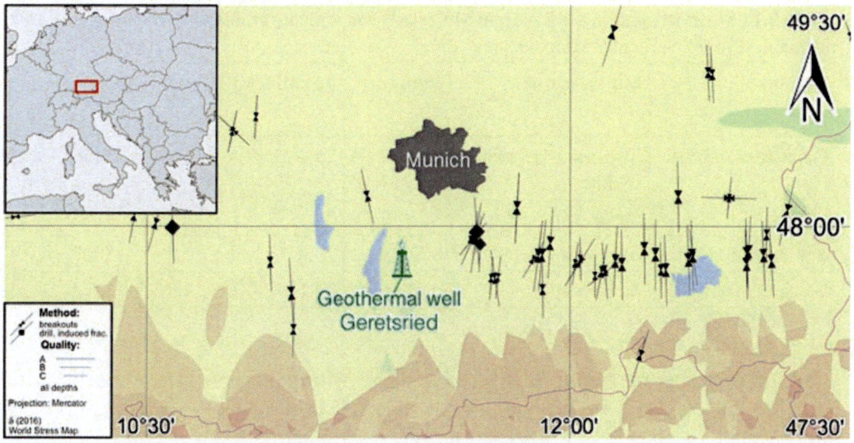

Fig. 2.8 Excerpt of the World Stress Map from the area of Southeast Germany with the main horizontal stresses, obtained by breakouts and drilling induced fractures, striking predominantly NS. (modified after Stockinger et al. [41], with data from Heidbach et al. [61])

hydraulic fracturing, geological indicators, and overcoring, accumulate to a total of roughly 43,000 entries [20]. Thirty-two thousand five hundred entries are classified into a quality ranking system from A-C, meaning that the direction of S_{Hmax} is certain within the range of ± 15°, ± 20°, and ± 25°, respectively.

Altered from an export of the CASMI (Create A Stress Map Interactively, after Heidbach and Höhne [60], with data from Heidbach et al. [61], Fig. 2.8 shows 55 S_{Hmax} stress orientations in Munich's southern area with the length of scalars according to their data quality from A-C, obtained by borehole breakouts and drilling-induced fractures. Section 2.4.2.3 describes both types of failure connected with the in situ stresses, plus the effect of core disking and geological indicators. In general, the scalars indicate that the orientation of S_{Hmax} strikes NS. Although there are no data points available around the area of Geretsried specifically (see Geothermal Well in Fig. 2.8), stress orientations to its east and west both show coinciding NS striking of S_{Hmax}. However, Budach et al. [62] state that most of the data from the WSM is from the Tertiary strata. Data on stresses from the Mesozoic sediments is individually examined by further authors.

2.4.2.2 Review on existing Stress Data for the Bavarian Molasse Basin

Besides information from the World Stress Map, many authors have contributed to estimating the stress gradients in the deep molasse basin for Mesozoic sediments. All researchers agree that the stress regime in the molasse sediments and the Jurassic carbonate rocks below is a strike-slip (SS) regime—likely close to a normal faulting (NF) regime. However, they disagree in terms of stress gradients, mainly since different methods and input parameters apply for their determination.

Table 2.1 Stress regimes in the Bavarian Molasse Basin with the gradients of the principal stresses, normalized to the vertical principal stress

Authors	Stress regime	Depth (m)	S_v (MPa/km)	Ratio, with $S_v = 1$		
				S_{Hmax}	S_{hmin}	p_f
Budach et al. [62]	Strike-slip/normal faulting	3300	25.2	1	0.50	0.38
	Strike-slip			2.1	0.76	
Seithel et al. [65]	Strike-slip	4000	23.0	1.2	0.67	0.4
Backers et al. [66]	Strike-slip	–	24.1	1.7	0.72	0.39

After a series of minor seismic events in Munich's southern suburbs, Unterhaching, Megies and Wassermann [63] derived—from focal mechanisms of three microseismic events—a predominant left-lateral strike-slip regime with sub-vertical dipping faults. The events are located in the crystalline basement below the Mesozoic strata, where fault orientations strike with N29E to S35W. Although these faults strike slightly steeper, this analysis still fits well with the indenter model of Ratschbacher et al. [64] (Sect. 1.4.1). Budach et al. [62] argue that the potential strike-slip stress regime is limited to the border of a NF stress regime. They narrow the coefficient of friction down to a range of 0.75–0.9, which resulted in a gradient for a normal faulting stress regime of $S_v = 25.2$, $S_{Hmax} = 25.2$, $S_{hmin} = 12.7$ and $p_f = 0.38$ Sv MPa/km and for a SS stress regime of $S_v = 25.2$, $S_{Hmax} = 52.2$, $S_{hmin} = 18.8$ and $p_f = 0.38$ Sv MPa/km (Table 2.1).

Ten kilometers south of the study site from Megies and Wassermann [63], Seithel et al. [65] derived from borehole breakouts and drilling-induced tensile fractures (DITFs) a SS stress regime in the three geothermal wells of Sauerlach where S_{Hmax} strikes NS. The stress magnitude for a depth of 4000 m, measured by Formation Integrity tests (FITs) and evaluated by the stress limitation method, determines a gradient of $S_v = 23$, $S_{Hmax} = 28$, $S_{hmin} = 15.5$ and $p_f = 0.4$ Sv MPa/km (Table 2.1). Additionally, two fracture systems were detected: the first is striking in a N-S to NNE-SSW direction and the second in an ENE-WSW direction while both dip by 70–90° [65].

Backers et al. [66] agreed with the SS stress regime and refined the considerations above by limiting both S_{hmin} from data of FITs, and the coefficient of friction to 0.4 and 0.45, which resulted in a magnitude of $S_v = 24.1$, $S_{Hmax} = 41.1$, $S_{hmin} = 17.3$ and $p_f = 0.39$ Sv MPa/km (Table 2.1).

2.4.2.3 Indicators and Quantifiers of in situ Stresses

The effects of high or differential in situ stresses often gleam with the occurrence of markers that indicate the orientation and magnitude of the principal stresses. This Chapter presents the four in situ stress markers, detected in the Geretsried geothermal wells.

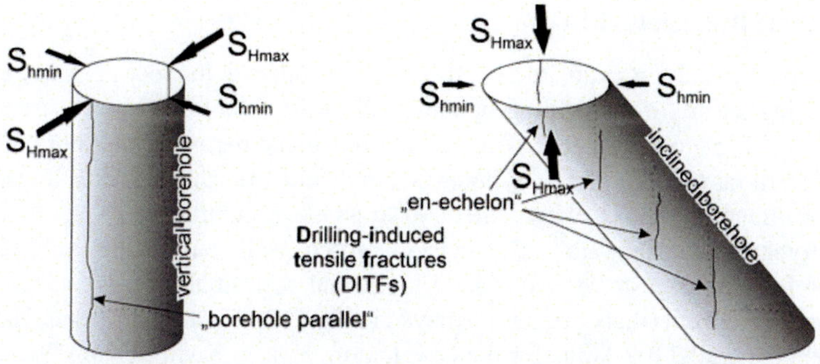

Fig. 2.9 Drilling-induced tensile fractures in a vertical borehole (right) and as en-echelon fractures in an oblique borehole (modified after Valley [67]: 85, Fig. 4.2)

Drilling Induced Tensile Fractures (DITFs)

Drilling induced tensile fractures occur when the $\sigma_{\theta,min}$ becomes smaller than the pressure within the well. The Mode I fracture initiates and always propagates parallel with the major principal stress and perpendicular with the least principal stress, allowing the determination of S_{Hmax} and S_{hmin}. Brudy and Zoback [45] differentiate these fractures clearly from artificially induced hydraulic fractures since they only penetrate the borehole wall for a shallow depth, and hence, the plastic deformation stabilizes the borehole. In a vertical borehole, these fractures appear in parallel, partly serrated lines that find themselves 180° to the opposite in the borehole. For inclined boreholes, these fractures appear at an oblique angle to the borehole axis and are named "en-echelon" fractures [45]. Brudy and Zoback [45] explicitly mention that the trace of a plane fault with a sinusoidal trace—which is not fully outlined—may be inadvertently mistaken with an en-echelon DITF or vice-versa. This can be a major source of error and should be checked in terms of mechanical plausibility and compared with the five properties below, which are characteristic for DITFs [45] (Fig. 2.9).

- Appear in pairs on the opposite sites of the borehole wall
- Parallel for vertical or oblique (en-echelon) for inclined boreholes
- Between 0.1 and 2.0 m in length
- Striking parallelly with the azimuth of S_{Hmax}
- Not perfectly straight, serrated lines

Thermal stresses, induced by a change in temperature, have a compressive effect upon cooling and act against the tensile motion. After Brudy and Zoback [45] the term, which must fall below zero to initiate DITFs, recapitulates to:

$$3\sigma_h - \sigma_H(+|\sigma_t|)(+p_f) - p_{w,max}^{frac} \leq 0$$

Borehole Breakouts (BBO)

The mechanical prerequisite for BBOs is introduced in Sect. 2.3.2.2. Besides opposing a risk for the drilling operation, they are a tool to measure stress orientations [51, 68, 69]. Figure 2.10 shows the schematic perimeter of a borehole, with the directions of S_{Hmax} and S_{hmin} being perpendicular to each other. Breakouts happen in the direction of S_{hmin}, where the hoop stress $\sigma_{\theta,max}$ culminates.

Borehole breakouts not only occur in vertical wells but also in inclined wells. Therefore, a change in the ratio of axial, tangential, and radial stresses lead to different forms of breakouts that occur, as displayed in Fig. 2.11. Figure 2.11 (a1 and b1) show borehole parallel breakouts for the case that $\sigma_\theta \geq \sigma_z \geq \sigma_r$, in case of Fig. 2.11 (a2 and b2) where $\sigma_z \geq \sigma_\theta \geq \sigma_r$ breakouts form in a spherical shape laterally around the borehole and in the third case, Fig. 2.11 (a3 and b3), $\sigma_z \geq \sigma_r \geq \sigma_\theta$ they follow a helical path as described by Fjær et al. [40].

Core Disking

Jaeger and Cook [70] first published the phenomenon of core disking, which describes the disintegration of drill cores, removed from a rock mass under high in situ stresses, into plates. The thickness of these plates or discs depends on the absolute stress

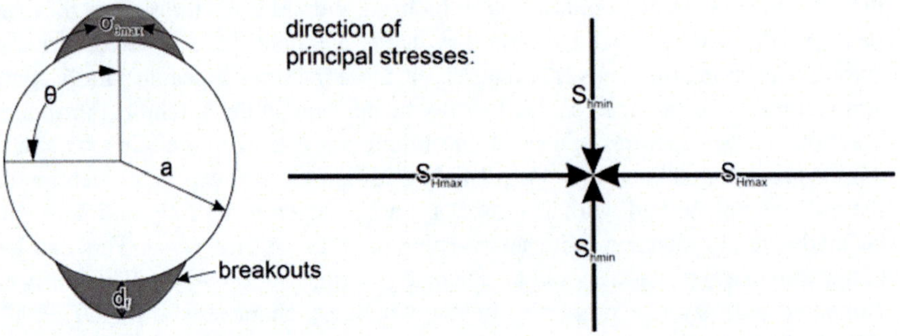

Fig. 2.10 Borehole breakouts at the point of maximum tangential stress ($\sigma_{\theta,max}$)

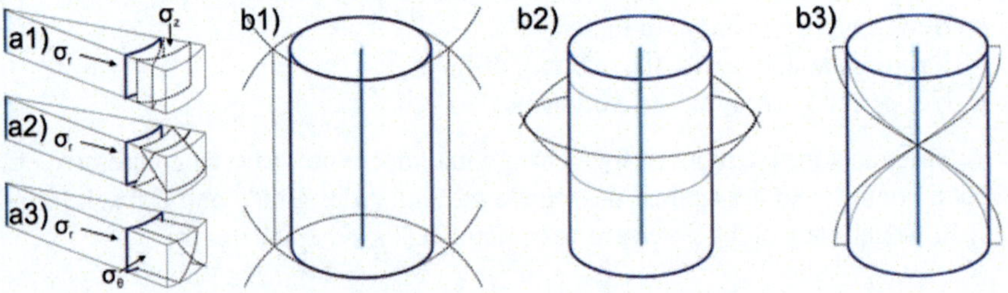

Fig. 2.11 Different forms of borehole breakout depending on the effective direction and ratio of radial stresses (σ_r), tangential stresses σ_θ and stresses parallel with the borehole σ_z (merged and modified after Fjær et al. [40]: 161, Figs. 4.16, 4.17 and 4.18)

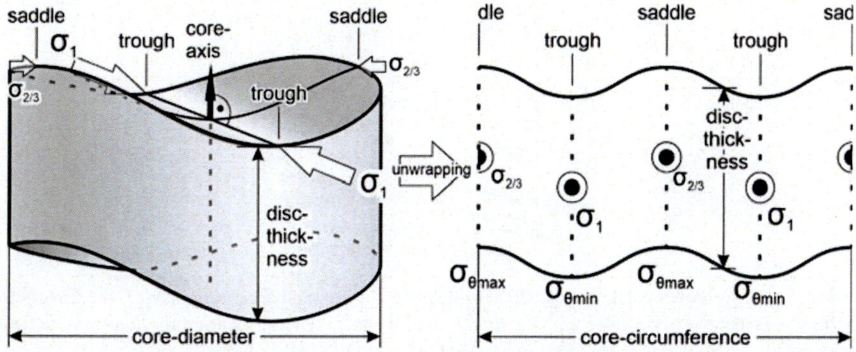

Fig. 2.12 Schematic saddle-shaped core disking (modified after Song [71]: 55, Fig. 2.15) phenomenon that is caused by high differential stress. The axis through the minima (troughs) is the direction of the major principal stress, the axis through the maxima (saddles) is a smaller principal stress. Unwrapping shows the position of the maximum tangential stress. The disc thickness is related to the absolute stresses

magnitudes. With increasing stress, the discs' thickness decreases [70]. Song [71] describes a particular feature of core disking, where the discs develop a horse-saddle shape. The curvature allows differentiating the direction of principal stresses (or radial stresses in case of an inclined well) (Fig. 2.12). Easily distinguishable from discontinuities, the circumference of the plane has two minima (troughs), which indicate the strike of σ_1, and two maxima indicating (saddles) the strike of σ_2 or σ_3, respectively (Fig. 2.12). Although not yet explicitly mentioned in the literature but consistently discussed in the previous Chapters, this pattern, when unwrapped, imitates the theoretical course of tangential stresses along the borehole wall.

Geological and Structural Stress Indicators

Geological and structural indicators are hardly applicable since wide surfaces with tectonic markers are not accessible in these depths. However, some microstructures on drill cores reveal structures that are a consequence of in situ stresses. Petit [72] published numerous descriptions of microstructures that indicate the sense of movements on planes in brittle rocks. These structures emerge when the shearing surface is insufficient to degrade the full amount of occurring shear stresses. These actual shear surfaces show striation in the direction of shearing, and fractures emerge to a slight angle against the shearing direction. After Petit [72] and Meschede [73], most commonly, these features are strongly lineated and stepwise fractured or have lunate breakouts as display in Fig. 2.13a and b. A second type is tensile fractures,

Fig. 2.13 Secondary fracture forms caused by shearing after Petit ([72]: 589, Fig. 2.1)

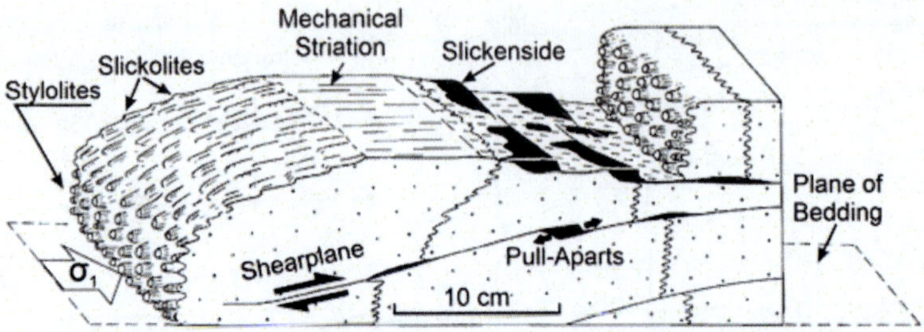

Fig. 2.14 Stylolites, slickolithes, slickensides and further phenomena indicating the direction of fossil or recent stresses, here with the direction of σ_1 and a dextral shearing motion (modified after Arthaud and Mattauer [74]: 739, Fig. 2.1) and Meschede ([73]: 121, Fig. 5.11)

as shown in Fig. 2.13c. These are secondary, crescent fractures. The concave side of the arc indicates the direction of shearing. Those secondary fractures are not healed by mineralization and often occur on bedding planes [72].

Stylolites are additional indicators for a paleo stress regime. Meschede [73] describes Stylolites as elongated, cone- or cylindrical-shaped bodies that can grow from millimeters to centimeters. Mostly found in carbonate rocks, they originate from pressure dissolution processes and concentrate along existing discontinuities, where material accumulates (e.g., bedding layers or faults). Horizons of stylolites are more or less perpendicular to the major principal stress σ_1. Slickolites, as shown in Fig. 2.14, are a slightly altered form of stylolites which appear on shearing planes with a sharp angle to σ_1. In some areas, recrystallization produces extremely smooth, mirror-like surfaces, so-called slickensides (Fig. 2.14).

Pull-Apart structures in between these stylolites can be compared to the fractured steps or lunate fractures from Fig. 2.14.

2.5 Application of Numerical Models

Although the analytical and empirical tools, as presented in Sect. 2.1, give a good approximation about the stability issues of cavities, they, on the one hand, completely neglect the most critical factor of a rock mass—the fracture network. On the other hand, they cannot be used to determine the absolute extend of the process of fracture initiation and propagation, as well as the full extent of the excavation damaged zone (EDZ). Knowledge about these properties in a rock mass becomes increasingly essential, particularly in the aspect of EGS utilization of deep geothermal boreholes. A state-of-the-art tool to evaluate fracture growth and further stability issues of rock mass are numerical models.

2.5.1 *Overview of Numerical Methods*

According to Jing and Hudson [75], three main governing methods exist to simulate the mechanical behavior of rocks and rock mass. These are:

- Continuum methods: the finite difference method (FDM), the finite element method (FEM), and the boundary element method (BEM)
- Discrete methods: the discrete element method (DEM), discrete fracture network (DFN) methods
- Hybrid continuum / discrete methods

Among the continuum methods, the FEM finds wide application in rock mechanical problems where fracture opening, or block detachment are of no significance [75]. The operating principle of this method is the discretization of an intact rock into elements. These elements deform when a force is applied—resulting in strain, which leads to failure when the elastic- or elasto-plastic deformation exceeds the material strength properties. However, the elements cannot disintegrate upon failure [76]. FEM can consider material heterogeneity, plastic deformation, complex boundary conditions, in situ stresses, and gravity [75] but also groundwater flow, heat, and radionuclide transport [77]. This approach's disadvantage is that the continuum's properties have to be adapted to the rock mass properties, using an empirically defined criterion, e.g., the Hoek–Brown-failure Criterion [78].

The DEM compensates for this lack of the FEM and transforms the continuum into a discontinuum by splitting the geometry into rigid (non-deformable) or deformable discrete blocks along predefined joints or discontinuities. These linear (2D) or planar (3D) elements allow finite displacement detachment and the recognition of new contacts [78]. However, the certainty about orientation, spacing, persistence, and strength, and stiffness parameters is the limiting factor in classifying these discontinuities. These parameters are hardly listed in the literature [76] and must be determined by extensive tests on discontinuities in outcrops or drill cores [79].

Hybrid methods are introduced by Jing and Hudson [75] as "*the FEM and DEM for the non-linear or fractured near-fields where explicit representations of fractures and/or non-linear mechanical behaviour, such as plasticity, are needed. This harmonizes the geometry of the required problem resolution with the numerical techniques available, thus providing an effective representation of the far-field to the near-field rock mass*". As a combination of the advantages of FEM and DEM, the simulation begins with a continuum. Upon meeting the failure criterion, the newly developed discontinuities become discrete elements [80]. ELFEN [81] and Y-Geo [80, 82] represent software applications of these hybrid methods. The Y-Geo Code constitutes the software used in this thesis—Irazu FEMDEM (Geomechanica).

2.5.2 Irazu FEMDEM (Geomechanica)

Mahabadi et al. [80] further developed the open-source code Y-2D Code from Munjiza et al. [82], and created the Y-Geo Code. The Software Irazu is the commercialized subsidiary of this code, which is distributed by the company Geomechanica Inc.

The continuum is discretized with three-noded triangular elements with four-noded cohesive (or crack) elements connecting each node of the neighboring triangular element (Fig. 2.15a). These three-noded elements represent the modeling domain's material properties and transform elastic deformation in constant-strain [83]. By exceeding the peak-strength of either tensional- or shear strength or in mixed mode, the strain is localized as a Fracture Process Zone (FPZ). Hence, the fracture propagation is controlled by the bonding stresses acting on the crack element (Fig. 2.15b). The response of these crack elements reacts accordingly to the cohesive tensile strength f_t and fracture energy G_{Ic}, resulting in Mode I type at failure, or after the Mohr–Coulomb criterion for f_s and fracture energy G_{IIc} for failure in Mode II. For mixed Mode I + II fracture, an elliptical coupling between opening (o) and slip (s) is applied (Fig. 2.15c). Once a cohesive element is broken, the residual friction is considered with the Mohr–Coulomb criterion for f_r (Fig. 2.15b) [83].

A Discrete Fracture Network (DFN) emulates existing discontinuities. In Irazu, it is implemented as a string of four-noded crack elements with different properties than the surrounding intact rock. These DFNs can either be employed as "cohesive" or "broken" elements. The same failure criterion applies for the cohesive DFN elements and the regular rock elements, while for the broken DFN, residual friction applies only [84]. The methodology section on Irazu in Sect. 4.4 lists and explains the source of all input parameters necessary to set up a model in Irazu.

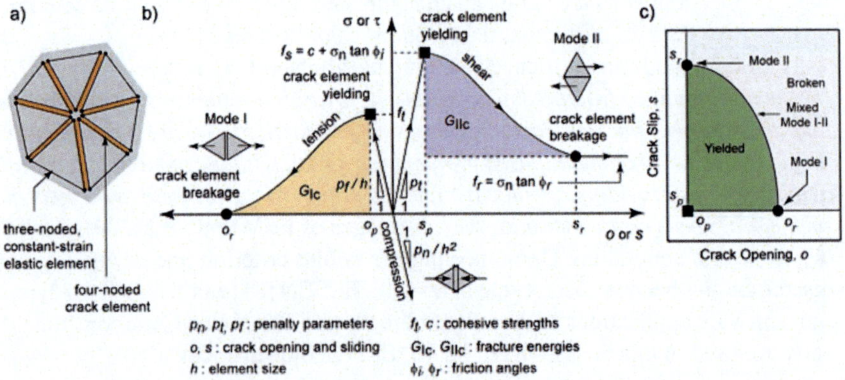

Fig. 2.15 Rock deformation and fracturing in FDEM. **a** Triangular elastic elements and four-noded crack elements represent the continuum. **b** Constitutive behaviour of the crack elements defined in terms of normal and tangential bonding stresses, σ and τ, versus crack relative displacements, o, and s (i.e., opening and slip). **c** Coupled relationship between o and s, for mixed-mode fracturing ([83]: 495, Fig. 2.1)

References

1. Landau LD, Lifshitz EM (1970) Theory of elasticity. vol 7. 2nd edn, Oxford, Pergamon Press
2. Fairhurst C (1971) Fundamental considerations relating to the strength of rocks. Veröff Inst. Bodenmechanik und Felsmechanik (Karlsruhe) 55:1–56
3. Griffith AA (1921) The Phenomena of rupture and flow in solids. Philosophical Trans Royal Society of London. Series A, Containing Papers of a Math Phys Character (1896–1934). 221(582):163–198
4. Griffith AA (1924) The theory of rupture. In: First international congress applications mechanics, pp 55–63
5. Coulomb CA (1776) Essai sur une application des regles maximis et minimis a quelques problems de statique, relatives a l'architecture. Acad Sci Paris Mem Phys 7:343–382
6. Mohr O (1900) Welche Umstände bedingen die Elastizitätsgrenze und den Bruch eines Materials? Zeitschrift des Vereins Deutscher Ingenieure 44:1524–1530
7. Labuz JF, Zang A (2012) Mohr–coulomb failure criterion. Rock Mech Rock Eng 45(6):975–979
8. Hoek E, Brown ET (1980) Empirical strength criterion for rock masses. J Geotech Engng Div 106:1013–1035
9. Hoek E, Brown ET (2018) The Hoek–Brown failure criterion and GSI—2018 edition. J Rock Mech Geotechn Eng 1–19
10. Jaeger JC, Cook NG, Zimmerman R (2007) In: Fundamentals of rock mechanics. 4th edn. Blackwell Publishing, pp 475
11. Hoek E, Martin C (2014) Fracture initiation and propagation in intact rock–a review. J Rock Mechan Geotechn Eng 6(4):287–300
12. Scholz CH (2019) In: The mechanics of earthquakes and faulting. 3rd edn. Cambridge, Cambridge University Press, pp 493
13. Martin CD, Chandler NA (1994) The progressive fracture of Lac du Bonnet granite. Int J Rock Mechan Mining Sci Geomech Abstracts 31(6):643–659
14. Mutschler T (2004) Neufassung der Empfehlung Nr. 1 des Arbeitskreises "Versuchstechnik Fels" der Deutschen Gesellschaft für Geotechnik e. V.: Einaxiale Druckversuche an zylindrischen Gesteinsprüfkörpern. Bautechnik 81(10):825–834
15. Bieniawski ZT (1967) Stability concept of brittle fracture propagation in rock. Eng Geol 2(3):149–162
16. Richart FE, Brandtzaeg A, Brown RL (1928) A study of the failure of concrete under combined compressive stresses. University of Illinois at Urbana Champaign. (College of Engineering), pp 103
17. Fossen H (2010) In: Structural geology. Cambridge, Cambridge University Press, pp 463
18. Cauchy A-L (1827) De la pression ou tension dans un corps solide. Ex. de math 2:42–56
19. Zang A, Stephansson O (2010 In: Stress field of the earth's crust. Dordrecht, Springer, pp 316
20. Heidbach O, Rajabi M, Cui X, Fuchs K, Müller B, Reinecker J, Reiter K, Tingay M, Wenzel F, Xie F, Ziegler MO, Zoback M-L, Zoback (2018) The World Stress map database release 2016: crustal stress pattern across scales. Tectonophysics 744:484–498
21. Anderson EM (1905) The dynamics of faulting. Trans Edinburgh Geol Soc 8(3):387–402
22. Anderson EM (1951) The dynamics of faulting and dyke formation with applications to Britain. Edinburgh (Oliver and Boyd), pp 206
23. Simpson RW (1997) Quantifying Anderson's fault types. J Geophys Res-Solid Earth 102(B8):17909–17919
24. Byerlee J (1978) Friction of rocks. Pure Appl Geophys 116:615–626
25. Zoback MD, Barton CA, Brudy M, Castillo DA, Finkbeiner T, Grollimund BR, Moos DB, Peska P, Ward CD, Wiprut DJ (2003) Determination of stress orientation and magnitude in deep wells. Int J Rock Mech Mining Sci 40(7–8):1049–1076
26. Alber M, Backers T, Bartmann K, Brenne S, Dinter S, Dresen G, Gipper P, Grühser C, Heft S, Meier T, Molenda M, Röske F, Stöckhert F (2015) Erforschung der Mechanismen und Simulation hydraulisch induzierter Risse in geklüfteten Gesteinen für die Optimierung des Aufschlusses geothermischer Lagerstätten. (Ruhr Universität Bochum, geomecon GmbH), pp 136

27. Terzaghi K (1923) Die Berechnung der Durchlässigkeitsziffer des Tones aus dem Verlauf der hydrodynamischen Spannungserscheinungen. Sitzungsberichte Akad Wiss Wien 132:125–138
28. Biot MA (1941) General theory of three-dimensional consolidation. J Appl Phys 12:155–164
29. Rice JR, Cleary MP (1976) Some basic stress diffusion solutions for fluid-saturated elastic porous media with compressible constituents. Rev Geophys Space Phys 14:227–241
30. Nur A, Byerlee J (1971) An exact effective stress law for elastic deformation of rock with fluids. J Geophy Res 76:6414–6419
31. Jaeger JC, Cook NG (1979) In: Fundamentals of rock mechanics. 3rd edn. New York, Chapman & Hall, pp 593
32. Zoback MD, Mastin L, Barton CA (1987) In situ stress measurements in deep boreholes using hydraulic fracturing, wellbore breakouts, and Stonely wave polarization. In: Rock stress and rock stress measurements, Stockholm, Centrek Publications, Lulea
33. Moos DB, Zoback MD (1990) Utilization of observations of well bore failure to constrain the orientation and magnitude of crustal stresses: application to continental deep sea drilling project and ocean drilling program boreholes. J Geophy Res 95:9305–9325
34. Kirsch (1898) Die Theorie der Elastizität und die Bedurfnisse der Festigkeitslehre. Zeitschrift des Vereines Deutscher Ingenieure 42:797–807
35. Hiramatsu Y, Oka Y (1962) Analysis of stress around a circular shaft or drift excavated in ground in a three dimensional stress state. J Mining Metallurgy Instit Japan 78:93–98
36. Hiramatsu Y, Oka Y (1968) Determination of the stress in rock unaffected by boreholes or drifts, from measured strains or deformations. Int J Rock Mech Mining Sci Geomech Abstracts 337–353 (Elsevier)
37. Haimson B, Fairhurst C (1967) Initiation and extension of hydraulic fractures in rocks. Soc Petroleum Eng J 7(03):310–318
38. Haimson BC, Fairhurst C (1969) In-Situ stress determination at great depth by means of hydraulic fracturing. In: Somerton WH (Ed) Rock mechanics—theory and practice. American Institute of Mining, Metallurgical, and Petroleum Engineers, pp 559–584
39. Peška P, Zoback MD (1995) Compressive and tensile failure of inclined well bores and determination of in situ stress and rock strength. J Geophys Res: Solid Earth 100(B7):12791–12811
40. Fjær E, Holt RM, Raaen AM, Horsrud P (2008) In: Petroleum related rock mechanics. vol 53. 2nd edn, Elsevier, pp 514
41. Stockinger G, Käsling H, Menschik F, Thuro K (2019) 3D rotation applied to in situ stress fields for 2D numerical modelling, borehole stability and drill core recovery in deep geothermal wells. In: Fontoura SD, Rocca RJ, Mendoza JP (eds) Rock mechanics for natural resources and infrastructure development—full papers. London, CRC Press, pp 3144–3151
42. . Haimson BC, Herrick CG (1989) Borehole breakouts and in situ stress. In: Rowley JC (ed) Drilling symposium 1989 [12th annual energy-sources technology conference and exhibition], vol 22. Houston (American Society of Mechanical Engineers, New York), pp 17–22
43. Feder G, Arwanitakis M (1976) Zur Gebirgsmechanik ausbruchsnaher Bereiche tiefliegender Hohlraumbauten. Berg- u. Huettenmaennische Monatshefte 121(4):103–117
44. Hubbert MK, Willis DG (1957) Mechanics of hydraulic fracturing. Petrol Trans AIME 210:153–168
45. Brudy M, Zoback MD (1999) Drilling-induced tensile wall-fractures: implications for determination of in-situ stress orientation and magnitude. Int J Rock Mech Mining Sci 36(2):191–215
46. Stock J, Healy J, Hickman S, Zoback M (1985) Hydraulic fracturing stress measurements at Yucca Mountain, Nevada, and relationship to the regional stress field. J Geophys Res: Solid Earth 90(B10):8691–8706
47. Moos D, Morin RH (1991) Observations of wellbore failure in the Toa Baja well - implications for the state of stress in the North Coast Tertiary Basin, Puerto Rico. Geophys Res Lett 18(3):505–508
48. Schmitt D, Zoback M (1989) Poroelastic effects in the determination of the maximum horizontal principal stress in hydraulic fracturing tests—a proposed breakdown equation employing a modified effective stress relation for tensile failure. Int J Rock Mech Mining Sci Geomech Abstracts 26(6):499–506

49. Shen B, Stephansson O, Rinne M (2002) Simulation of borehole breakouts using FRACOD2D. Oil Gas Sci Technol 57(5):579–590
50. Martin CD, Christiansson R (2009) Estimating the potential for spalling around a deep nuclear waste repository in crystalline rock. Int J Rock Mech Mining Sci 46(2):219–228
51. Zoback MD, Moos D, Mastin L, Anderson RN (1985) Well bore breakouts and in situ stress. J Geophys Res: Solid Earth 90(B7):5523–5530
52. Martin CD (1997) The 17th Canadian geotechnical colloquium: the effect of cohesion loss and stress path on brittle rock strength. Can Geotech J 34:698–725
53. Diederichs MS (2007) The 2003 CGS Geocolloquium adress: damage and spalling prediction criteria for deep tunneling. Can Geotech J 44:1082–1116
54. Perras MA, Diederichs MS (2016) Predicting excavation damage zone depths in brittle rocks. J Rock Mech Geotech Eng 8(1):60–74
55. Martin CD, Kaiser PK, McCreath DR (1999) Hoek-Brown parameters for predicting the depth of brittle failure around tunnels. Can Geotech J 36(1):136–151
56. Tondera D, Klapperich H, Blöcher G, Moeck I, Steiger T, Bems C, Hild S (2013) Geothermie-Forschungsprojekt" Allgäu 2.0"-Forschungsansätze, Laboruntersuchungen & Planungsschritte. In: Thuro K (ed.): 19. Tagung für Ingenieurgeologie mit Forum für junge Ingenieurgeologen und Fachausstellung, Technische Universität München, Fachsektion Ingenieurgeologie, Deutsche Gesellschaft für Geotechnik, 13.-16. März 2013, München (TUM), pp 285–293
57. Homuth S, Götz AE, Sass I (2014) Lithofacies and depth dependency of thermo-and petrophysical rock parameters of the Upper Jurassic geothermal carbonate reservoirs of the Molasse Basin. Zeitschrift der Deutschen Gesellschaft für Geowissenschaften 165(3):469–486
58. Hedtmann N, Alber M (2017) Investigation of water-permeability and ultrasonic wave velocities of German Malm aquifer rocks for hydro-geothermal energy. Proc Eng 191:127–133
59. Mraz E, Bohnsack D, Stockinger G, Käsling H, Zosseder K, Thuro K (2018a) Die Bedeutung von Analogaufschlüssen des Oberjura für die Interpretation der Lithologie der geothermalen Tiefbohrung Geretsried. Jahresberichte und Mitteilungen des Oberrheinischen Geologischen Vereins 100:517–548
60. Heidbach O, Höhne J (2008) CASMI—a visualization tool for the world stress map database. Comput Geosci 34:783–791
61. Heidbach O, Rajabi M, Reiter K, Ziegler MO, WSM Team (2016) World stress map database relese 2016. *V 1.1*. GFZ Data Services. https://doi.org/10.5880/WSM.2016.001
62. Budach I, Moeck I, Lüschen E, Wolfgramm M (2018) Temporal evolution of fault systems in the Upper Jurassic of the Central German Molasse Basin: case study Unterhaching. Int J Earth Sci 107(2):635–653
63. Megies T, Wassermann J (2014) Microseismicity observed at a non-pressure-stimulated geothermal power plant. Geothermics 52:36–49
64. Ratschbacher L, Frisch W, Linzer H-G, Merle O (1991) Lateral extrusion in the eastern Alps, Part 2: structural analysis. Tectonics 10(2):257–271
65. Seithel R, Steiner U, Müller B, Hecht C, Kohl T (2015) Local stress anomaly in the Bavarian Molasse Basin. Geothermal Energy 3(1):22
66. Backers T, Meier T, Gipper P, Munsch P, Bücken D, Nokar K (2017) Abschlussbericht zum Teilprojekt B: Struktur- und Spannungsfeld im Verbundprojekt MAFA: Parametrisierung von Fazies, Diagenese, Struktur- und Spannungsfeld sowie Optimierung der Testabläufe im Malm zur Verringerung des Erfolgsrisikos. Berlin (geomecon GmbH), pp 44
67. Valley BC (2007) The relation between natural fracturing and stress heterogeneities in deep-seated crystalline rocks at Soultz-sous-Forêts (France). Dissertation, Group of Engineering Geology, ETH Zurich, Zurich, pp 245
68. Bell JS, Gough DI (1979) Northeast-soutwest compressive stress in Alberta: evidence from oil wells. Earth and Planet Sci Lett 45:475–482
69. Zoback ML, Zoback M (1980) State of stress in the conterminous United States. J Geophys Res: Solid Earth 85(B11):6113–6156
70. Jaeger J, Cook N (1963) Pinching-off and disking of rocks. J Geophys Res 68(6):1759–1765

71. Song I (1998) Borehole breakouts and core disking in Westerly granite: mechanisms of formation and relationship to in situ stress. Dissertation, University of Wisconsin, Madison, pp 207
72. Petit JP (1987) Criteria for the sense of movement on fault surfaces in brittle rocks. J Struct Geol 9:597–608
73. Meschede M (1994) *Methoden der Strukturgeologie : ein Leitfaden zur Aufnahme und Auswertung strukturgeologischer Daten im Gelände und im Labor.* Stuttgart (Enke;), pp 169
74. Arthaud F, Mattauer M (1969) Exemples de stylolites d'origine tectonique dans le Languedoc, leurs relations avec la tectonique cassante. Bulletin de la Société Géologique de France S7-XI(5):738–744
75. Jing L, Hudson JA (2002) Numerical methods in rock mechanics. Int J Rock Mech Mining Sci 39(4):409–427
76. Stead D, Eberhardt E, Coggan JS (2006) Developments in the characterization of complex rock slope deformation and failure using numerical modelling techniques. Eng Geol 83(1):217–235
77. Chan T, Scheier N (1987) Finite-element simulation of groundwater flow and heat and radionuclide transport in a plutonic rock mass. In: 6th ISRM Congress, international society for rock mechanics and rock engineering
78. Barla G, Barla M (2000) Continuum and discontinuum modelling in tunnel engineering. Rudarsko-geološko-naftni zbornik 12:45–57
79. Barton N, Choubey V (1977) The shear strength of rock joints in theory and practice. Rock Mech 10(1):1–54
80. Mahabadi OK, Lisjak A, Munjiza A, Grasselli G (2012) Y-Geo: new combined finite-discrete element numerical code for geomechanical applications. Int J Geomech 12(6):676–688
81. Rockfield Software Ltd (2004) ELFEN 2D/3D numerical modelling package. Swansea, UK, Rockfield Software Ltd
82. Munjiza A, Bangash T, John NWM (2004) The combined finite–discrete element method for structural failure and collapse. Eng Fracture Mech 71(4):469–483
83. Lisjak A, Figi D, Grasselli G (2014) Fracture development around deep underground excavations: insights from FDEM modelling. J Rock Mech Geotech Eng 6(6):493–505
84. Lisjak A, Tatone B, Mahabadi OK, Kaifosh P (2019b) Irazu 2D theory manual. Toronto (Geomechanica), pp 70
85. ISRM (1999) Draft ISRM suggested method for the complete stress-strain curve for intact rock in uniaxial compression. Int J Rock Mech Mining Sci 36:279–289
86. ASTM D7012-14 (2014) Compressive strength and elastic moduli of intact rock core specimens under varying states of stress and temperatures. ASTM International, West Conshohocken, PA www.astm.org

Chapter 3
The BMWi Project "Dolomitkluft" and the Study Site

Initiated in 2015, the joint project "Dolomitkluft", funded by the Federal Ministry for Economic Affairs and Energy (BMWi), started in May 2016. The project partners comprise the companies Enex Geothermieprojekt Geretsried Nord GmbH & Co. KG., Enex Power Germany GmbH, Geothermie Neubrandenburg (GTN), and G.E.O.S. Ingenieurgesellschaft mbH and the research partners: Leibniz Institute for Applied Geophysics (LIAG) and the Technical University of Munich who pursue the goal to develop, test, and analyze a discontinuity-governed dolomitic aquifer. Inseparably, "Dolomitkluft" is connected with the well site Gut Breitenbach, around 600 m asl, which belongs to the municipality of the city Geretsried as part of the industrial area of Gelting. The geothermal site's concession area belongs to the company Enex Geothermieprojekt Geretsried Nord GmbH & Co. KG (ENEX). The first well GEN-1, realized in 2013, was the longest and deepest geothermal well in Europe with a length of 6036 m MD and a depth of 4840 m TVD. Although temperatures in the borehole deepest of around 165 °C exceeded expectations, the flow rate of 10 l/s fell short of predictions made at 150–160 l/s, and hence the well was of no commercial use [15]. In 2017, tied to the project BMWi Project "Dolomitkluft", a new approach was made, accessing a fractured rock formation rather than a matrix/porosity driven aquifer, which also required a new well path trajectory. This well, named GEN-1ST-A1, was deflected from GEN-1 in a depth of 4200 m as a sidetrack. The length of the sidetrack comprises 5963 m MD and a final depth of 4735 m TVD. In addition to conventional drilling, the project also included funding for drill cores, which the project partner Leibniz Institute for Applied Geophysics (LIAG) contributed to geomechanically characterize the rock mass. Ultimately, from 300 m initially planned coring length, 20 m could have been recovered in seven core runs.

G. M. Stockinger, *Fracturing in Deep Boreholes*, Springer Theses,
https://doi.org/10.1007/978-3-030-94569-5_3

3.1 Technical Execution of the Wells GEN-1 and GEN-1ST-A1

The first well GEN-1 started on January 13th, 2013, and took 228 days to complete. The borehole was divided into 5 sections, including the open hole section. Starting with a 26″ inch hole, rejuvenating to an open hole diameter of 6 1/8″. The open hole section of GEN-1 starts in a depth of 4497 m MD (4337 m TVD) and ends at 6036 m MD (4852 m TVD), covering a length of more than 1500 m and a vertical distance of 515 m.

The second well, the sidetrack GEN-1ST-A1, started on April 18th, 2017, and took 198 days to complete. After a long delay due to a drill bit loss, milling was conducted until a whipstock, placed in 4253 m MD, allowed to proceed drilling with an 8 1/2″ convectional drill bit. Finally, open hole drilling was conducted with a 6″ drill bit. The coring operations, which are subjected to in Sect. 3.3.2.1, slowed down the drilling process over extended periods. The open hole section of GEN-1ST-A1 starts in 4484 m MD (4346 m TVD) and ends at 5963 m MD (4735 m TVD). Hence, the well extends over a length of 1480 m and a vertical distance of 390 m.

3.2 Well Trajectories of GEN-1 and GEN-1ST-A1

Figure 3.1 shows the trajectories of both the wells, adopted from [1]. To emphasize the deflection of the well, each trajectory is colored according to the inclination of the drill path. The purple color indicates vertical inclination, with a shift to blue, green, yellow, and finally red, representing an almost horizontal inclining well. Both wells share a vertical borehole from the start, which inclines to a maximum of 40° and strikes in an azimuthal direction of approx. 60°. In a depth of 4200 m, after the

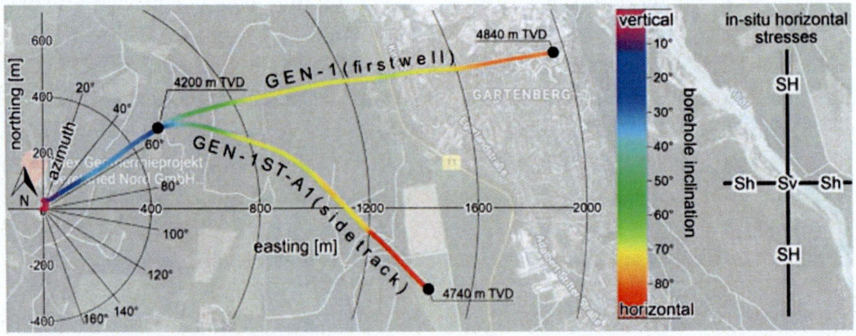

Fig. 3.1 Drill trajectory of the first geothermal well GEN-1 and its sidetrack GEN-1ST-A1. The drill paths are colored according to their inclination. An inclination of 0° (purple) represents a vertical well, and an inclination of 90° describes a horizontal well (red). The color transition implies the increasing inclination of the boreholes (modified after [1], background: maps.google.de)

sidetrack (GEN-1ST-A1) branches off, the first well (GEN-1) constantly strikes 80° in an ENE direction and increases its inclination to 80° over a length of 1700 m. In divergence, the sidetrack takes a sharp turn to a SE direction, changing its azimuthal direction from 80° to 135° over a length of 1350 m, finally reaching an inclination of 85°.

3.3 Data from "Dolomitkluft"

Although the project "Dolomitkluft" focused on the evaluation of the sidetrack GEN-1ST-A1, data from the main well GEN-1 was made available from the project coordinator and site owner ENEX in order to get better insights into the geological situation. Furthermore, a comparison of geological attributes and mechanical indicators and aspects between two wells reused data from GEN-1.

The geomechanical investigations examine facies and lithology, HMI logs, and, if available, drill cores and the location of fault zones for both wells. This thesis deals with the carbonate reservoir and therefore, only considers the sections of the open hole. Figure 3.2 gives a detailed overview of the data available and used in this work's further course.

3.3.1 GEN-1

The green line in the middle column of Fig. 3.2 shows the 3D trajectory of the well GEN-1. To the right in Fig. 3.2 plots the facies, the lithology, and the HMI log. The next paragraph contains the facies classification from [2] and the lithological description from [3]. ENEX kindly provided the data from the HMI log along with the accompanying report from the company Erdwerk [4].

According to the facies classification, three specific areas can be distinguished (compare Fig. 3.2). Firstly, the Purbeck facies, which marks the transition from Cretaceous sediments to Jurassic sediments, is lithologically dolostone. Secondly, there are two separate regions of platform carbonates in 4600 to 4800 m MD and 5050 to the end of the well in 6036 m MD. Both differ lithologically in mineral content and color. While the upper part consists of light limestone, the lower part transits from a dark limestone to a Dolostone in 5730 m MD back to dark limestone in 6000 m MD. According to [2], these sediments can be compared to the "Treuchtlingen Formation" with thick-bedded to massive sediments, partly dolomized (comp. Section 1.4.2.2). Thirdly, a platform and basin-dominated depositional environment separates the platform carbonates. Lithologically, these micritic limestones—despite their dark color—are comparable with the Solnhofener Limestone.

Schubert [4] from the commissioned company Erdwerk examined the image logs. From 203 interpreted bedding planes, he derived a mean dip of 11°, with maxima up to 20° in the direction of SW-SE—which Schubert ([4]: 15) claimed is a typical value

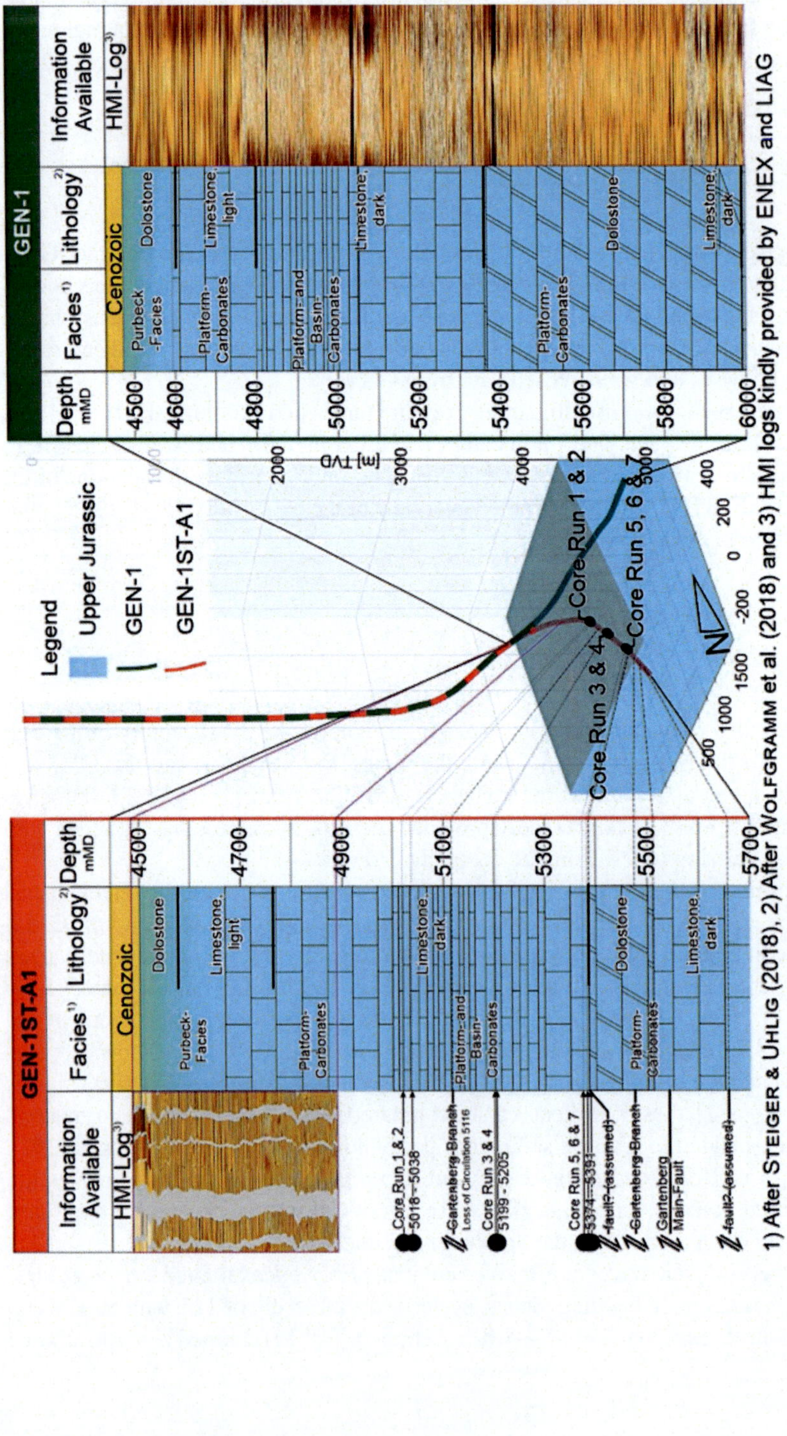

Fig. 3.2 Overview of the geothermal wells in Geretsried. The drill trajectories are illustrated in the middle column: the main well GEN-1 in green, the sidetrack GEN-1ST-A1 in red. Both share the same path up to 4200 m TVD. From there, the left and right column illustrate the available data for each well. Right: For GEN-1 data is available for the whole drill string on (1) facies from [2], on (2) lithology from [3], and (3) an HMI log, kindly provided by ENEX. Left: For GEN-1ST-A1, data is available for the whole drill string for (1) facies from [2], for (2) lithology from [3], complemented by rock cores (black dots) and faults (oppositional arrows). Data for (3) the HMI log kindly provided by ENEX and LIAG is available up to 4888 m MD

for the Mesozoic sediments of the Weißjura-Group. He argues that a slight change in the depositional environment changes the dip direction 4500 m MD to 4770 m MD from SE to SW in 4770 m MD to 5650 m MD until it changes back in a SE direction. This change in direction in 4770 m MD fits perfectly with the results from [2], who interpreted the transition from platform to basin carbonates. The change in direction may also allow conclusions on an actual basin geometry.

Schubert ([4]: 16f) found 70 surfaces in the image log that he identified as discontinuities—but due to the bad image quality, he argues that even more joints or faults exist. The mean orientation of these discontinuities strikes 170° N–S. However, he also mentions that due to the E-W trajectory of GEN-1, E–W striking discontinuities may be statistically underrepresented. All discontinuities incline steeply, almost vertical with a mean dip of 78°. A dense frequency of faults is located between 4770 m MD and 5650 m MD—and seems again connected with the appearance of the depositional basin structure.

Schubert ([4]: 17) also found Borehole Breakouts (BBOs), which he mapped due to their symmetric appearance along the borehole. Although karstified structures did not allow a safe distinction from the breakout structures, he determined a N–S striking S_{Hmax} direction from mainly E–W oriented BBOs. In particular, BBOs accumulate between 5850 and 6000 m MD, where he assumes a fault zone.

However, a second look at the logs shows that these BBOs are not as obvious as it may seem at first sight. Given reasons emphasize why this thesis reinvestigates the BBOs:

- E–W oriented BBOs are less likely in an E–W oriented, nearly horizontal well
- BBOs appear in the top and bottom, some at the sidewalls and some on the hole circumference
- the possibility of structural failure due to the fracture network was not considered
- the resistivity log—not used in the interpretation—apparently corresponds to the BBOs
- Presumably, BBOs might concentrate at the transition of facies as shown in Fig. 3.2.

The newly interpreted BBOs from GEN-1 are part of the results and discussion section of this dissertation (Sects. 5.3.2 and 6.5.3.1).

3.3.2 GEN-1ST-A1

The red line in the middle column of Fig. 3.2 shows the 3D trajectory of the well GEN-1ST-A1. The left side of Fig. 3.2 shows the depth allocation of different types of facies and lithologies and their description from [2, 3]. ENEX and LIAG kindly provided the HMI log, presented in Sect. 5.3.2.

In addition, drill cores, obtained from various depths of the reservoir, allow a deeper insight into the geology and geomechanical properties (Fig. 3.2, black dots).

Table 3.1 Located faults along the geothermal well GEN-1ST-A1 after [3], complemented by results from [5]

Depth m MD	Faults after [3] *after [5]	Orientation after [5]		Remark
		Orientation	Dip	
5116	Gartenberg-Branch-Fault 2a (North*)	E10S	60–65	Total mud loss
5391	Fault-Assumed?	No data		
5475	Gartenberg-Branch-Fault	E10S	60–65	
5544	Gartenberg-Main-Fault (South*)	E10N		
5658	En-echelon with Gartenberg-Main-Fault?			

Section 5.2 gives an overview of the drill cores and presents the investigations carried out on these cores.

The Purbeck facies starts with dolostone and changes to light-colored limestone to the bottom, where it transits to the facies of platform carbonates. A change of color from light to dark limestone indicates an anoxic milieu although the depositional environment remains the same [2]. This environment changes at around 5000 m MD where, slightly deeper than in GEN-1, the platform-basin carbonate facies starts. This facies is well-developed up to 5300 m MD, where the platform carbonates reoccur. In a range from 5390 m MD to 5550 m MD the dark limestone of the platform carbonates is intercepted by Dolostone. Equivalent to GEN-1, [2] relates the platform carbonates with the "Treuchtlichen Formation" and the platform-basin carbonates with the "Solnhofen-Formation".

The path of GEN-1ST-A1 intersects several faults. Table 3.1 lists the faults according to their depths and mud loss from [3] as well as their orientation from [5].

3.3.2.1 Rock Cores

The introduction to this Chapter briefly notes that the initial planning intended to recover 300 m of drill core within one week, starting at 5020 m MD. The company "ALS Corpro" (now Reservoir Group) was commissioned for the core drilling operation. The most recent technology, a so-called "Triple Tube TSS", a three-barrel system, was used to recover the cores. The three parts of this system comprise the outer barrel, the inner barrel, and the liner. The outer barrel, which mainly ensures the stability of the drill string, carries the core head, which has a parabolic profile, 12 blades, and an inner diameter of 3.5 inches, and an outer diameter of 6 inches (Fig. 3.3a). The inner barrel is inserted into the outer barrel and can rotate freely, eliminating any damage from rotation as no torque is transmitted to the rock core material. An aluminum half-moon liner (displayed in Fig. 3.3d) coated with Teflon completes the inner barrel setup. This type of liner has a lower friction coefficient than, e.g., a steel inner barrel and thus eases the core entry into the string, theoretically reducing the occurrence of core jamming [16]. Besides, the half-moon liner

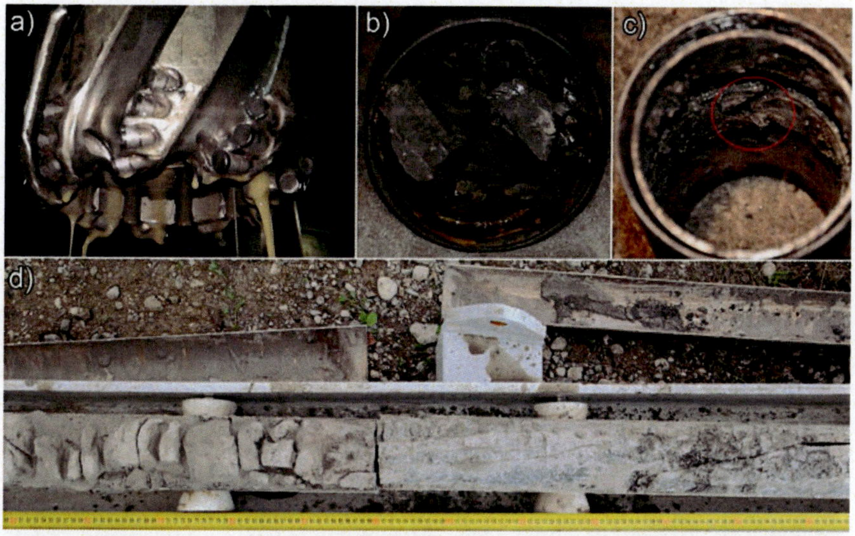

Fig. 3.3 Impressions of the drilling; **a** drill bit used for coring (Photo kindly provided by Lukas Rainer; **b** jammed core barrel with disintegrated rock core material; **c** piece of rock core jammed in the core catcher (red circle), **d** rock cores after removal of the halfmoon liner, covered in drill mud

allows the visual examination of the core right at the drill site without causing any further damage to the core.

In contrast to 300 m of cores expected, only 18.73 m of cores were recovered in seven individual core runs. Table 3.2 shows an overview of these core runs. None of the core runs finished successfully since core jams led to the abort of every single

Table 3.2 Summarized results from the coring operation, sorted in the chronological core runs, with the core boxes, the total length cored, the recovered length, the total gain, and the loss for each core run

Core run-no	MD (m)	Core box no	Length cored (m)	MD of length recovered (m)	Gain (m)	Loss (m)
1	5018.18–5035.92	1	17.74	5018.18–5019.3	1.12	−16.62
2	5035.80–5037.35	1	1.55	5035.80–5036.51	0.71	−0.84
Drilling was continued conventionally from 5037.35 m MD to 5199.0 m MD						
3	5199.0–5204.0	2–4	5.0	5199.0–5203.0	4.0	−1.0
4	5204.1–5205.37	4	1.27	5204.1–5205.1	1.0	−0.27
Drilling was continued conventionally from 5205.37 m MD to 5374.0 m MD						
5	5374.0–5378.0	5	4.0	5374.0–5375.0	1.0	−3.0
6	5378.0–5388.0	5–9	10.0	5378.0–5387.0	9.0	−1.0
7	5388.0–5390.9	10	2.9	5388.0–5389.9	1.9	−1.0
Total			42.46		18.73	23.73

core run. The middle top part of Fig. 3.3b shows a jammed core barrel, where the rock is completely fractured and wedged. In Core Run 1, this jamming was not detected, and hence, from a total length of 17.74 m, only 1.12 m of core were recovered while 16.62 m were milled and probably washed out by the drilling mud. In CR 1, a piece of rock stuck in the core catcher presumably caused the core jam (Fig. 3.3c). The core drilling was interrupted between runs 2 and 3 and 4 and 5 by conventional drilling but jamming continuously occurred throughout all further core runs.

The half-moon liner was detached from the inner barrel, and the upper half was removed for visual inspection of the recovered cores. To distinguish "top" and "bottom", the core was marked with two parallel lines of red and black waterproof pens along the axis. With a look from bottom to top, the red line is always on the right, and the black line is, correspondingly, to the left. This is also marked at the very ends of each core box, as seen in Fig. 3.4. Finally, the core was cut into pieces of 1 m length and stored in the core boxes.

Figure 3.4 shows the drill cores from the seven core runs, as described above. These cores are subject to multiple examinations, such as ultrasonic measurements, computer tomography scanning, and further static strength testing. Static strength testing and the resonance US method require a specific specimen geometry, a cylindrical specimen with a length to diameter ratio of at least 2:1 (compare Sect. 4.2). The cores in their original state did not meet this requirement. Hence, the cores had to be overcored with a smaller diameter, covered in Sect. 4.2.1.2. Figure 3.4 illustrates the overcored segments with either 2 cm (orange) or 5 cm (yellow) core drill bits.

Ultrasonic testing encompasses the determination of the P-wave and S-wave velocity of the rock and allows the derivation of dynamic elastic parameters from the rock's physical resonance, which is part of the methodology section (Sect. 4.2.2). The P-wave velocity was measured parallel to the core's axis (axial) along a great length of the whole core run, as shown by the blue line in Fig. 3.4. Although the axial P-wave determination covers nearly 80%, it was not possible in areas where the core was fully destroyed or disturbed. The radial measurement of the P-wave velocity was limited to cores with an entire lateral surface, which restricted these measurements to the cores of the core runs 1, 2, 6, and partly 7 (Fig. 3.4, red line). The determination of the S-wave velocity requires a plane surface, which limited the measurements to the segments of the core runs 1, 2, and 3 (Fig. 3.4, green line). In most cases, the cores from run 4 to 7 fell apart before completing a planar end surface, which applied for both the axial and lateral preparation of the cores.

In addition, six segments of the cores were subject to long-term US testing. Figure 3.4 marks these segments with the labels LT-US 1 to 6. While the whole string of cores was only measured once, testing of the long-term segments was conducted over a period of several hours, starting at the moment of recovery from the core barrel. Section 5.2.4.2 deals with the results of long-term US testing.

A general geological description of the rock cores can be found in [2, 3]. In this thesis, Sect. 5.2.1 contains the geomechanical description of the rock cores, including elements of [6] and information beyond standard descriptions.

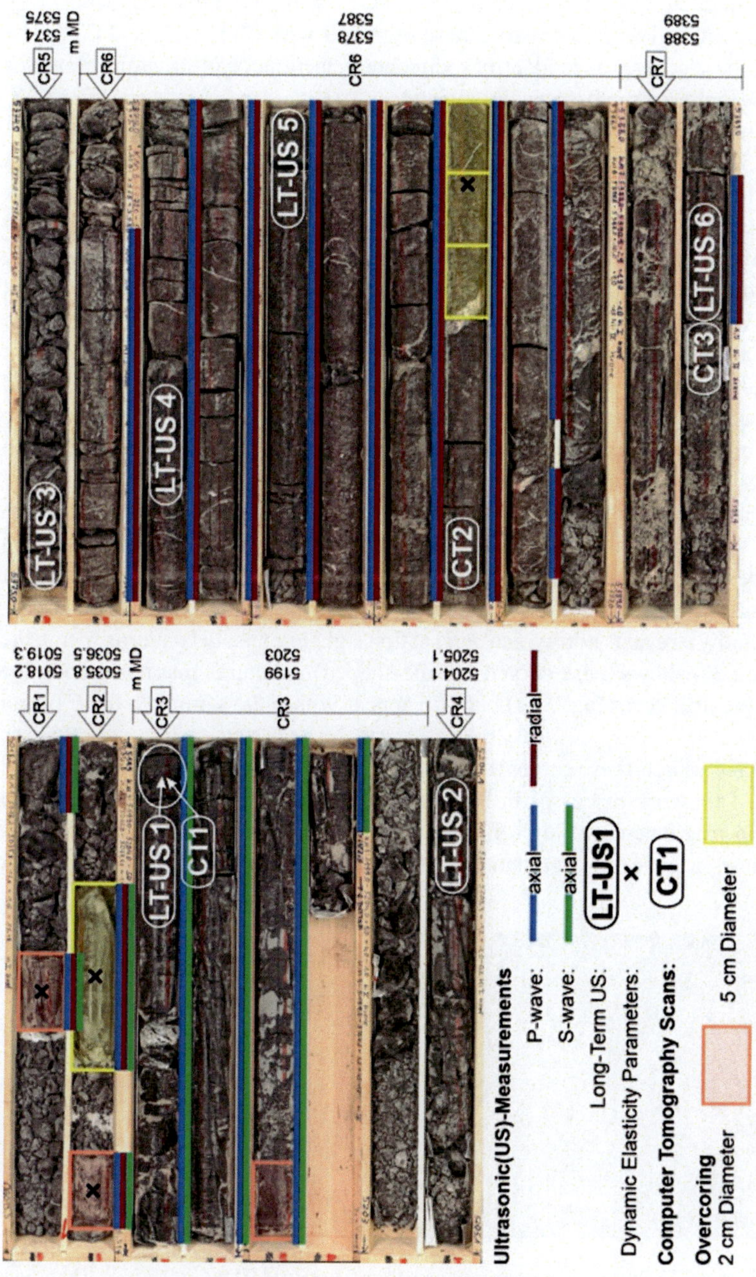

Fig. 3.4 Photos of the rock cores listed after core runs allocated to their depth (modified after [7]). Due to an initial error, the cores' top is at the right side of each core box. The legend marks the measurements: Ultrasonic (US) as blue, green, and red lines, long-term US with **LT-US**, Dynamic properties with an **x**, computed tomography with **CT** and overcored pieces with orange and yellow squares

3.3.2.2 Computed Tomography (CT) Scans on Rock Cores

Koch [8] conducted computed tomography (CT) scans on three segments from different core runs. Figure 3.4 marks these segments with CT1, CT2, and CT3. The purpose of these scans was to identify sedimentary, tectonic, and structural elements in the cores that are macroscopically invisible.

CT1 is recorded on a 10 cm long segment from the beginning of core run 3 from 5199.9 to 5200.0 m MD. Koch [8] describes the core as a micritic limestone with conchoidal fractures indicating a mechanically brittle behavior. He further interprets a stress-related, irregular, and jagged fracture pattern as a sign for isotropic texture (Fig. 3.5, red arrows). Besides the stress-related fractures, he locates two naturally occurring discontinuity sets perpendicular to each other (Fig. 3.5, yellow and green arrows). The scan in Fig. 3.5a, b, shows a stepwise gradation of dark grey in the middle to a light grey outside. Although [8] does not explain this phenomenon, it may indicate a stronger disruption on the outside than inside. Hence, drilling-induced damage may have activated outer lying discontinuities, while inner discontinuities are not triggered and are mechanically still active. Section 5.2.4.2 presents the long-term US (LT-US1) testing results obtained on the same specimen before CT scanning.

CT2 is a record of an almost 18 cm long segment from core run 6 from a depth between 5384.5 and 5385.0 m MD. From a macroscopic, external description, Koch [8] describes that several stylolites with different spatial orientation and two thicker discontinuities traverse the core. Although the CT-scan does not reflect these structures internally, it reveals non-typical, and at first sight unexpected patterns within the core. Figure 3.6 shows these curved (saddle-shaped) structures that trace along the mantel of the drill core (Fig. 3.6a) but also appear within the sample in the CT-scan (Fig. 3.6c).

Koch [8] interprets these curves as a trace of primary bedding, which was disturbed by coring. This work discusses in Sect. 6.4.2 that these structures are due to core disking and originate from high differential in situ stresses. The top view cross-section shows a concentric gradation from a dark grey in the middle to a lighter

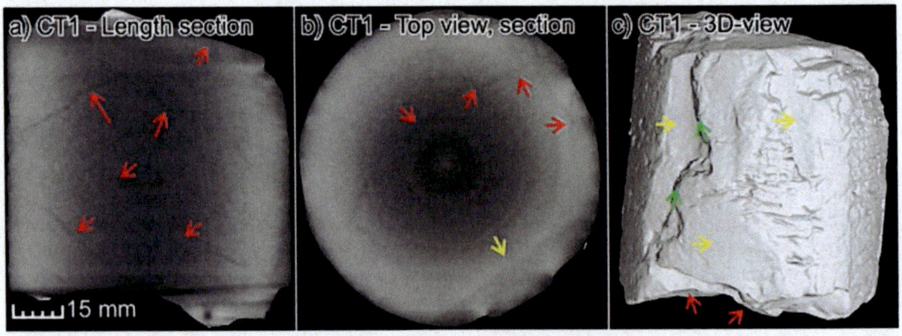

Fig. 3.5 CT scan CT1 from a 10 cm long rock core from 5199.9 m MD to 5200 m MD. The arrows mark possible discontinuity planes, detected by the scan. **a**, **b** are cross-sections through the core, **c** shows a 3D model

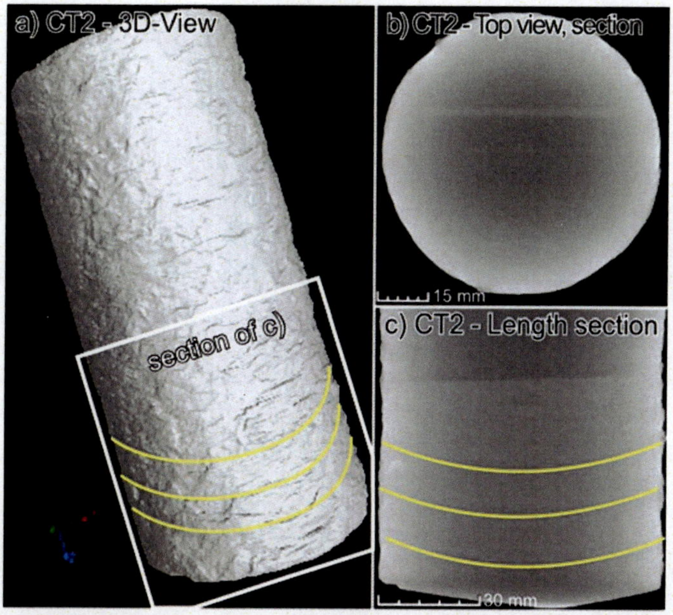

Fig. 3.6 CT scan CT2 from a 30 cm long rock core from 5384.5 m MD to 5385.0 m MD. The yellow curved lines trace a closely spaced set of cracks, identified as saddle-shaped core disking in Sect. 5.3.1.1. **b**, **c** are cross-sections through the core, **a** shows a 3D model

grey in the outer rims, indicating that the curvature is uniformly in all directions. In contrast to CT 1, the greyish darkening from the middle to the outside is very steady, and no further disturbances in the scan are detected.

CT3 is a scan of a 9 cm long segment from Core Run 7 from a depth between 5389.2 and 5389.3 m MD.

Koch [8] describes the sample as a strongly tectonically fractured, fine-crystalline, and homogeneous dolostone with several narrow and thicker healed fractures. In the CT-scan, he identifies two nearly parallel fracture sets (Fig. 3.7b, red arrows) intersected by at least two additional sets. These additional sets are marked with blue and yellow arrows in Fig. 3.7b, c. Figure 3.7, a shows the heavily fractured habitus, as described by Koch [8].

3.3.3 Testing at the Well Site

The set-up of a temporary laboratory (Fig. 3.8) at the well site enabled long-term ultrasound (US) and acoustic emission (AE) measurements on the rock cores. Supplementary acoustic emission (AE) measurements should record and quantify the damage to the rock during stress relaxation of the drill core by seismic micro-events caused by fracturing. Wieser [9] conducted these experiments successfully on magmatic

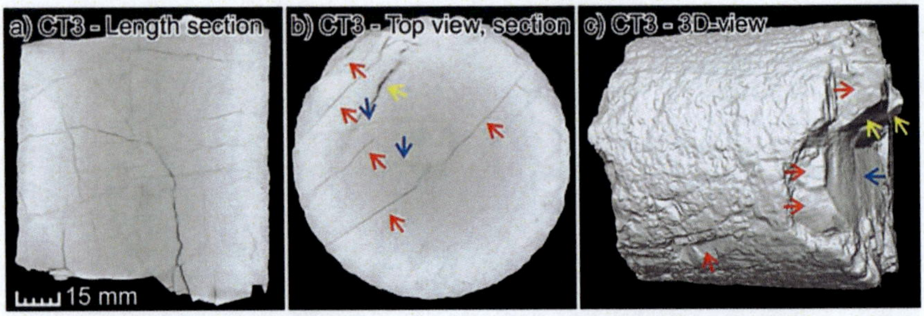

Fig. 3.7 CT scan CT3 from a 10 cm long rock core from 5389.2 m MD to 5389.3 m MD. The arrows mark possible discontinuity planes, detected by the scan. **a**, **b** are cross-sections through the core, **c** shows a 3D model

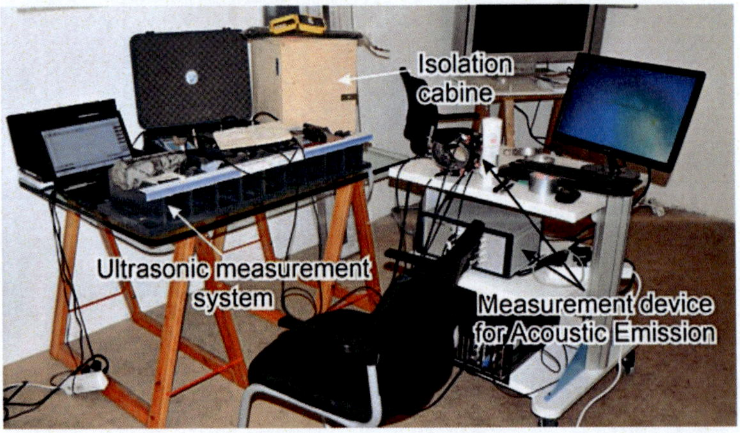

Fig. 3.8 Test setup at the drill rig at Gut Breitenbach to determine the Ultrasonic-velocity and detect Acoustic Emission events from the fresh recovered rock cores (from [7]: 67, Abbildung 69)

rocks of the Pyhäsalmi Iron Mine (Finland) at a depth of approx. 1400 m, also in a differential stress field. Measurements over time showed acoustic events lasting up to 150 h and a continuous decrease of the sound wave velocity during the first 200 h after coring [10]. Stockinger [11] deduced from these previous works that acoustic events and a decrease of the P-wave velocity over time could also be measured in the carbonate drill cores of Geretsried—and at best to derive magnitude and orientation of the in situ stresses. However, despite a custom-designed isolation cabin and a self-built corset with attached AE sensors that allowed a perfect mounting of the sensors to the drill cores, no AE events could have been recorded. Previous laboratory testing showed AE events with the same method, which rules out technical errors. Stockinger [11] assume that the reason for the lack of AE events is a short relaxation process, i.e., the initiation and propagation of fractures, during or shortly after coring. Therefore, this process is already completed before recovery, while the cores are still in the barrel. Stockinger [11] claim that this might indicate a mechanically brittle behavior of the examined carbonates.

3.3.4 *HMI Logs of GEN-1 and GEN-1ST-A1*

High-Resolution Micro Imager logs (HMI) are available for both wells GEN-1 and GEN-1ST-A1. The company Weatherford performed the logging with a wireline, microresistivity imaging tool in water based drill mud (WBM) [17]. An electrical device, equipped with six pads, measures resistivity in a conductive borehole fluid with WBM. Prensky [12] describes that the resistivity measurement is a function of porosity, pore fluid, pore geometry, cementation, and clay content under the influence of mineralogy. For each pad, 25 sensors measure the resistivity as a function of azimuth and depth. In contrast to acoustic imaging methods, which only represent the borehole wall and the casing, resistivity imaging penetrates the surface beyond the borehole wall and resolves smaller features [12].

The HMI log of GEN-1 covers the whole open hole section from 4500 m MD to 6000 m MD. Logging was terminated prematurely in GEN-1ST-A1 by the operator, and thus, the HMI log only covers part of the Purbeck and a small section of the platform carbonates (compare Fig. 3.2) up to 4888 m MD. In contrast to the HMI log of GEN-1, the image quality and resolution of the HMI of GEN-1ST-A1 also allows the tracing of planar features in this short passage, although one recording pad out of six failed. In absolute ratios, the HMI log of GEN-1 has a vertical resolution of 24 Pixel/m while the HMI log of GEN-1ST-A1 counts nearly 200 Pixel/m, equivalent to a microresistivity resolution of 5 mm.

The application "Vinland Software Suite®", used to evaluate the HMI logs, was kindly provided free of charge by Eriksfiord (http://www.eriksfiord.com). This software package, primarily for geomechanical borehole stability analysis, is also a vast image analysis tool to pick planar and areal features. A combined import of the HMI logs as png-files with the associated las-files provides true dip and azimuth, apparent dip and azimuth, and data on the tool's orientation as well as the corresponding depth of planar features. For BBOs and Unilateral structures, the information contains the direction of the center of the element in the borehole coordinate system and the global coordinate system, the height and the width, and the associated depth.

Enex Geothermieprojekt Geretsried Nord GmbH & Co. KG (ENEX) kindly provided the HMI log from the main well GEN-1. This work adopts the information on interpreted fractures from [4] since the log does not allow an explicit demarcation of line-shaped (1-D) elements (planar features, e.g., discontinuities) due to its low resolution. However, this limitation does not apply for areal elements like borehole breakouts (BBOs), or newly introduced in this work, structures that occur only unilaterally in the borehole. The mapping of these Unilateral Patterns follows two criteria: firstly, they are evident on the unwrapped HMI log as dark structures on the well's wall, and secondly, they lack the 180°-symmetry of borehole breakouts. These Unilateral Patterns (UPs) map as unclassified features with only the information on failure width and height and the center of failure in a clockwise angle from the top of the borehole. Figure 3.9 shows a 25 m long excerpt of the GEN-1 HMI log, where the pixelated habitus is visible. The highside (top of the borehole) is left and unwraps the image in a clockwise direction. The BBOs are mapped automatically by the software with pairs of turquoise squares that oppose each other. The Unilateral Patterns (UPs) are picked by single ellipsoid elements (purple ellipsoids in Fig. 3.9).

The HMI log of GEN-1ST-A1, recorded from 4484.0 m MD to 4888.3 m MD, allowed the mapping of BBOs and UPs. The identification of specific joint sets was possible due to its good resolution. Figure 3.10 shows an excerpt from the HMI log

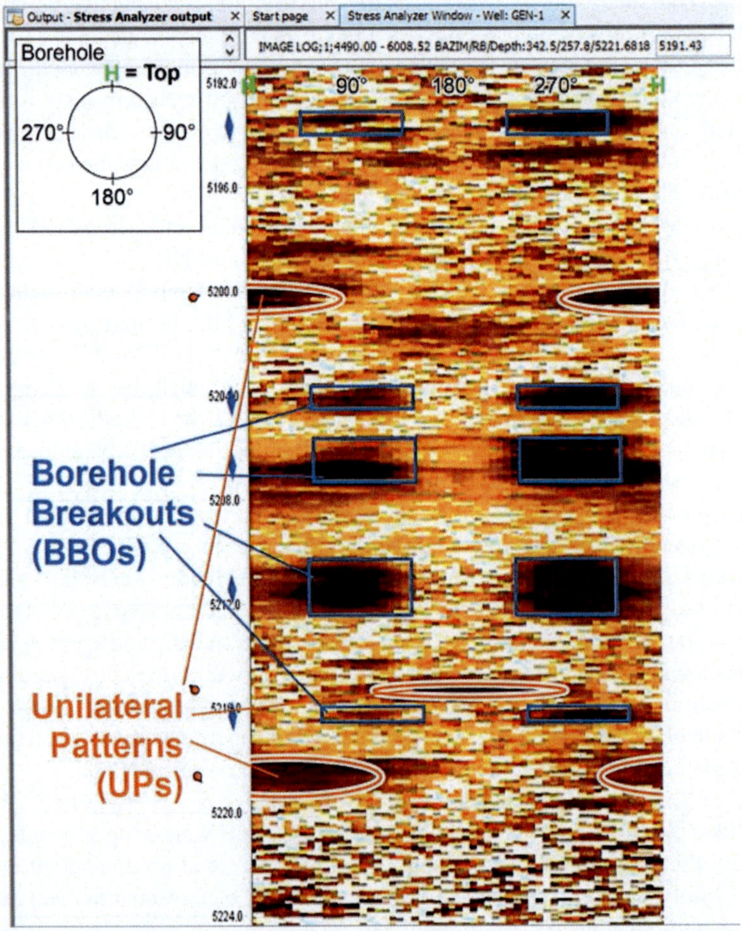

Fig. 3.9 Excerpt of an HMI log from GEN-1 in 5192 m MD to 5224 m MD, unwrapped in a highside coordinate system. From the left, the top or highside (green H), the borehole is unwrapped clockwise to the right wall (90°), followed by the bottom (180°), the left wall (270°) and closing the circle back to the top. The HMI log shows laterally symmetric appearing borehole breakouts (blue) and asymmetric Unilateral Patterns (orange). Image by courtesy of Robert Straubinger (ENEX)

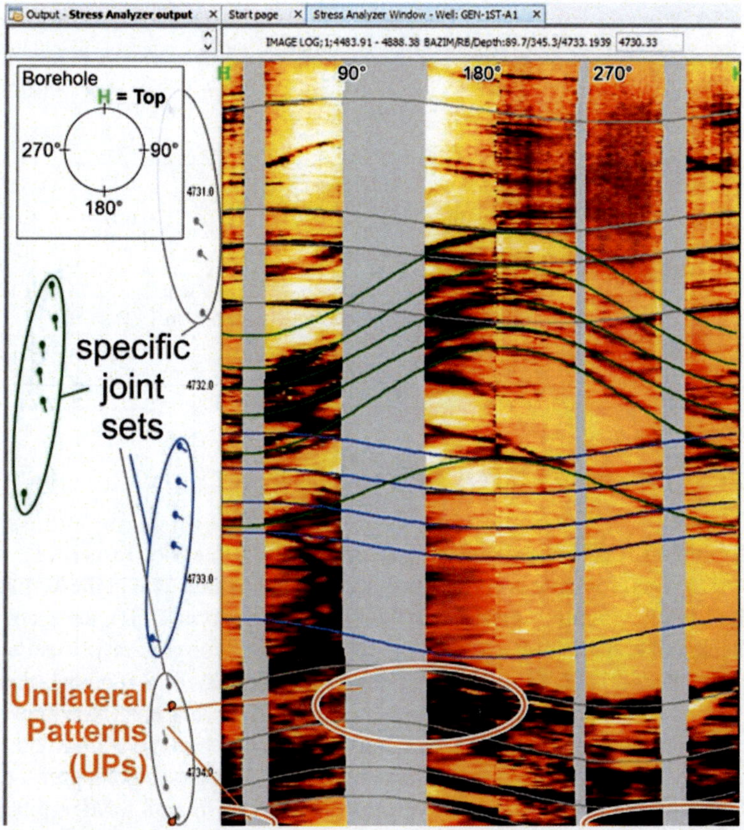

Fig. 3.10 Excerpt of an HMI log from GEN-1ST-A1 in 4731.5 m MD to 4734.2 m MD, unwrapped in a highside coordinate system. From the left, the top or highside (green H), the borehole is unwrapped clockwise to the right wall (90°), followed by the bottom (180°), the left wall (270°) and closing the circle back to the top. The HMI log shows laterally symmetric appearing borehole breakouts (blue) and asymmetric Unilateral Patterns (orange). Tadpoles at the left of the image show the orientation of the mapped discontinuity traces. Image by courtesy of Robert Straubinger (ENEX) and Inga Moeck (LIAG)

of GEN-1ST-A1. The left column is divided into nine vertical divisions and contains tadpoles that belong to the individual planar feature. Points in the divisions show the true dip, and the tick shows the azimuthal direction of the feature. Section 4.1.2.1 explains the evaluation methodology, and Sect. 5.3.2 presents the results for the BBOs and the UPs, and Sect. 5.4.2 for the planar elements.

Table 3.3 Stress regimes with their gradients of the principal stresses, normalized to the vertical principal stress from research reports or publications linked with the project Dolomitkluft

Authors	Stress regime	Ratio, with $S_v = 1 = 24.9$ MPa/km in 4375 m TVD		
		S_{Hmax}	S_{hmin}	Pp
Wolfgramm [3]	Strike slip to normal faulting	1	0.52	0.38
	Pure strike slip	2.1	0.76	
	Strike slip to reverse faulting	3.2	1	
		Ratio, with $S_v = 1 = 24.0$ MPa/km in 4500 m TVD		
Moeck [13]	Strike slip	1.6	0.88	0.4
	Strike slip to reverse faulting	2.2	1	

3.3.5 *Findings on Stress Regimes*

Wolfgramm [3] performed and evaluated Formation Integrity Tests (FIT), which is part of the final research report. Two FITs were conducted, one in the Tertiary strata of the Lithothamnium Limestone in 4260 m MD (4150 m TVD), the second in the Purbeckian Strata in 4480 m MD (4338 m TVD). Wolfgramm [3] use a vertical gradient of 24.3 MPa/km, which equals a bulk density of 2.5 g/cm^3, and a mud density of 1.32 g/cm^3. Although neither Leak-off-Pressure nor Hydrofrac pressure was reached, they could put a lower limit for S_{hmin} of 13.8 MPa/km in the Tertiary Strata and 14.2 MPa/km in the Purbeckian Strata. For the actual aquifer, based on acidification conducted with an overpressure of more than 8.5 MPa, totaling in a radial pressure 51.7 MPa, no fracturing was achieved either, leading to a minimal S_{hmin} of 14.6 MPa/km. Wolfgramm [3] uses a varying value for hydrostatic pressure gradient between 9.2 and 9.4 MPa/km. A drawdown test reduced the radial pressure by 10.5 MPa, resulting in a minimum pressure of 30 MPa within the borehole.

Depending on the prevailing stress regime, Wolfgramm [3] suggest a strike-slip regime delimited to a normal- or reverse stress state in 4375 m TVD with the stress ratios listed in Table 3.3. Moeck [13] come to similar results, using the frictional faulting limitations with a friction coefficient $\mu = 0.8$ in 4500 m TVD.

3.3.6 *Temperature of the Reservoir*

The estimated reservoir temperature is at least 159 °C. This agrees with the temperature in the main well GEN-1 of 169 °C, which ends slightly deeper. However, measuring data shows that the temperature drops to 130 °C during drilling operation [14]. Accordingly, the temperature difference in the bottom hole is at least 30 °C.

References

1. Stockinger G, Käsling H, Menschik F,Thuro K (2019) 3D rotation applied to in situ stress fields for 2D numerical modelling, borehole stability and drill core recovery in deep geothermal wells. In: Fontoura SD, Rocca RJ, Mendoza JP (eds) Rock mechanics for natural resources and infrastructure development–full papers. CRC Press, London, pp 3144–3151
2. Steiger T, Uhlig S (2018) Bio- und Lithostratigraphie der Geothermie-Bohrung Geretsried GEN- 1 mit Sidetrack GEN-1ST-A1. (Anlage 1). In: Thuro K, Zosseder K, Bohnsack D, Heine F, Konrad F, Mraz E, Stockinger G (eds) Dolomitkluft - Erschließung, Test und Analyse des ersten kluftdominierten Dolomitaquifers im tiefen Malm des Molassebeckens zur Erhöhung der Erfolgsaussichten. Teilprojekt B: Geomechanische und hydro-geologische Parametrisierung und Modellierung, München (Technische Universität München - Ingenieurfakultät Bau Geo Umwelt - Lehrstuhl für Ingenieurgeologie)
3. Wolfgramm M, Thiem S, Zimmermann J, Budach I, Buse C, Kabus F (2018) Forschungsbericht GTN: Erschließung, Test und Analyse des ersten kluftdominierten Dolomitaquifers im tiefen Malm des Molassebeckens, 93p. Berlin (Geothermie Neubrandenburg GmbH)
4. Schubert A (2013) Geothermieprojekt Geretsried-Nord - Kurzbericht Reservoircharakterisierung Geretsried GEN 1, 34p (unpublished report)
5. Shipilin V, Tanner DC, von Hartmann H, Moeck I (2019) Temporal evolution of faults in the Southern German Molasse Basin: a case study of Wolfratshausen, Germany. In: 81st EAGE Conference and Exhibition 2019, pp 1–5
6. DIN EN ISO 14689 (2018–05) Geotechnische Erkundung und Untersuchung – Benennung, Beschreibung und Klassifizierung von Fels (ISO 14689:2017); (Deutsche Fassung EN ISO 14689:2018). Beuth Verlag GmbH, Berlin, 2018–05
7. Thuro K, Zosseder K, Bohnsack D, Heine F, Konrad F, Mraz E, Stockinger G (2019) Dolomitkluft - Erschließung, Test und Analyse des ersten kluftdominierten Dolomitaquifers im tiefen Malm des Molassebeckens zur Erhöhung der Erfolgsaussichten. Teilprojekt B: Geomechanische und hydro-geologische Parametrisierung und Modellierung, 413p. München (Technische Universität München - Ingenieurfakultät Bau Geo Umwelt - Lehrstuhl für Ingenieurgeologie)
8. Koch R (2018) CT-Analyse von drei ausgewählten Kernproben der Bohrung GEN-1ST-A1. (Anlage 2.3). In: Thuro K, Zosseder K, Bohnsack D, Heine F, Konrad F, Mraz E, Stockinger G (eds) Dolomitkluft - Erschließung, Test und Analyse des ersten kluftdominierten Dolomitaquifers im tiefen Malm des Molassebeckens zur Erhöhung der Erfolgsaussichten. Teilprojekt B: Geomechanische und hydro-geologische Parametrisierung und Modellierung, München (Technische Universität München - Ingenieurfakultät Bau Geo Umwelt - Lehrstuhl für Ingenieurgeologie)
9. Wieser C (2016) Quantifying the effect of stress changes on the deformation and cracking behavior of solid rock using acoustic emission techniques, 166p. Dissertation, Chair of Engineering Geology, Technical University of Munich, Munich
10. Wieser C, Käsling H, Thuro K, Nuijten G (2015) Monitoring microcrack evolution during core relaxation from high insitu stresses by acoustic emission at Pyhäsalmi Mine, Finland. In: 13th ISRM international congress of rock mechanics, Paper 623 (International Society for Rock Mechanics and Rock Engineering)
11. Stockinger G, Bohnsack D, Moeck I, Käsling H, Thuro K (2019a) Möglichkeiten und Grenzen der Erhebung geomechanischer Parameter an tiefen Bohrkernen. In: Fachsektionstage Geotechnik 2019, 6, Würzburg, Deutschland
12. Prensky SE (1999) Advances in borehole imaging technology and applications. In: Lovell MA, Williamson G, Harvey PK (eds) Borehole Imaging: applications and case histories, vol 159. Geological Society, London, pp 1–43
13. Moeck IS, Dussel M, Buness H, Wawerzinek B (2019) Verbundprojekt Dolomitkluft - Erschließung, Test und Analyse des ersten kluftdominierten Dolomitaquifers im tiefen Malm des Molassebeckens; Teilprojekt E: Spannungsfeldanalyse und Charakterisierung der Störungs- und Kluftzonen, 61p. Hannover (Leibniz-Institut für Angewandte Geophysik)

14. Kahnt R, Gabriel A, Schlegel M (2019) Verbundprojekt Dolomitkluft: Erschließung, Test und Analyse des ersten kluftdominierten Dolomitaquifers im tiefen Malm des Molassebeckens, Teilprojekt C, 47p. Halsbrücke (G.E.O.S. Ingenieurgesellschaft mbH)
15. www-05: https://www.tiefegeothermie.de/projekte/geretsried. Accessed on: 03.03.2020
16. www-06: https://www.worldoil.com/news/2012/12/10/corpro-s-core-barrel-technology-debuts-in-the-gulf-of-mexico. Accessed on: 02.02.2021
17. www-07: https://www.weatherford.com/en/documents/technical-specification-sheet/products-and-services/formation-evaluation/wireline-services/compact-microimager/. Accessed on: 11.11.2020

Chapter 4
Sampling and Methodology

This thesis unites different methods to get a holistic view of the rock mechanical processes—during and post-drilling—around the geothermal boreholes of Geretsried. The following Chapter presents the applied methodology throughout this work. Starting with the collection of rocks from outcrops and drill cores, on warding on the parametrization methods, and settling with a new method to determine the stresses acting along a borehole. Finally, limitations and assumptions concerning this work complete this Chapter.

4.1 Sampling

This Chapter outlines the sampling of analog rocks from outcrops of the Franconian Alb, the Swabian Alb, and the Helvetian facies and describes the recording, preparation, examination, and analysis of the fragile rock cores recovered from the well GEN-1ST-A1.

4.1.1 Sampling from Outcrops

Subject to laboratory testing and parametrization, as presented in Sect. 4.2, are eleven different carbonate rocks. Analysis of the cuttings from the well GEN-1, conducted by Mraz et al. [1], led to the selection of eight specific sampling sites located with pins in Fig. 4.2. Figure 4.2a allocates the sampled outcrops in the Franconian and Swabian Alb. These quarries are in operation, and mining of these carbonate rocks accounts for different purposes. The limestones “Bankkalk” (BK) and “Basisoolith” (BO) of the Swabian Alb and the “Obere Krumme Lage” (OKL) from Painten serve as flux limestone for concrete. The “Pfraundorfer Dolomit” (PFD), the “Wachenzeller Dolomit” (WAD), the “Kelheimer Auerkalk” (KAK), and the “Dietfurter Kalkstein”

G. M. Stockinger, *Fracturing in Deep Boreholes*, Springer Theses,
https://doi.org/10.1007/978-3-030-94569-5_4

and "Dietfurter Dolomit" (DK and DD) find application as quarried blocks in architectural design, landscaping, or carving. The well-known "Solnhofener Plattenkalk" (SPK), mostly due to its thin-bedded properties and a lack of freezing resistance, is used as indoor flooring tiles. Finally, the "Wasserzeller Dolomit" (WD), which is a Dedolomit according to Mraz et al. [1], is recovered as hydraulic blocks.

Kindly supported by these quarries' operators, blocks of rocks were recovered and transported to the laboratory for subsequent processing. The blocks' dimension demanded a sufficient thickness to recover a cylinder of at least 10 cm in length. Directional features, like bedding or natural discontinuities, were not considered during sampling. However, the carbonates' intended purpose brings some challenges to sampling, which deserve attention when analyzing the later laboratory results.

A huge challenge for sampling proposes the flux limestone since the rock mass is either of bad quality or is extracted destructively. Figure 4.1c shows the quarry in Painten, where laminations of the OKL < 1 mm interstratify thicker beds of limestone. These thin beds are prone to weathering and further expose larger limestone beds to weathering, causing trouble finding blocks representing the true rock properties. Despite these issues, two blocks of an unaltered OKL rock could have been collected after removing some of the altered rock. Similar applies to BK and BO, where the limestone is mined subsurface by drill and blast, causing trouble finding blocks that were not damaged by the blasting. Nevertheless, two blocks each that were

Fig. 4.1 Exemplary quarries from the Franconian Alb. Upper left: the massive Pfraundorfer Dolostone (PFD) with clearly visible karstification and secondary residual fillings in Kinding, Lower left: the massive Kelheimer Auerkalk (KAK) from Kelheim. Right: well stratified carbonates of the Obere Krumme Lage (OKL) with various bedding thicknesses in Painten

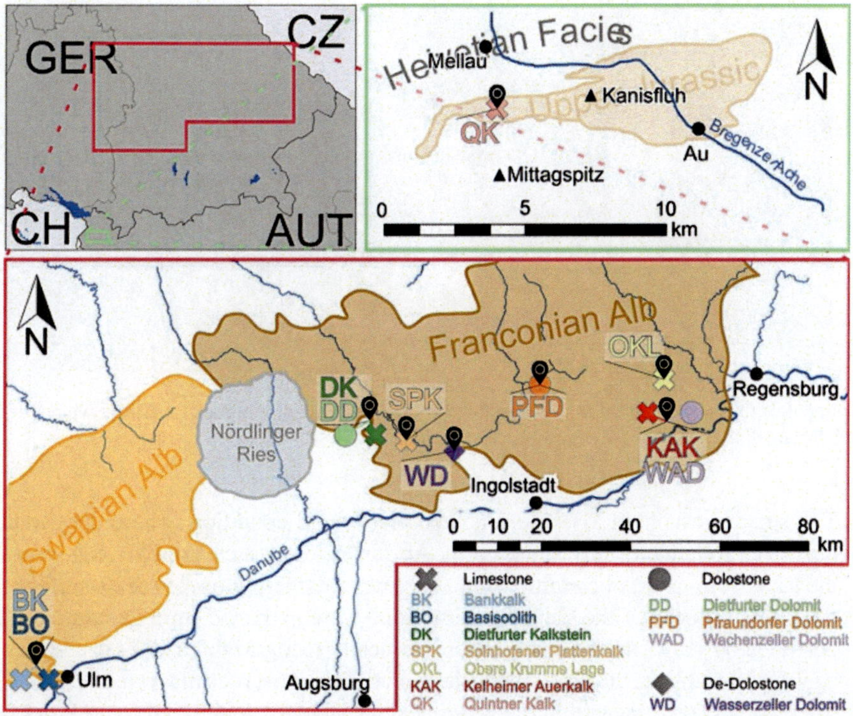

Fig. 4.2 Sampling locations of all analog rocks examined in this thesis. The top left overview shows the sampling areas framed in pink for southern Germany and framed in green for western Austria. The legend at the right bottom shows lithology of each rock. A cross marker represents limestone, a circle dolostone and a diamond Dedolomit (modified and supplemented after Stockinger et al. [2])

macroscopically in an undamaged condition could be obtained of a fresh heading wall.

For carbonate rocks, processed as larger blocks, the mining method is less destructive. In the quarries of Kinding or Kelheim (Fig. 4.1 a–b), where the carbonate bedrock is outcropping unstratified, the blocks are extracted by sawing. In a well-stratified rock mass, such as the DK with a layer thickness of up to 1.5 m, the bedding plane is used as a natural discontinuity, and the rock is extracted from the rock mass by the plug-and-feathers method. Both methods do not lead to further artificially damaging the rock. Therefore, the sampling of unweathered blocks is well feasible. Blocks from PFD and WD were sampled directly in the quarries, while blocks from DD, DK, KAK, WAD, and SPK were provided by the operators formatted to cuboids 20 × 20 × 30 cm.

In contrast to the Alb plateaus' carbonates, where the unweathered, autochthonous, almost undisturbed stratified limestones are easily accessible in quarries, the limestone of the Quinten formation is difficult to access—hence difficult to sample. One outcrop of the Quintner limestone that is also accessible by car is in Vorarlberg, Western Austria, at the WGS84 coordinates 47.334414, 9.858797.

Fig. 4.3 Outcrop of the Quintner Limestone (QK). The bedding of the dark limestone, which strikes EW and dips with 60° to the north, is clearly visible

The strata presents in a severely faulted and folded condition, where the bedding dips with an angle of 60° to the north. At this location, a cavern-like ledge shields the rock from external conditions and confines weathering to the rock's surface. In total, three blocks of the Quintner Limestone were extracted from this outcrop, all pervaded by very thin and open, macroscopically recognizable cracks or fissures—again supporting the dislocated deposition character of the rock mass (see Sect. 1.4.2). These small fractures within the bedding slaps were exploited to remove blocks of sufficient dimension from the rock mass.

The further processing of these rocks, among preparation of the specimens and tests conducted follows in Sect. 4.2.

4.1.2 Acquisition and Evaluation of Drill Cores and Well Logging Data

Besides analog rocks, in situ rock cores provide information on the rock and the rock mass. In their original state, the cores, as presented in Sect. 3.3.2, deliver the in situ orientation of spatial features (e.g., discontinuity planes or saddle-shaped core disking structures) with a description of their surface conditions. Complementarily, the HMI logs also provide the orientation of discontinuities and stress-related features such as BBOs and UPs (Sect. 3.3.4, Figs. 3.9 and 3.10). Obtaining these features from HMI logs and drill cores is identical. A geotechnical classification of the cores completes the acquisition of the rock cores' data and provides the basis for further examinations in the laboratory (Sect. 4.2).

After the 360° core scans, employees of the Technical University of Munich (TUM) brought the cores to Munich for further processing, preparation, and subsequent non-destructive testing (see Sect. 4.2).

4.1.2.1 Tracing and Determining the Orientation of Discontinuities on HMI Logs and Drill Cores

When a planar object, e.g., a discontinuity or a joint, intersects a cylindrical object, the overlapping part is a cylindrical section (Fig. 4.4a). These sections can be circles when the plane intersects the borehole at a right angle, or—when these objects are oblique to one another—the overlapping section becomes an ellipse. In case the cylindrical and planar objects are parallel, a pair of opposite lines traces the cylindrical surface.

HMI logs and 360° core image scans are available as image-files (Figs. 3.9, 3.10, and Fig. 4.5) from unwrapped boreholes or rock cores. Figure 4.4a–c shows the unwrapping schematically. Traces of discontinuities recorded in these images can be either a straight line (Fig. 4.4 , J1 and J2) or a sinusoidal line, depending on the intersection angle (Fig. 4.4, J3).

If the trace is a straight line, the discontinuity is perpendicular or parallel to the cylinder's axis. For all cases in between, the discontinuities shows as a sinusoidal pattern on the image (Fig. 4.4c). Without knowing the cylindrical object's orientation, the sinusoidal wave provides a normalized dip direction and dip. The maximum and minimum turning point of the wave indicates the dip direction, the two inflection points show the strike direction, and the amplitude indicates the dip, depending on the length-to-diameter ratio of the cylinder.

In general, this preliminary data on the orientation of planar features from the images is called the **apparent dip** and **apparent azimuth**. In case the cylindrical object is a vertical borehole, this data is also the **true dip** and **true azimuth**, which can be traced by the cardinal points that orient the logs (Fig. 4.4c). For inclined boreholes, a so-called **high side** local coordinate system is commonly used, as Fig. 4.4c shows

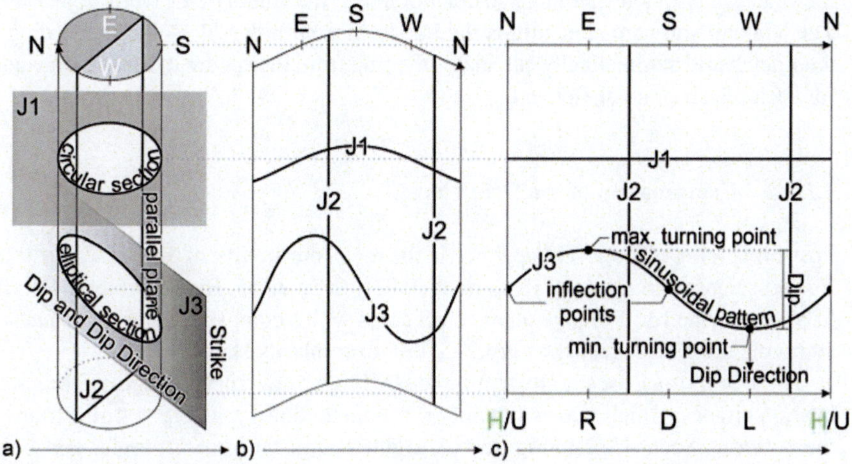

Fig. 4.4 Exemplary illustration of unwrapping a cylindrical borehole with the steps: **a** vertical borehole with intersecting joints J1, J2, J3., **b** intermediate stage of unwrapping—the joints imply their unwrapped form, **c** final planar form with the true discontinuity traces

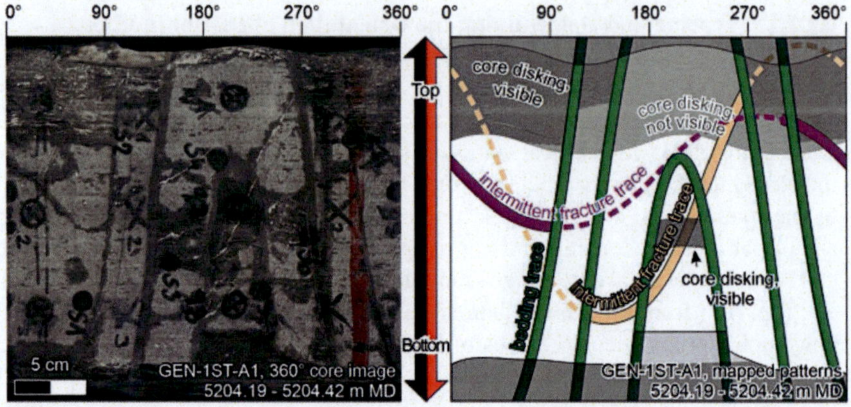

Fig. 4.5 Left: 360° core image from the well GEN-1ST-A1 out of 5204.19—5204.42 m MD. Black joint traces protrude from the greyish surface. The crosses and circles on the surface are the position of the long-term ultrasonic measurements (Photo: M. Dussel, LIAG). Right: The mapped traces from the 360° core image show 3 different discontinuity sets (green, purple, and beige) and tightly spaced saddle-shaped core disking traces in two areas

in the bottom row, with the green "H". This local coordinate system divides the image into quarters, starting at the top of the borehole with "up" (H/U), followed by a clockwise rotation in the direction of drilling to "right" (R), "down" (D), "left" (L), and closing the loop with "H/U" again. Since a deflecting tool was sacrificed to minimize the risk of core jamming, the drill cores lack orientational markers, and the method was only fully applicable for the HMI logs. Text-based ".las"-files, recorded during the drilling operation, provide information for GEN-1 and GEN-1ST-A1 on the values of borehole azimuth and deviation over the length of the wells trajectories. The Vinland software suite aligns the image files with the ".las"-files according to their depth and automatically calculates the true from the apparent orientation values for each planar or areal element.

4.1.2.2 Examination of the Drill Cores

Post-processing of the cores happened in the rock core archive of the Federal Institute for Geosciences and Natural Resources (BGR) in Spandau, Berlin. At the BGR, the LIAG performed 360° image scans on all cores with a completely preserved surface, archiving the cores visually as Fig. 4.5, left, exemplarily shows.

These 360° image scans are, similar to HMI logs, unwrapped images of the cores' lateral surfaces, which allows to apply same methodology, as Fig. 4.4 illustrates in the previous Sect. 4.1.2.1. Figure 4.5, right, shows the green, purple, and beige discontinuity and intermittent fracture traces, mapped as sinusoidal pattern on the unwrapped 360° core image (Fig. 4.5, left). In addition, the saddle-shaped core disking show twice as frequent as the discontinuity traces.

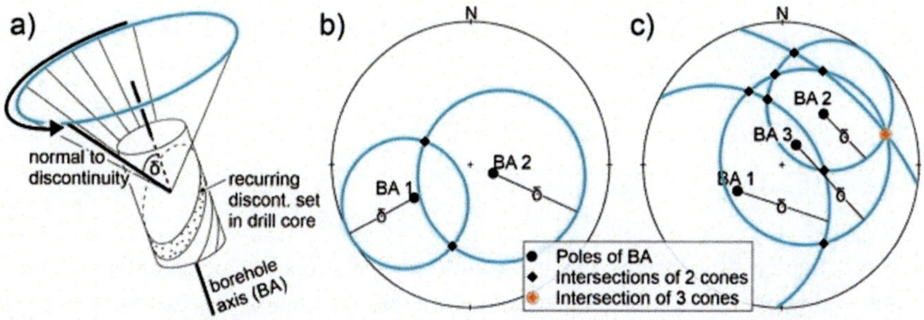

Fig. 4.6 Reestablishing the initial stratification with the steps **a** identification of a recurring discontinuity set in the drill core [3]: 73, Fig. 4.3), **b** plotting the pole of two borehole axes (BA) with their corresponding cones gives two possible surfaces, **c** a third point reduces the max. possible surfaces to one (modified after Ragan [4]: 511, Fig. 20.7)

4.1.2.3 Back-Rotation of Unoriented Rock Cores

While conventional methods, such as orientated rock cores or matching cores with an image log, fail to determine the rock's original stratification, a method applies that preconditions an interpretation of the cores' structures.

The prerequisites for this method include borehole azimuth and deviation (obtained from ".las"-files), recurring planes of a single discontinuity set, and a continuous change of the well trajectory, labeled as borehole axis (BA) in Fig. 4.6. Once one discontinuity set is identified, their apparent inclination angles δ with the core's / borehole's axis are individually measured (see Fig. 4.6a). For each inclination value, the according pole of borehole azimuth and deviation plots in a stereographic projection. The inclination angle δ—which increases from a perpendicular to a parallel bearing of the discontinuity—is drawn as a cone around the corresponding borehole section's pole. Poles of potential planes lie at the intersections of the cones. There are two or four possible planes for two borehole axis poles BA 1 and BA 2 in Fig. 4.6 (b and c). A third value, as BA 3 schematically shows in Fig. 4.6c, reduces the possibilities to one pole only, shown by an orange, encircled diamond marker, and allows a distinct conclusion on the discontinuity set's true orientation.

4.1.2.4 Drill Core Description

While Wolfgramm et al. [5] managed the geological and lithological description of the rock cores, this thesis complements the description according to the geotechnical aspects after the Suggested Methods of the International Society for Rock Mechanics [6, 7] and the international valid German standard [8]. Particular markers, such as saddle-shaped structures from core disking or tectonic indicators (Sect. 2.4.2.3) that are not comprised by these standards, are explicitly mentioned, and allocated. A preliminary geomechanical rock core description took place at the drill site right after coring, and again in the laboratory to determine changes in the cores' condition.

4.2 Laboratory Work

This Chapter summarizes the methodological approaches of sample preparation for analog rocks and drill cores and the subsequent non-destructive testing and destructive testing procedures performed on them.

4.2.1 Sample Preparation

The series of laboratory experiments within this thesis requires two different kinds of specimen geometry. Firstly, an elongated cylindrical body with a length to diameter ratio of at least 2:1 for Ultrasonic Testing and the determination of static elastic and strength parameters by uniaxial compressive strength tests. Secondly, a disk-shaped cylindrical body for tensile strength testing.

4.2.1.1 Preparation of the Analog Samples

The preparation of the specimens from the analog samples happened according to [9]. This standard regulates the specimen preparation very precisely, leading to a better quality and comparability of the different rock samples.

4.2.1.2 Overcoring of Drill Cores

Overcoring of the drill cores was inevitable since the cores neither fulfilled the required length to diameter ratio, nor had a complete lateral surface, or a plane end surface. However, due to the bad core quality, overcoring posed some challenges achieving a satisfying specimen quality after [9]. Hence, Stockinger et al. [10] developed an alternate approach that suggests a gentle method to obtain a maximum of testable specimens from severely damaged drill cores. 5 cm drill bits were intended overcoring drill cores with a fully preserved lateral surface. Since most of the rock cores' surface was unstable, and rock flakes kept spalling off, it was impossible to clamp these cores without destroying them or triggering any discontinuities at first hand. Furthermore, practice shows that hollow cylinder walls, thinner than 2 cm, cannot ablate the occurring torsional forces and break.

Figure 4.7a shows the designed device with the drill bit attached to the drilling machine. To its right side, Fig. 4.7 shows the gradual approach from placing the samples into the device to the recovery of the finished overcored specimens.

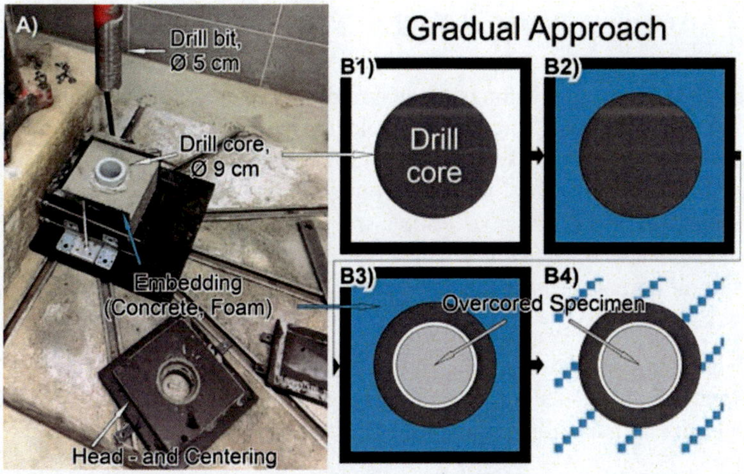

Fig. 4.7 Overcoring apparatus (A), where strongly shattered rock cores are embedded into a fine matrix of concrete or foam (B2), overcored (B3) and afterwards recovered (modified after Thuro et al. [11]: 102, Abbildung 73)

B1: An approximately 15 cm long piece of the rock core, wrapped with cling film to prevent contamination, is placed in the middle of the box.

B2: Concrete or construction foam (both was applied successfully) fills out the free space. The Head- and Centering plate with a hole for the drill bit closes the box.

B3: Overcoring starts after the hardening of the filling material

B4: After removing the sidewalls and the head- and centering plate, the sample's hollow cylinder remains and the overcored specimen is retrieved from within the drill bit.

This method solves both problems mentioned above: Firstly, by embedding the test specimen in a fine matrix, the test specimen's lateral surface is stabilized so that the core cannot move, and secondly, torsional forces can distribute evenly over the extended area.

After successfully overcoring several dummy samples and one real drill core, three failed attempts—where the cores disintegrated in plates during drilling—led to this method's cancelation. Hence, the drill bit size was decreased to 2 cm drill bit. Stockinger et al. [10] further introduced a lower limit of an axial P-wave velocity of at least 4 km/s and a macroscopically stable condition as additional criteria to pick and choose drill cores from the core runs.

4.2.2 Non-destructive Testing

Ultrasonic (US) testing belongs to the non-destructive testing methods and actively measures elastic waves' propagation velocity within materials. "Active" methods require an external impulse source and a receiver, whereas, e.g., Acoustic Emission testing, as a "passive" method, requires only a receiver [12]. Specimens that are subject to non-destructive testing stay intact and can accordingly be tested destructively afterward (Sect. 4.2.3).

Multiple purposes can be satisfied with US testing without damaging or destroying the rock.

- Determining elastic properties of the rock [13, 14]
- Monitoring and detection of significant properties, including discontinuities [15]
- Detection of anisotropic rock properties and stress relaxation [16]

The theory of linear elasticity—when stress and strain relate linearly (see Sect. 2.1.1)—describes an elastic wave's propagation in a medium. Two of six elastic moduli are sufficient to describe an isotropic material, all of which correlate [14]. Besides the compressional wave modulus, the bulk compressional modulus, the shear modulus, and the Lamé parameter, the two most popular moduli in rock mechanics with their definitions after Schön [14] are:

- the Young's modulus E, as the ratio of stress to strain in a uniaxial stress state
- the Poisson's ratio ν, as the (negative) ratio of lateral strain to axial strain in a uniaxial stress state

Both moduli cannot be determined from US testing only but also from Uniaxial Compression Strength (UCS) testing (see Sect. 2.1.3). Although the definitions for both are identical, different frequencies, strain, and stress amplitudes distinguish both methods [17]. Hence, the elastic parameters are supplemented with the annotation "dynamic" when obtained by US testing, e.g., E_{dyn}, and with "static" (E_{stat}) when obtained by UCS testing.

The two dynamic moduli correspond to the velocity of two elastic body waves—the P-wave (also: compressional or longitudinal wave) and S-wave (also: shear or transversal wave). Both correlate with the bulk density ρ of the material and express in terms of Young's modulus E and Poisson's ratio ν as:

$$v_P = \sqrt{\frac{E}{\rho} * \frac{1-\nu}{(1+\nu)*(1-2\nu)}}$$

$$v_S = \sqrt{\frac{E}{\rho} * \frac{1}{2*(1+\nu)}}$$

The ratio of both elastic waves can be expressed with the Poisson's ratio only. For a minimum value for the Poisson's ratio of 0 for isotropic and homogenous material,

Table 4.1 Compilation of P- and S-wave velocities, densities and elastic properties of carbonate minerals and rocks

Rock or mineral	v_P (m/s)	v_S (m/s)	ρ (g/cm³)	Young's modulus (GPa)	Poisson's ratio (-)	References
Calcite	6640	3440	2.71	82.9	0.32	Mavko et al. [18]
Dolomite	7340	3960	2.87	113.9	0.30	
Limestone	3700–6300	2000–3700	–	–	–	Schön [14]
	4000–7000	–	–	–	–	Guéguen and Palciauskas [19]
	–				0.1–0.35	Gercek [20]
Dolostone	3550–6450	2300–3400	–	–	–	Schön [14]
	5000–7500	–	–	–	–	Guéguen and Palciauskas [19]
	–				0.1–0.35	Gercek [20]

Schön [14] calculates that a ratio of $v_P/v_S > \sqrt{2}$ is representative for real isotropic rocks.

Table 4.1 lists typical P- and S-waves' velocities from selected carbonate minerals and rocks, their true (or particle) density, and the derived calculated elastic properties.

4.2.2.1 Controls on P-wave and S-wave Velocity

According to the P- and S-waves' previous definition, their velocity is governed by the Poisson's ratio of the rock-forming minerals. Further factors such as porosity, microstructures (e.g., cracks), or secondary minerals either speed up or slow down the elastic wave. Hence a conclusion on the rocks' lithology by only knowing the Poisson's ratio and v_P is not possible. However, an increasing P-wave velocity correlates with the increasing density of sedimentary rocks [19].

Wyllie et al. [21] describes that **porosity** has a huge influence on v_P in sedimentary rocks. For a rock that is saturated with fluid, this relation can be expressed with the linear equation:

$$\frac{1}{v_P} = \frac{\phi}{v_f} + \frac{1-\phi}{v_m}$$

where v_f is the elastic wave velocity in the fluid, v_m the wave velocity of the rock and φ is the rocks' porosity. Hence, v_P decreases with increasing porosity. Since all experiments are conducted on dry samples, the effect of **saturation** is neglected for all results.

Existing **Microstructures** decrease the elastic wave speed. The effect of anisotropy—as a form of oriented and aligned structures, has a huge impact on v_P in three dimensions and is explicitly described in Sect. 4.2.2.3 [22].

Pressure does not directly influence the elastic wave velocity. However, with an increase in differential pressure (effective pressure, Sect. 2.2.3), microstructures close, and the elastic waves' velocity increases [19]. Increasing **temperature** decreases the velocity of both v_P and v_S. This effect shows in specimens from the Solnhofen Limestone a decrease in velocity from 6.03 km/s at 25 °C to 5.78 km/s at 200 °C [19].

Secondary minerals that contain iron also increase the velocity. Whether a secondary mineral decrease or increases the velocity depends on the individual material property [19].

4.2.2.2 Attenuation of Elastic Waves in Rocks

Since rocks do not deform perfectly elastic, but more likely in a viscoelastic behavior, a small amount of elastic waves energy dissipates during propagation. This effect drains energy and decelerates an elastic wave [14, 19]. Hence, evaluation and discussion of elastic wave velocities of samples of different scales must consider a possible effect of attenuation.

Winkler and Murphy III [23] describes three major mechanisms that are responsible for this loss of energy:

- **Frictional sliding between grains**: This effect requires large strain rates from seismic waves triggered by earthquakes or explosions. Strain rates below 10^{-6}—as is the case for laboratory US testing (since the stress is infinitesimally small)—do not influence the velocity noticeably.
- **Scattering**: When an elastic wave hits a heterogeneity (density difference, pores, cracks, grain boundaries), it is scattered, and its energy loss is reducing the velocity of the first pulse. Winkler and Murphy III [23] finds this effect, particularly in sandstone, which combines multiple heterogeneities. Silica or carbonate grain cement reduces this effect significantly.
- **Saturation**: Attenuation of P-wave velocity is weak in saturated and slightly stronger in partially saturated rocks due to a viscous dissipation of acoustic energy. For S-waves, attenuation is at a maximum since Newtonian fluids cannot transfer shear waves.

In the context of this work, US testing was conducted on carbonate rocks only with a maximum testing length of 45 cm and a minimum testing length of 4 cm. Due to the short measuring distance and the low strain rates, frictional sliding is ignored. Scattering at pores, cracks, or grain boundaries may apply for carbonates. However, the homogenous composition from either calcite or dolomite minerals besides a dense grain binding, makes this effect negligible. Finally, the saturation aspect is not applicable since US testing was performed on dry samples only. In summary, results from lateral and radial measurements on elongated specimens in this work are comparable and do not consider an elastic wave attenuation. Hence, differences in wave velocities that go beyond the maximum measuring error are interpreted as an anisotropy of the rock.

4.2.2.3 Velocity Anisotropy

Simplified, anisotropy is a coalescence of structures, microcracks, unconformities, or discontinuities aligned in a specific direction [23]. Rock-forming minerals like calcite, quartz, or plagioclase, and other lithological attributes like grain size, grain alignment, or porosity mostly have natural anisotropic properties. Still, Guéguen and Palciauskas [19] claims that a random distribution of those attributes makes rocks, in general, behave isotropic. However, orientation-dependent, elastic wave velocities under isotropic stress conditions apply for sedimentary rocks with aligned platy clay minerals, for metamorphic rocks with a preferred structure, or where an anisotropic fracture network exists [23]. Winkler and Murphy III [23] also describes that in a differential stress field, natural cracks close in one direction, while cracks in other directions stay open or open up, causing unequal elastic wave velocities in a general isotropic material. Schön [24] describes a way of quantifying the anisotropy by the ratio of maximum to minimum wave velocity. In the 2nd edition of [14], this quantification is missing. Hence, this work uses a different approach for quantifying anisotropy, which is simply the ratio of the highest axial v_P to the least lateral v_P. Thuro et al. [11] introduced this quotient, which indicates the anisotropy's extent, as "(An)isotropy Index". Hence, a perfect isotropic material would have a quotient of 1. However, due to a possible measuring error of $\pm$ 3% [13], a material is still classified as isotropic if the "Anisotropy Index" quotient is 1.00 $\pm$ 0.05. Mathematically, the maximum measuring error (1.03) divided by the minimum error (0.97) results in 1.06. A rounded variance of $\pm$ 0.05 keeps the value user-friendly and conservative. As discussed in Sect. 4.2.2.2, wave attenuation must not be considered. All indices beyond 1 $\pm$ 0.05 indicate anisotropic behavior that can be further classified by the determination of v_P in all three orthogonal dimensions:

1. If v_P is equal in all directions, the material is isotropic.
2. If v_P is alike in two dimensions but differs from the third, it is transversely isotropic.
3. If v_P is different in all three dimensions, the material is anisotropic.

For the S-wave, cases 1 and 2 depend on the polarization of the S-wave impulse. For the third case, there exist two different v_S [19].

4.2.2.4 Experimental Setup for Ultrasonic Testing

This work adopts the experimental setup of Wieser [16], who established US testing for P-wave velocity on air-dried rock specimens with a diameter of 5 and 8 cm in an axial and a lateral direction. The testing procedure of Wieser [16] used planar

P-wave transmitters and receivers with an ultrasonic gel as a coupling medium. The P-wave velocity was measured twice in the axial direction and each 45° in the lateral direction over three to four steps over the lateral surface. In this work, the previous experimental setup experienced some major changes and improvements:

(I) New, multifrequency, piezoelectric transducers with a probe tip remove the limitation of sample dimensions and better connect to the specimens.
(II) The resonance frequency technique with the new transducers allows the determination of elastic properties and the derivation of an S-wave velocity.
(III) Planar S-wave transducers allow true S-wave velocity measurement.

The experimental setup for (I)–(III) is—except a change of transducers—identical and schematically shown in Fig. 4.8. The ultrasonic sound generator (USG40) emits an impulse with a frequency of 20 kHz to both the US-transmitter and the Oscilloscope (ScopeMeter/PicoScope) (Fig. 4.8). The trigger signal marks t0 (Fig. 4.8). The P-wave velocity results from the specimen's length divided by the arrival of the first impulse at t1 (Fig. 4.8, Setup B-1). This procedure follows the guidelines of the standard [25]. The measurements were conducted with the Software "LightHouse UMPC" by Geotron Elektronik [www-08].

(II) A method to measure the rock specimens' dynamic properties is by measuring their resonance frequency after [28]. When P-waves pass through an elongated object, e.g., through a rod, the body contracts and dilatates with a distinctive frequency in a lateral direction (Fig. 4.8, B-2, in [13]. The yielding material's natural frequency,

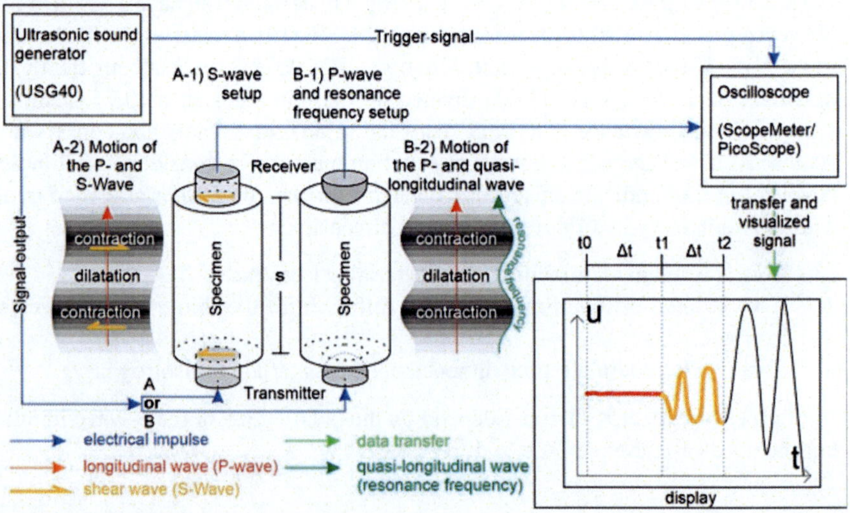

Fig. 4.8 Experimental setup for US testing. The initial signal comes from the ultrasonic sound generator and travels through either A, where it is transferred as a S- and P-wave or B, where it passes the specimen as P-wave and proceeds to the oscilloscope. Simultaneously marks a trigger signal the time t0. From the runtime difference calculates the elastic wave velocity (modified after Menschik [27]: Fig. 3.9)

measured by the multifrequency transducers, is multiplied by twice the specimen's length and results in the velocity of a so-called quasi-longitudinal wave v_D. The dynamic Young's modulus is calculated from $E_{dyn} = (v_D)^2 \times \rho$, where ρ is the bulk density [13]. This method eases the effort of determining the elastic properties to the simple effort of measuring the P-wave velocity since the body's natural frequency is measured simultaneously. The mere demand for the specimen's geometry is an elongated ratio of 2:1 to 4:1 and planar circular or square end surfaces, parallel to each other, forming a cylindrical or prismatic body [13]. The measurements were conducted and evaluated with the Software "LightHouse DW" by Geotron Elektronik [www-09].

(III) The true S-wave velocity measurement uses a planar transducer transmitter, which triggers the piezometric movement in a linear anisotropic motion, sending a polarized shear wave through the specimen. The receiver is aligned parallelly with the motion of the transmitter. The anisotropic impulse with a specific frequency f of 80 Hz allows distinguishing the arrival time t1 of the longitudinal wave from the arrival time t2 of the transversal wave (Fig. 4.8). According to the P-wave, the difference of t2-t0 determines the shear wave velocity v_S. The specimens' geometry demands are two planar end surfaces and a length of at least one wavelength. The minimum wavelength calculates from the relationship of frequency f, wavelength λ, and elastic wave velocity $c = f \times \lambda$. For instance, if the materials v_P is 6000 m/s, the corresponding minimum length must be 7.5 cm with these transducers [www-10].

Sections 5.1.1 and 5.2.2 present the results from the methods (I) and (II); results from the method (III) apply in Sect. 5.2.4.1, directly on core measurements.

4.2.2.5 Acoustic Impedance

The acoustic impedance (AI), commonly abbreviated with Z, is the resistance of a medium against the propagation of an elastic wave. The AI is defined as the product of the elastic wave velocity, for v_P and v_S, with the materials bulk density ρ [31, 14, 32]:

$$Z_P = \rho * v_P$$

$$Z_S = \rho * v_S$$

Since the reflection and transmission of an elastic wave between two materials depend on the acoustic impedance, it can also distinguish between lithologies. Odegaard and Avseth [33] show in a cross plot of v_P/v_S (or the Poisson's ratio) versus AI that a growing ratio of v_P/v_S, and a steep decrease of AI indicates an increase in porosity. An increase in shale content is indicated in the same direction but with a slighter decrease of AI. A decline of fluid (gas) saturation leads to a decrease in v_P/v_S-ratio and AI. According to Hossain and MacGregor [34], a decline in v_P/v_S ratio and higher AI indicates strengthened cementation, which mainly applies to clastic sediments and complements carbonate rocks by Gegenhuber and Pupos [35]. Carbonate rocks form two clouds in Fig. 4.9, one for saturated (brine) carbonates and a second one for dry (gas, air-filled) carbonates. Saturated carbonates show higher AI in a range between 13,000 and 19,000 (g/cm^3)/(m/s), with a v_P/v_S-ratio of 2.0 to 2.5 (approx. range of Poisson's ratio from 0.3 to 0.4). Gas carbonates cover the same range but with slightly lesser AI between 11,000 and 17,000 (g/cm^3)/(m/s) and a v_P/v_S-ratio of 1.5–2.0 (approx. range of Poisson's ratio from 0.1 to 0.3). Hence, plotting the US testing results in Fig. 4.9 below gives important information about cementation, porosity, saturation, shale content, and the analog samples' comparability with the in situ rock cores.

AI provides an empirical relationship with the rock's Uniaxial Compressive Strength (σ_u) and shear strength (τ_S). Since none of the drill cores is suitable for static strength (destructive) testing, this is the only method to derive strength parameters from the drill cores. Müller et al. [36] states that linear correlations between Z_P and σ_u and between Z_S and τ_S exist. These equations express as follows:

$$\sigma_u = 16.4831 * Z_p - 31.9662$$

$$\tau_s = 6.1603 * Z_s - 11.3206$$

Sections 5.1.4 and 5.2.3 contains the strength parameters derived from the US testing results for the analog rocks and the drill cores.

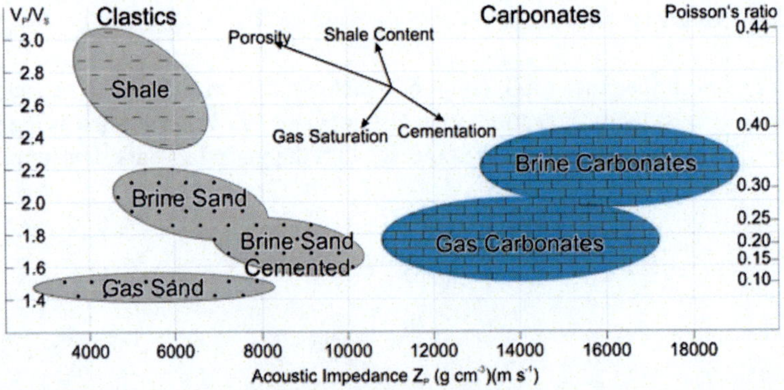

Fig. 4.9 Partitioning of different lithologies according to their v_P/v_S-ratio versus the acoustic impedance of the P-wave (Z_P). This allows a distinction between clastic and carbonate sediments and gives an estimation of the direction to which porosity, shale content, gas saturation and cementation increase (overhauled from Schön [14]:259, Fig. 6.57), based on publications from Avseth and Ødegaard [33], Hossain and MacGregor [34], and Gegenhuber and Pupos [35]

4.2.3 *Destructive Testing*

The Uniaxial Compression Test and the Brazilian Disk Test act as destructive testing methods in this thesis. Rock strength parameters are stresses calculated by the force at material failure applied over specimens' area. The values of rock strength find application in calculating borehole stability (Sect. 5.5.4.1), estimating the depth of the excavation damaged zone (EDZ) and for finite-discrete element simulations as input parameters.

4.2.3.1 Uniaxial Compression Testing

The Uniaxial Compression test determines the strength of the rock without confining the lateral expansion of the specimen. The test is conducted after Mutschler [37], "Option 6". "Option 6" includes, besides the uniaxial compressive strength (σ_u), the continuous record of the axial and lateral strain with cycles. The specimens' geometry is according to Sect. 4.2.1.1. A uniaxial compression test delivers the following data (compare to Fig. 4.11).

- Uniaxial Compressive Strength σ_u [MPa]: peak strength of the rock
- Average Young's modulus V_{stat} [GPa]: the slope of the initial linear part of the stress–strain curve
- Static Young' modulus E_{stat} [GPa]: the slope of the unloading cycle of the stress–strain curve
- Poisson's ratio ν_{stat} [-]: the ratio of change of axial strain to change of lateral strain within the limits of V_{stat}

The employed hydraulic testing machine has a maximum loading capacity of 2000 kN with an accuracy of class 1, according to [38]. The axial strain controls the rate of force applied. The axial displacement averages measurements from three parallelly connected linear variable differential transformers (LVDTs) between the loading platens. A radial strain gauge tightly attached around the circumference in the middle of the specimen records the lateral deformation, hence the circumferential strain. The software "TestXpert"® by Zwick/Roell unites data on force, axial and lateral deformation and automatically calculates all parameters.

It is notable that the ISRM suggested methods for a complete stress–strain curve of the uniaxial compression test [39] only determines the Young's modulus in the initial stress–strain slope and does not intend an unloading and reloading cycle. Mutschler [37] claims that the measurement of the rock's actual elastic properties requires loading cycles. This work uses the terms average Young's modulus V_{stat} (Verformungsmodul in German) as the elastic property suggested by the ISRM [39] and static Young's modulus E_{stat} as the real elastic property of the rock. The range of the unloading cycle is rock-specific and shares its lower and upper stress-limits with V_{stat}. Usually, the range for both parameters is 20–30% of σ_u for the lower limit and

70 -80% of σ_u for the upper limit for the examined carbonates. If applicable, loading applied perpendicular to a visible anisotropy.

4.2.3.2 Tensile Strength of Rock

The tensile strength of the examined carbonate rocks was determined with the Brazilian Disc Test after Lepique [40]. This testing method determines the tensile strength of cylindrical rock discs (l/d = 1:2) indirectly by linearly loading two opposite sides of the rock disc until failure.

A curved steel loading jaw, which is clamped in a hydraulic testing machine with a maximum loading capacity of 200 kN, applies the load to the cylindrical rock disc. The software "TestXpert"® by Zwick/Roell uses the diameter d and the length l of the specimen and the force of failure F to calculate the tensile strength by the equation:

$$\sigma_t = \frac{2 * F}{\pi * d * l}$$

For each test, the loading direction was either isotropic (orientation case after Lepique [40]: SPZ 1) or perpendicular to a visible anisotropy (SPZ 2), aiming for the highest strength possible.

4.3 Stress Rotation

For the stability of boreholes, beyond other geomechanical issues, it is sufficient to look at a 2D cross-section only, which is always a slice of an actual 3D rock-mechanical or geological question; Besides, solving 2D problems is less time- and (computational-)resource-consuming. However, a transfer from 3D spatial information to two dimensions can often pose problems. For example, when a 2D cross-section should implement a 3D fracture network, the fracture traces' apparent strike and dip depend on the 2D plane's orientation. The same applies to differential stresses, as they emerge in the Molasse Basin (see Sect. 2.3.3). If the 2D cross-section is parallel to any of the principal stresses, these stresses apply unchanged (Fig. 4.10).

If a 2D cross-section of a model is oblique to any direction of the principal stresses, the stress acting at the point of the 2D section must be rotated to determine the actual effective stress tensor. More specifically, Fig. 4.11 schematically shows a borehole striking and dipping with an angle non-perpendicular to the principal stresses. Hence, there would be two options to model a 2D cross-section of the borehole:

Option: 1 Constructing the circular borehole diameter as an elliptical shape, which would allow the unaltered use of the in situ stress field in the model. A red frame illustrates a WE-oriented 2D cross-section with the elliptical borehole in yellow (Fig. 4.11).

Option: 2 Rotating the stresses to act perpendicular to the circular circumference of the borehole. Shear stresses emerging from the rotation.

Ultimately, the decision fell for Option 2 for the following reasons:

- Stress rotation is a common tool in reservoir geomechanics [41].
- The calculation of an ellipses' eccentricity and tilt means the same effort as the stress rotation.
- Meshing artifacts are less likely for circular and uniform objects than for tilted ellipsoids.

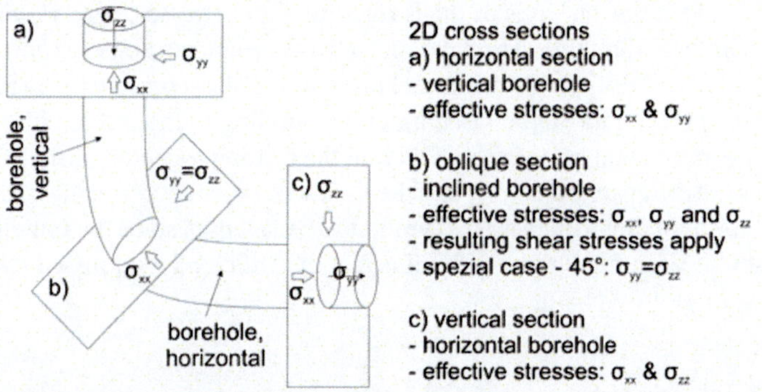

Fig. 4.10 Simplified inclined borehole, with the possible 2D cross-section with the effective stresses to the right. **a** 2D cross-section through a vertical borehole, **b** 2D cross-section through an oblique borehole, **c** 2D cross-section through a horizontal borehole.

Fig. 4.11 Illustration of the fundamental problem: An oblique borehole crosses a 2D cross-section (red square) perpendicular to principal stresses. The trace of the borehole on the cross-section is an ellipse

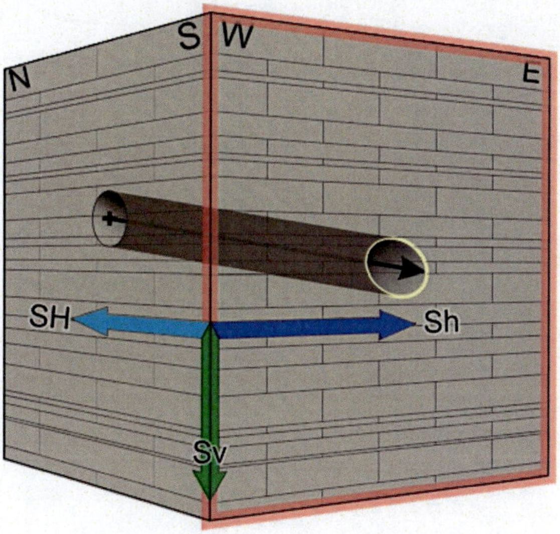

4.3.1 General Approach

The general approach follows Goldstein's methodology [42], which describes this rotation as an x-convention in a right-handed coordinate system. The process of stress rotation is shown with its single steps in Fig. 4.12. The initial consideration is based on the common acceptance of the tectonic principal stresses, where Sv is the principal stress acting vertically and perpendicular to the horizontal stresses. In the top left of Fig. 4.12, all principal stresses act orthogonally to another (compare Sect. 2.2). Hence, the principal stresses can be substituted according to the stress matrix in Sect. 2.2 with a 3D cartesian coordinate system x, y, and z. The first rotation (Fig.4.12, r1) aligns one axis of the former principal stresses, here y', with the drill path's azimuthal direction. The rotation happens around an angle φ (phi) around Sv, which equals the drill path's azimuthal direction. The second rotation (Fig.4.12, r2) tilts the z-axis with an angle θ (theta), corresponding to the drill path's inclination, around the horizontal axis perpendicular to the drilling trajectory, here x'. Hence, y'' becomes the drill path trajectory, and the x'- and z'-axes act perpendicular to the drill path. A third rotation (Fig.4.12 ,r3) around ψ (psi) lapses since the rotation around a drill path, in simplified terms, a 2D line can be sufficiently expressed by the angles φ and θ.

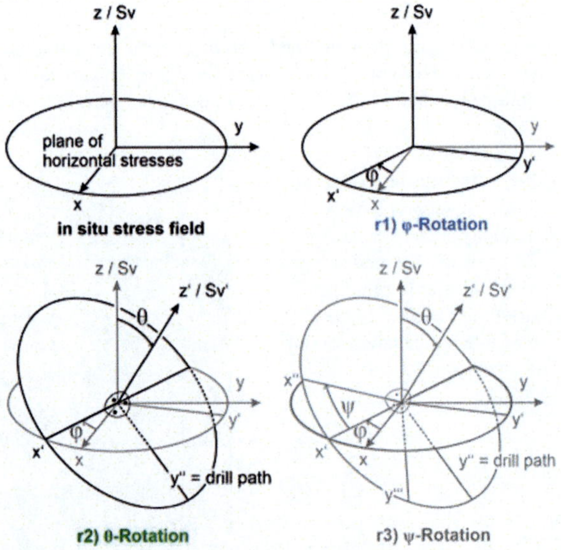

Fig. 4.12 Sequence of the stress rotation based on Goldstein [42], starting with the in-situ stresses acting perpendicular to each other, following the first rotation (r1) in the horizontal plane, followed by r2, tilting the vertical stresses onto the axis y'', which is our drill path. The third step r3 is shown for the sake of completeness, but no rotation is performed around this axis (from Stockinger et al. [43])

4.3.2 Mathematical Stress Rotation

Mathematically, a single rotational matrix [R] composes of the single rotational steps r1, r2, and r3 after Sect. 4.3.1, Fig. 4.12:

$$R = r1 * r2 * r3 = \begin{bmatrix} \cos\varphi & \sin\varphi & 0 \\ \sin\varphi & \cos\varphi & 0 \\ 0 & 0 & 1 \end{bmatrix} * \begin{bmatrix} 1 & 0 & 0 \\ 0 & \cos\theta & \sin\theta \\ 0 & -\sin\theta & \cos\theta \end{bmatrix} * \begin{bmatrix} 1 & 0 & 0 \\ 0 & 1 & 0 \\ 0 & 0 & 1 \end{bmatrix}$$

The matrix [R] is the amount of full rotation that rotates the in situ stresses. The in situ stress matrix [σ] is, as in Sect. 2.2.2, defined by x and y as horizontal stresses and z as the vertical stress. The vector transformation, hence, multiplies the in situ stress matrix with [R] and its transposed matrix [R^T] due to the symmetry of the stress tensor (comp. Sect. 2.2.1). The resulting stress matrix [σ'] composes from the adapted principal stresses σ_{xx}', σ_{yy}', σ_{zz}' and newly occurring shear stresses $\tau_{xy} = \tau_{yx}$, $\tau_{xz} = \tau_{zx}$, $\tau_{yz} = \tau_{zy}$. One of the principal stresses is parallel to the borehole trajectory.

$$[\sigma'] = R*[\sigma]*R^T = R * \begin{bmatrix} \sigma_{xx} & 0 & 0 \\ 0 & \sigma_{yy} & 0 \\ 0 & 0 & \sigma_{zz} \end{bmatrix} * R^T = \begin{bmatrix} \sigma'_{xx} & \tau_{xy} & \tau_{xz} \\ \tau_{yx} & \sigma'_{yy} & \tau_{yz} \\ \tau_{zx} & \tau_{zy} & \sigma'_{zz} \end{bmatrix}$$

$$SH\ and\ Sh\ subsitute\ for\ either\ \sigma_{xx}\ or\ \sigma_{yy},\ Sv\ subsitutes\ for\ \sigma_{zz}$$

4.3.3 Regulations Concerning the Stress Rotation

The previous Chapters define the horizontal stresses as x and y in the cartesian coordinate system. However, only the x-axis controls the rotation r2 (Sect. 4.3.2) and therefore requires further definition to guarantee a correct stress rotation, especially when multiple different trajectories apply that need to be processed by code (see next Chapter).

Hence, the x-axis as one horizontal stress defines the direction to the geographic north. Independent from magnitude or gradient, the x-axis can represent SH or Sh, allowing to connect the cartesian coordinate system with a geographic reference, and azimuth (φ) and inclination (θ) of the drill path can be employed reasonably. However, this would limit the stress field that either SH and Sh always strike absolute NS and WE, respectively. Although a NS striking of SH is, exceptionally, the case for the stress field in the Bavarian Molasse Basin, globally, principal stress directions differ vastly in their orientation [44].

Consequently, Stockinger et al. [43] defined that the horizontal stress with the smallest deviation from N (> 315° and < 45°) substitutes σ_{xx}. Thus, Stockinger et al. [43] introduce a fourth angle that corrects the rotational matrix regarding the strike of the stress field to account for all stress regimes that do not strike NS or EW. This angle χ (chi) results from the difference of the geographic north to the direction of the strike of the stress field. The angle χ is limited to a range between $-45°$ and $45°$. If this angle is supposed to be larger, the basic in situ stress matrix needs to be adjusted.

Figure 4.13 shows three exemplary cases. For case 1, the stress tensor (green) strikes NNW-SSE with a strike of 155°. Hence, if the rotation applies to a drill path with an azimuth from 0° to 360°, the correcting angle χ is 25°, and the maximum horizontal stress takes place for σ_{xx} in the in situ stress matrix. In case 2, the maximum horizontal stress strikes with a 115° WNW-ESE. Thus, σ_{xx} substitutes S_{hmin}. The angle χ is $-25°$ for a counter-clockwise rotation. Case 3 represents an in situ stress field that strikes 45° in NE-SW direction. The decision of whether SH or Sh replaces σ_{xx} depends on the direction of rotation. Therefore, for a counter-clockwise rotation (χ is negative), σ_{xx} is S_{Hmax} or S_{hmin} for a clockwise rotation.

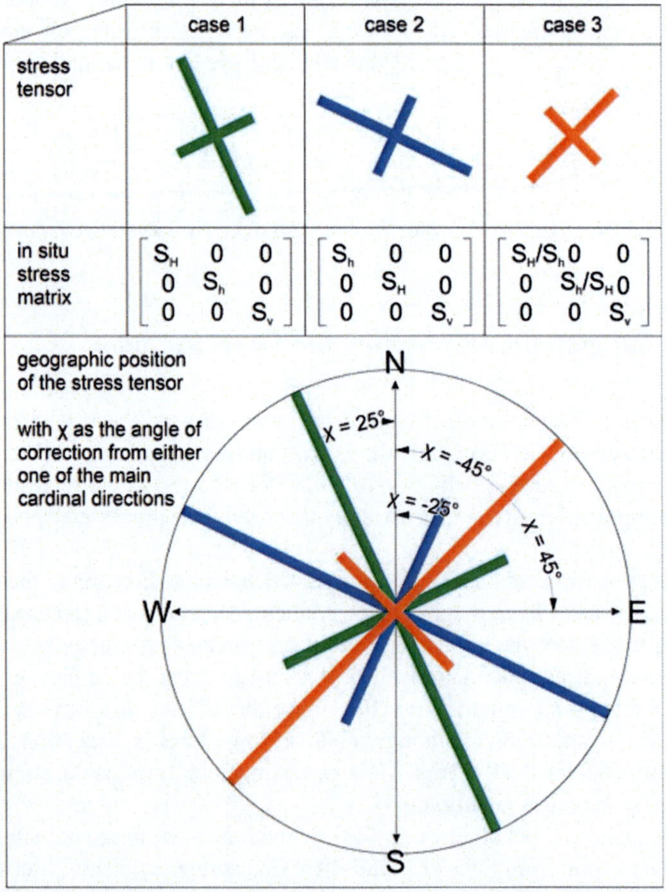

Fig. 4.13 Determination of the correction angle χ for various oriented stress fields (case 1 to 3) in relation to the geographic position of the stress tensor

4.3.4 Step by Step Rotation and Implementation in ©Python

The equation introduced in Sect. 4.3.1.2 applies for each section of the drill path with a unique value for azimuth (φ), inclination (θ), deviation to the north ($\chi\,\chi$), and the stress gradient as a 3 × 3 matrix. Hence, all six parameters (neglecting ψ, since its matrix determinant is constantly 1) are within a calculation for one single section. A code, written with the programming language ©Python, including the package "NumPy" (Version 1.19.9) by Harris et al. [45], performs these calculations. The code runs the following steps:

1. At first, the in situ stress field's gradient in [MPa/m] and the stress matrix is defined:

```
# Gradients of the in situ stress field
Sigmaxx = 0.0411 #MPa/m
Sigmayy = 0.0173 #MPa/m
Sigmazz = 0.0241 #MPa/m
Sigmapp = Sigmazz*0.39 #MPa/m
Sigma = np.matrix([
[Sigmaxx-Sigmapp,0,0], #Sigmax-xx; x-axis
[0,Sigmayy-Sigmapp,0], #Sigma-yy; y-axis
[0,0,Sigmazz-Sigmapp] #Sigmax-zz; z-axis
])
```

The stress gradients (Sigmaxx, Sigmayy, and Sigmazz) are defined as the elements σ_{xx}, σ_{yy}, and σ_{zz} of the 3 × 3 absolute in situ stress matrix. Sigmapp is the pore/fluid pressure as a fraction of the vertical stress Sigmazz, which affects each principal stress equally. Sigma is the actual 3 × 3 effective stress matrix.

2. Secondly, the strike of the stress field in relation to the magnetic north is defined with the angle χ:

```
### Strike of the horizontal principal stresses, for strike parallel to NS or
WE = 0, a max. angle of +-45° is possible
### for a clockwise rotation, use positive values; for counter-clockwise
rotation, use negative values
### angle chi:
if chi >= -45.0 and chi <= 45.0:
chi = chi
elif chi < -45.0 or chi > 45.0:
print("The stress field cannot be rotated by more than 45 degree, please
perform the changes in the in situ stress matrix")
sys.exit(1)
```

The angle χ is allowed in a range between -45° to 45°, including the outer limits. For values below or above, the code aborts the calculations and demands an adaption of the in situ stress matrix, i.e., an exchange of Sigmaxx and Sigmayy.

3. Thirdly, the code performs the single rotations r1, r2, r3

r1: Rotation around the z-axis with the angle φ (phi), according to Fig. 4.12, r1:

```
### angle phi, theta und psi as input for the 3D-rotation
#-----------------------------------------------
# phi: azimuth of the borehole 0 (N) - 360 (N) degree
if theta <= 3 and theta >= -3:
phi = 0
elif theta > 3 or theta < -3:
if phi <= 45:
phi = -1*phi+chi #for rotation when azimuthal direction is in the
1st quadrant <45°
elif phi > 45 and phi <= 135:
phi = 90-phi+chi #for rotation when azimuth is in the 1st and
4th quadrant
elif phi > 135 and phi <= 225:
phi = 180-phi+chi #for rotation when azimuthal direction is in the
3rd and 4th quadrant
elif phi > 225 and phi <= 315:
phi = 270-phi+chi #for rotation when azimuthal direction is in the 2nd
and 3rd quadrant
elif phi > 315 and phi <= 360:
phi = 360-phi+chi #for rotation when azimuthal direction is in the
2nd quadrant >315°
```

The code snippet sets the horizontal rotation to zero if the drill path's inclination deviates by not more than 3° from the vertical. Therefore, a slight deviation of the borehole from the vertical is neglected and, measuring errors or artifacts are suppressed. If the borehole deviates more than 3° from the vertical, the rotation performs regularly after Sects. 4.3.1 and 4.3.2 with consideration of the angle χ.

Figure 4.14 shows the operation principle according to the cartesian coordinate system concerning the geographic north. Each quadrant's angle bisectors (roman numeral) show that the stresses become equal when rotated by 45°.

r2: Rotation around the z-axis with the angle θ (theta), Fig. 4.12, r2:

```
# theta: inclination of the drill path 0 = vertical, 90 = horizontal,
limited to 90°
if theta <=45:
theta = theta*(-1)
elif theta > 45 and theta < 90:
theta = theta-90
elif theta >= 90:
print("The inclination is limited in a range of 0 to not more than 90 degree)
and excludes 90!"
sys.exit(1)
```

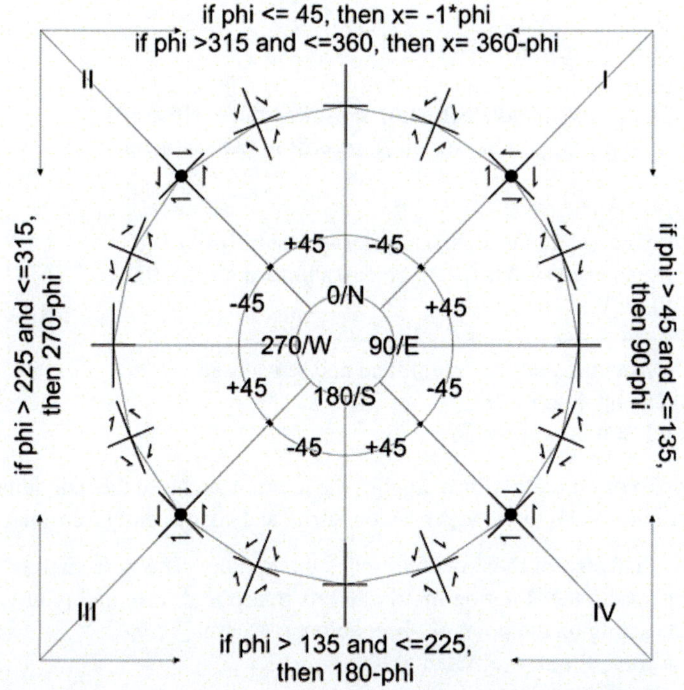

Fig. 4.14 Illustration of the rotation around the z-axis with the angle φ (phi) with regard to the geographic coordinate system (inner circle) and the applied stress tensor (crosses on the outer circle)

For an inclination of < 45°, the stresses rotate counter-clockwise. Until then, the borehole changes from vertical to an inclination of 45°. Up to this point, the stresses acting on the 2D borehole cross-section are the horizontal stresses (Fig. 4.10a–b). When exceeding this inclination, the vertical stress is one stress acting on the 2D plane (Fig. 4.10b–c), stated by the first "elif" statement. The second "elif" statement limits the inclination to an excluded angle of 90° since a borehole can only move between vertical and horizontal. Otherwise, the code cancels the calculation.

r3: No rotation a round ψ, Fig. 4.12, r3:

```
# psi: no rotation around this axis!
psi = 0
```

No rotation equals an angle of zero. The matrix' determinant is 1, as step 4, r3, shows.

4. Applying the angles and combining all rotation matrices:

```
# Rotation matrices r1, r2 and r3
r1 = np.matrix([
[np.cos(np.radians(phi)), np.sin(np.radians(phi)), 0],
[-np.sin(np.radians(phi)), np.cos(np.radians(phi)), 0],
[0, 0 ,1]
```

```
])
r2 = np.matrix([
[1, 0, 0],
[0, np.cos(np.radians(theta)), np.sin(np.radians(theta))],
[0, -np.sin(np.radians(theta)), np.cos(np.radians(theta))]
])
r3 = np.matrix([
[np.cos(np.radians(psi)), np.sin(np.radians(psi)), 0],
[-np.sin(np.radians(psi)), np.cos(np.radians(psi)), 0],
[0, 0 ,1]
#-----------------------------------------------
# Rotation matrices multiplied and transposed
R = np.dot(r1,r2,r3)
RT = np.transpose(R)
```

This part of the code snippet applies the rotation angles to the individual matrices and multiplies them to the aggregated matrix, and concurrently transposes it.

5. The final step of the code multiplies the initially set 3 × 3 matrix "Sigma" in step 1 with the Rotation matrix and its transposed form and returns the stress state acting on the point of observation as "Sigma_Prime":

```
Sigma_Prime = R*Sigma*RT
print(Sigma_Prime)
print(Sigma_Prime)
```

Both inclined wells assemble from several sections with individual pairs of value for azimuth and inclination, as Fig. 3.1 shows. GEN-1 counts 211 sections and GEN-1ST-A1, 254, respectively. Each section is a line in an Excel-sheet with rows of floats for depth, φ, θ, ψ, and χ. For each line, the code must run once. Hence, a new python code applies the calculation script for each line. As a result, the python code generates a comma-separated file, listing the stresses according to their position in the matrix with the consecutive sequence of rows and plots a stress gradient versus depth diagram. Appendix 1 contains the full codes for the single calculation and the application with an Excel-file.

4.3.5 *Advantages to previously Applied Methods*

The methodology of stress rotation is state-of-the-art in rock mechanics and petroleum engineering [41, 46]. Nevertheless, it was either used to solve borehole breakout observations by an inversion method [46] or to estimate tendencies and orientations for potential breakouts or tensile fractures [41]. Neither of these methods directly relates to the stresses acting on a specific section of the borehole, nor do they cover all possible stress regimes. Also, determining absolute stresses at a certain point is not intended, which is an important input parameter in modern modeling codes.

This method removes these limitations. Firstly, from the 3 × 3 matrix, stresses acting on a plane can easily be derived, required for 2D models. Secondly, a clearly defined and flexible stress matrix allows the representation of any arbitrary stress regime. Further adaption of the codes could also take oblique stress fields, as they occur around salt domes, into consideration. Thirdly, the stresses adapt for the entire drill trajectory, which allows a quick estimation and detection of potentially hazardous areas—concerning breakouts, drill induced fractures—or potentially viable areas for hydraulic fracturing even before the trajectory is executed.

4.4 Numerical Modelling with Irazu (©Geomechanica)

Section 2.5.2 presents the mechanical and hydraulic fundamentals of Irazu and the advantages of this simulation code over others. This Chapter presents the model setup and the required input parameters with their sources. Finally, this Chapter concludes with the fundamental ideas of how the aquifer is presented best in terms of stresses, lithology, anisotropy, and structural elements. Lisjak et al. [47] further verify the software's applicability for a cylindrical hole in an infinite elastic medium.

4.4.1 Model Setup

From the experiences gained from Lisjak et al. [11], the model setup experienced some major changes:

1. Downsizing of the 2D cross-section
2. Refining the mesh
3. Changing absolute stresses with effective stresses
4. Adapting the material parameters
5. Embracing Discrete Fracture Networks (DFNs)

4.4.1.1 Geometry and Meshing

Fracture propagation after [11] showed that—under significantly larger in situ stresses—fractures do not propagate by more than 5 m into the rock. Hence, the edge length of the former square 2D cross-section, from formerly 20 m, compacts to 12 m (Fig. 4.15, left). In this square's center, a second surface embeds a borehole with a diameter of 0.1524 m (Fig. 4.15, right). A mesh implemented with the software Gmsh (https://gmsh.info/) discretizes the geometry. The global mesh size is 0.15 m and reduces twice towards the center, once in a 3 m square to 0.02 m and again in a 1 m square to 0.005 m. This double step should smooth the transition from a coarse

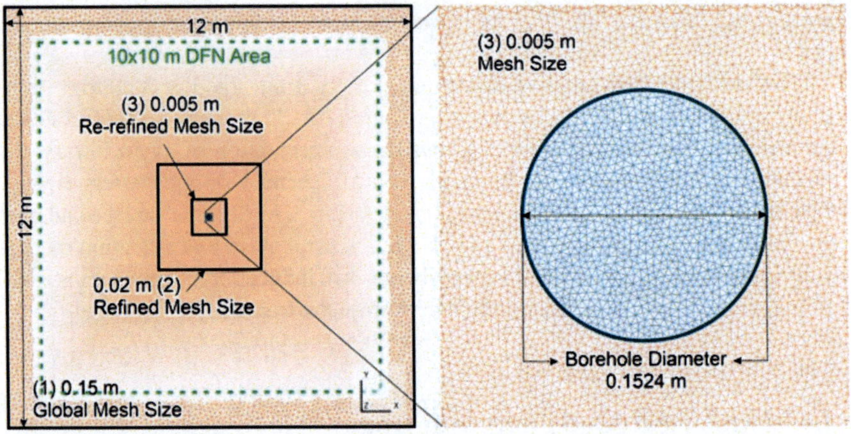

Fig. 4.15 Left: Dimensions of the numerical model with mesh refinement zones and the area where Discrete Fracture Networks apply. Right: Mesh size of 0.005 m in relation to the borehole diameter of 0.1524 m

mesh to a fine mesh. A finer mesh should imply the micritic or micro-crystalline character of the rock. The green, dashed square in Fig. 4.15 defines the area for randomly generated Discrete Fracture Networks (DFNs).

4.4.1.2 Boundary Conditions and in situ Stress Conditioning

Once the creation of the model's geometry and the mesh's generation is complete, boundary conditions reproduce the rock or rock mass's initial state. Since the observations focus on the borehole and its proximity surroundings, mechanical pins (Fig. 4.16a, black triangles) at the edge lengths guarantee no movement or deformation. Along the borehole, circumference acts as a fluid pressure boundary condition that does not apply initially (Fig. 4.16b, Step I). Once the simulation excavates the borehole, the full fluid pressure applies (Fig. 4.16b, Step II), which is the maximum difference between the pore pressure and the mud weight. Consecutively, a steady reduction of the hydraulic pressure simulates returning to the natural water table (Fig. 4.16b, Step III).

The in situ stresses derive from the stress rotation after Sect. 4.3. The stresses apply in the form of σ_{xx}, σ_{yy}, and τ_{xy} in the software-explicit coordinate systems, where σ_{xx} is horizontal, σ_{yy} is vertical, and τ_{xy} is the shear stress acting on the plane. In this thesis, the 2D cross-sections are almost vertical. Thus, σ_{xx} remains, σ_{yy} becomes σ_{zz} (from the rotated stress matrix), and consequently, τ_{xy} changes to τ_{xz}. In contrast to Thuro et al. [11], where the absolute stress applies, this thesis implements effective stress.

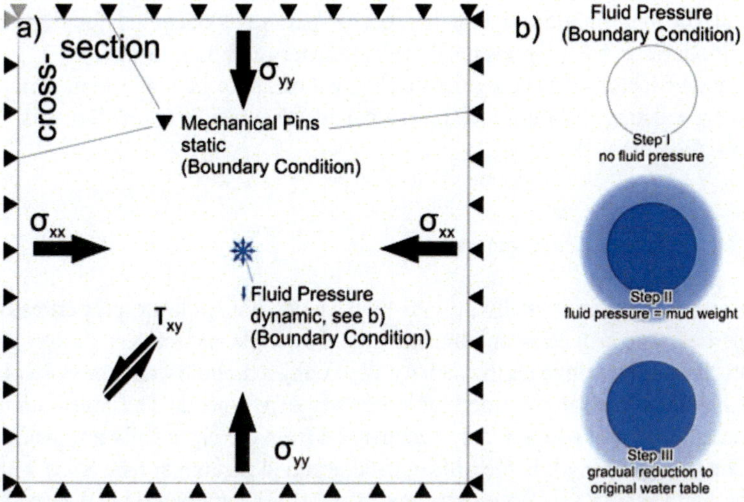

Fig. 4.16 Boundary conditions and in situ stresses that apply in the numerical simulation. **a** mechanical pins allow no movement on the outer boundaries. In situ stresses apply in direction of σ_{xx}, σ_{yy}, and τ_{xy}. **b** Fluid Pressure applies on the circumference of the borehole

The in situ stresses apply in a preliminary finite element run. This FEM-run is the basis for the finite-discrete element run. Section 4.4.3 summarizes all steps to create a fully functioning model.

4.4.1.3 Lithologies and Discrete Fracture Networks (DFN)

Different lithologies and existing fractures are the major change to the previous models, published by Thuro et al. [11]. Different lithologies classify as surfaces with specific thicknesses and apparent dips over the model's whole distance, while DFNs are line features with apparent dips only. Different material parameters assign to surfaces (Sect. 4.4.2.1), while stiffness and strength properties define the DFN line features (Sect. 4.4.2.2). In the case of Scenario 2 in Sect. 4.4.4.2, a DFN set supplements the lithology. Hence, two kinds of surfaces that pretend to be the bedding with different properties are parallel separated by a DFN, indicating a tighter spacing of bedding planes.

4.4.2 Input Parameters

Material parameters define the mechanical behavior of the rock under the preset boundary conditions. Due to the complex mechanical calculation steps, with elastic and plastic deformation, a vast number of parameters account for the material's definition. Sect. 4.4.2.1 presents these parameters and their origin.

Another factor influencing the mechanical behavior beyond the material parameters are existing discontinuities, a so-called "Discrete Fracture Network" (DFN). They define the rock mass.

4.4.2.1 Material Parameters

The material parameters divide into three major units: elastic properties, strength properties, and penalties subdivided into the single parameters (Table 4.2). The elastic properties comprise the density, the Young's modulus, and the Poisson's ratio, which were all self-determined by laboratory experiments. The fourth component, the viscous damping factor, stays unchanged from the basic software configuration.

The tensile strength is the only self-determined parameter (see Sect. 5.1.3) from all strength properties. Cohesion, one of the shear parameters, is derived from Z_S (Sect. 5.2.3) and after Jaeger et al. [50]. The applied friction coefficient of 0.6, as the second shear parameter, is the default parameter and agrees with Byerlee [48]. The software allows an estimation of Mode I and Mode II fracture energy after Whittaker

Table 4.2 Mandatory material input parameters for Irazu with their unit and source

	Parameters		Unit	Source
Elastic properties	Density	ρ	kg/m^3	Laboratory results
	Young's modulus (static)	E_{stat}	GPa	
	Poisson's ratio	ν	–	
	Viscous damping	μ	kg/(m s)	Basic software configuration
Strength properties	Tensile strength	σ_t	MPa	Laboratory results
	Cohesion	c	MPa	Laboratory results and literature ()
	Mode I fracture energy	G_{IC}	J/m^2 N/m	Software recommendation, derived from σ_u
	Mode II fracture energy	G_{IIC}		
	Friction coefficient	μ	–	Basic software value, reconciled with [48]
Penalties	Normal contact penalty	p_n	GPa m	Basic software configuration
	Tangential contact penalty	p_t	GPa/m	Artificially increased to account for the small mesh size after personal communication by email with [49]
	Fracture penalty	p_f	GPa	Basic software configuration

[51], who use an empirical relationship with the tensile strength. Mode I to Mode II Fracture Energy is used in a ratio of 1:10.

From the last unit, the penalties control the interaction between discrete elements. Due to high strain, discrete bodies may computationally overlap in the normal direction or by shearing. However, these discrete bodies are supposed to be impenetrable. Hence, a repulsive force must counter these deformations, which are the penalty forces. For the normal contact penalty p_n, they act in the normal direction, and for the tangential contact penalty p_t, they act in the tangential direction [52]. For the normal contact penalty and the fracture penalty, the values are an order higher than the Young's modulus. Mahabadi [49] recommended, after personal communication, increasing the tangential contact penalty by 2.5 orders to account for the small mesh size.

4.4.2.2 DFN Parameters

DFN lines are either broken or cohesive within the model. Broken DFN sets only have a friction coefficient and penalty values for p_n and p_t. Weakened joint sets that still allow the transfer of tensile Cohesive DFN are sets of weakened joints that can still withstand some tensile force. Hence, additional strength properties apply, such as the tensile strength, cohesion, Mode I, and II fracture energy.

4.4.3 *Follow-Up Steps to the Final Model*

Before the actual finite-discrete element run starts, a preliminary FEM run implements the in situ stresses in the model.

4.4.3.1 FEM Run

The in situ stress implementation runs for 2,000,000 (2 M) time steps as a finite element simulation. A "plot-selection-over-time" diagram indicates whether the in situ stresses are in equilibrium for each model, which was the case for all conducted models.

4.4.3.2 FEMDEM Run

The FEMDEM run calls the created FEM files and starts the consecutive finite-discrete element simulation, which runs for 500,000 (five hundred thousand) time steps. The software outputs the results every 1000 time steps. For the initial stage, the borehole does not yet exist. The borehole excavation happens with the "core modulus reduction method", i.e., a constant decrease of the Young's modulus of the excavating material over given time steps. Here, the reduction of the Young's modulus of the borehole surface starts at 50 k time steps at the initial value and decreases the value by two orders of magnitude before the software removes the material completely at a time step of 200,001. This method guarantees a smooth and artifact-free removal of the borehole material [52]. With the time step of excavation, the hydraulic boundary condition mobilizes the radial stress at 4.5 MPa. With a time step size of 30 k, the hydraulic pressure decreases by 0.5 MPa, reaching 0 MPa at a global time step size of 450 k

4.4.4 Scenarios

The results of the numerical models in Sect. 5.6.2 come from two different modeling scenarios, which developed in the course of this thesis.

4.4.4.1 Scenario 1

Scenario 1 consists of 12 simulations with one single lithology. For GEN-1 and GEN-1ST-A1 with their unique stress parameters, a model runs for each lithology: "weak limestone", "limestone", and "dolostone" (Fig. 4.17), totaling six models. Afterward, individual DFNs supplement the lithologies, creating another six models. Noticeably, the unstratified lithology "dolostone" waives DFN-1.

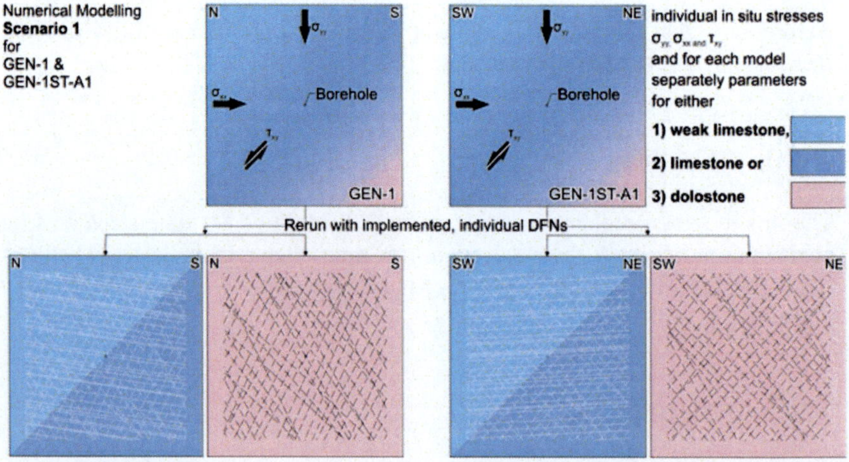

Fig. 4.17 Model setup of scenario 1: three lithologies each GEN-1 and GEN-1ST-A1 with individual in situ stresses, followed by a rerun of the same models with individual Discrete Fractured Networks (DFNs)

4.4.4.2 Scenario 2

Scenario 2 consists of the three independently considered lithologies from the previous Sect. 4.4.4.1. This setup, discussed in Sect. 6.5.1.2, exemplifies an advancing excavation of an inclined borehole through alternating carbonate layers, aiming to identify sequentially unique rock mass behavior. Figure 4.18 illustrates, for the wells GEN-1 and GEN-1ST-A1, from bottom to top, an unstratified dolostone covered by three alternating sequences of 60 cm "limestone" and 30 cm "weak limestone", completed by a layer of "limestone" in the hanging wall. The individual sub-scenarios are:

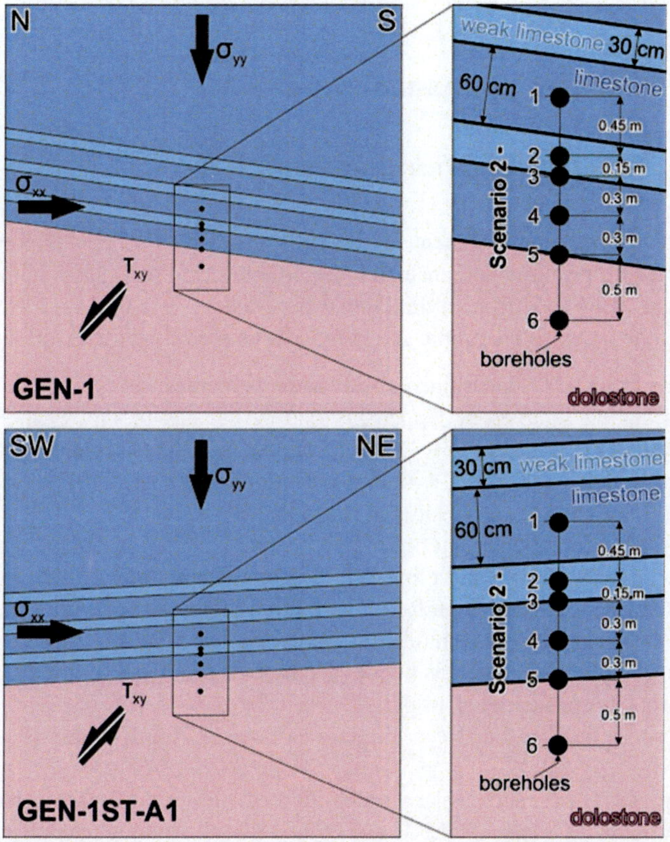

Fig. 4.18 Model setup of scenario 2 with three lithologies in one numerical model for each GEN-1 and GEN-1ST-A1; Each borehole position is a separate model, simulating unique rock mass behavior when drilling through lithologies with different rock mechanical parameters. The enlarged area to the right shows the ascending substages (SC2-1, SC2-2, SC2-3, SC2-4, SC2-5, SC2-6) with the different positions of the borehole

SC2-1: Borehole in the "limestone" between two layers of "weak limestone".
SC2-2: Borehole 45 cm deeper, only in "weak limestone".
SC2-3: Borehole 15 cm deeper, half "weak limestone" and half "limestone".
SC2-4: Borehole 30 cm deeper, only in "limestone", above "weak limestone", below "dolostone".
SC2-5: Borehole 30 cm deeper, half "limestone" and half "dolostone".
SC2-6: Borehole situated in "dolostone" only, half a meter below "limestone".

In total, six substages for GEN-1 and GEN-1ST-A1 build 12 models in scenario 2. The substages in Fig. 4.18 are labeled according to the previous listing. Furthermore, a DFN supplements the scenarios SC2-3 and SC2-5, resulting in four extra simulations with the labels SC2-3_DFN and SC2-5_DFN.

4.5 Limitations and Assumptions

4.5.1 Anisotropy and Inhomogeneities

The sedimentation environment of the Upper Jurassic Carbonates predefines the rock's anisotropic structure since bedding, lithological properties/inhomogeneities, faults, and stresses enforce a directional dependency in either way. However, we consider the rock itself isotropic and homogenous according to the following points:

- Thuro et al. [11] determined the P-wave velocities on cylindrical specimens from analog samples of Upper Jurassic carbonates (comp. Sect. 5.1.5) with length/diameter ratio = 2:1. Results from eleven rocks showed that the axial to radial velocity ratio was—with one exception—1.00 ± 0.03. The rock structure itself, lithologically, is considered isotropic.
- The same effect applies in terms of porosity. Although some analog samples show high porosity of 10% and more [1], the homogeneity of P-wave velocity along the specimen indicates a uniform distribution of pore space within the rocks.
- The hybrid FEM/DEM modeling approach used in this thesis implements anisotropy features, such as bedding, faults, or joints, as geometry or as DFN. The rock is considered isotropic.
- The stress field in the Upper Jurassic carbonates is anisotropic (Sects. 2.4.2, , 3.3.5).
- Vuggy structures, such as larger karstified cavities in the reservoir rock, did not show in the HMI logs of GEN-1 and GEN-1ST-A1. Although Schulz and Thomas [53] characterize karstification structures in Munich's greater area, their extent is eager to decrease to the south. Strasser [54] and Koschel [55] discuss that the cretaceous strata overlying the carbonate strata diminish karstification. Hence, analytical and numerical approaches neglect the influence of karstification.

4.5.2 Poroelasticity

According to Detournay and Chang [56], the requirement for poroelastic mechanical response is free-moving fluid in porous rock. The rock cores from the geothermal well of Geretsried have a mean porosity of 1.35 ± 0.5% [11]. One exception was one sample with a porosity of 8% [11], where the permeability was at 0.09 mD at 3 MPa effective pore pressure, decreasing with rising effective pressure to 0.04 mD [11]. Hence, the rock is impermeable. Still, fluid pressure is present but is assumed not distributing freely in porous space but more likely through fractures or fault zones. Wolfgramm et al. [5] described that mud loss mainly concentrated at fault zones in GEN-1ST-A1. The presence of fluid pressure needs to consider effective stress instead of absolute stress. The fluid pressure gradient in the Upper Jurassic reservoir, used for calculating the effective stress, is individually between 0.38 and 0.4 times the vertical stress (compare to Sects. 2.4.2.2 and 3.3.5).

4.5.3 Overpressure

The pore pressure in the tertiary strata, in the hanging wall of the Upper Jurassic carbonates, is under overpressure with a factor of 1.2–1.6 g/cm^3 in the Rupelian and 1.2–1.4 g/cm^3 in the Upper Cretaceous [57,, 58]. A maximum equivalent mud weight (EMW) of 1.94 g/cm^3 applied while drilling through those units. A maximum EMW of 1.07 g/cm^3 in GEN-1 and 1.05 g/cm^3 in GEN-1ST-A1 in the Upper Jurassic carbonates is sufficient for a successful drilling operation [59–61]. Although Drews et al. [59] mentions that underbalanced drilling in GEN-1 indicates a slightly overpressured Upper Jurassic reservoir, this work uses the fluid pressure gradients from the stress regimes in Sects 2.4.2.2 and 3.3.5, that matches the underpressure described by Lemcke [62] in the Upper Jurassic carbonate aquifer. A smaller formation pressure than mud weight leads to larger (effective) stresses in the numerical models and enables the simulation of a successive decline of the radial pressure in the borehole to a drop of the support pressure to zero.

4.5.4 Well Logging Data

Resistivity, density, and gamma ray logs help to determine differences in lithology (clean/shaley areas) and petrophysical parameters (e.g., porosity, recoverable/movable hydrocarbons) [63] and are available for both wells. However, the assumption of a homogenous limestone or dolostone without shaley or clayey sections from Sect. 4.5.1 and the negation of porosity from Sect. 4.5.2 excluded these logs for further evaluations on geomechanical purposes.

A caliper log, which mechanically measures the borehole's shape in several diametral directions, would deliver valuable geomechanical information. The results from a multifinger caliper [www-11] may show a varying shape of the borehole or a deviation from the deployed drill bit size, evaluating borehole stability and measuring deformations (e.g., borehole breakouts, Sect. 2.4.2.3). Unfortunately, the logging data for GEN-1 did not include a caliper log. For GEN-1ST-A1, a 6-arm caliper log supplemented the HMI log over its distance. However, problems with assigning the pads (arms of the caliper) reliably to the local well coordinate system disqualified this data.

4.5.5 Temperature Effects

Although Kahnt et al. [65] describes a decrease in temperature by 30° to the borehole bottom, thermal stresses will not be quantified. Generally, by a cooling of the rock mass, compressional stresses increase, and tensional stresses decrease. However, the applied numerical models do not consider thermal stresses, which they will in future versions, and therefore, this thesis will renounce thermal stresses due to the comparability of empirical and numerical methods.

References

1. Mraz E, Bohnsack D, Stockinger G, Käsling H, Zosseder SK, Thuro K (2018) Die Bedeutung von Analogaufschlüssen des Oberjura für die Interpretation der Lithologie der geothermalen Tiefbohrung Geretsried. *Jahresberichte und Mitteilungen des Oberrheinischen Geologischen Vereins* 100:517–548
2. Stockinger G, Mraz E, Menschik F, Thuro K (2018) Geomechanical model for a higher certainty in finding fluid bearing regions in non-porous carbonate reservoirs. In: Shakoor A, Cato K (eds) IAEG/AEG annual meeting proceedings, San Francisco, California, Cham, Springer International Publishing, pp 193–198
3. Lisle RJ, Leyshon PR (2004) Stereographic projection techniques for geologists and civil engineers. 2nd edn. Cambridge, Cambridge University Press, pp 124
4. Ragan DM (2009) Structural geology—an introduction to geometrical techniques. 4th edn. Cambridge, Cambridge University Press, pp 624
5. Wolfgramm M, Thiem S, Zimmermann J, Budach I, Buse C, Kabus F (2018) Forschungsbericht GTN: Erschließung, Test und Analyse des ersten kluftdominierten Dolomitaquifers im tiefen Malm des Molassebeckens. Berlin, Geothermie Neubrandenburg GmbH, pp 93
6. ISRM (1978) Suggested methods for the quantitative description of discontinuities in rock masses. Int J Rock Mech Mining Sci Geomech Abstracts 15:319–368
7. ISRM (1981) Basic geotechnical description of rock masses. Int J Rock Mech Mining Sci Geomechan Abstracts 18:85–110
8. DIN EN ISO 14689 (2018–05) Geotechnische Erkundung und Untersuchung – Benennung, Beschreibung und Klassifizierung von Fels (ISO 14689:2017); (Deutsche Fassung EN ISO14689:2018). Beuth Verlag GmbH, Berlin, 2018–05

9. ASTM D4543-19 (2019) Standard practices for preparing rock core as cylindrical test specimens and verifying conformance to dimensional and shape tolerances. ASTM International, West Conshohocken, PA. https://www.astm.org
10. Stockinger G, Bohnsack D, Moeck I, Käsling H, Thuro K (2019) Möglichkeiten und Grenzen der Erhebung geomechanischer Parameter an tiefen Bohrkernen. In: Fachsektionstage Geotechnik, vol 6. Würzburg, Deutschland
11. Thuro K, Zosseder K, Bohnsack D, Heine F, Konrad F, Mraz E, Stockinger G (2019) Dolomitkluft—Erschließung, Test und Analyse des ersten kluftdominierten Dolomitaquifers im tiefen Malm des Molassebeckens zur Erhöhung der Erfolgsaussichten. Teilprojekt B: Geomechanische und hydro-geologische Parametrisierung und Modellierung. München (Technische Universität München - Ingenieurfakultät Bau Geo Umwelt - Lehrstuhl für Ingenieurgeologie;), pp 413
12. Grosse CU, Ohtsu M (2010) In: Acoustic emission testing. Berlin, Springer, pp 396
13. Rentsch W, Krompholz G (1961) Zur Bestimmung elastischer Konstanten durch Schallgeschwindigkeitsmessungen. Bergakademie: Zeitschrift für Bergbau, Hüttenwesen und verwandte Wissenschaften, Sonderdruck aus Heft 7–8/1961:492–504
14. Schön JH (2015) In: Physical properties of rocks: Fundamentals and principles of petrophysics. vol 65. 2nd edn, Amsterdam, Elsevier, pp 512
15. Workman GL, Kishoni D, Moore PO (2007) Nondestructive testing handbook, ultrasonic testing. vol 7. 3rd edn, The American Society for Nondestructive Testing, pp 600
16. Wieser C (2016) Quantifying the effect of stress changes on the deformation and cracking behavior of solid rock using acoustic emission techniques. Dissertation, Chair of Engineering Geology, Technical University of Munich, Munich, pp 166
17. Alber M (2014): Experimentally-based relation between dynamic and static elastic moduli of various sandstones and a quartzite. In: Alejano R, Ollala C, Jiménez R, Perucho Á (eds) Rock engineering and rock mechanics: structures in and on rock masses, London, Taylor & Francis Group, pp 77–82
18. Mavko G, Mukerji T, Dvorkin J (2009) In: The rock physics handbook: tools for seismic analysis of porous media. 2nd edn., Cambridge, Cambridge University Press, pp 511
19. Guéguen Y, Palciauskas V (1994) In: Introduction to the physics of rocks. Princeton (Princeton Univ. Press), pp 294
20. Gercek H (2007) Poisson's ratio values for rocks. Int J Rock Mech Mining Sci 44(1):1–13
21. Wyllie MRJ, Gregory AR, Gardner LW (1956) Elastic wave velocities in heterogeneous and porous media. Geophysics 21:41–70
22. Hudson JA (1981) Wave speed and attenuation of elastic waves in material containing cracks. Geophy JR Astron Soc 64:133–150
23. Winkler KW, Murphy III WF (1995) Acoustic velocity and attenuation in porous rocks. In: Ahrens JT (ed) Rock physics and phase relations: a handbook of physical constants, vol 3. pp 20–34
24. Schön JH (1996) In: Physical properties of rocks: fundamentals and principles of petrophysics. London, Pergamon Press, pp 583
25. DIN EN 14579 (2005–01) Prüfverfahren für Naturstein - Bestimmung derGeschwindigkeit der Schallausbreitung; Deutsche Fassung EN 14579:2004. Beuth Verlag GmbH, Berlin, 2005–01
26. www-08: https://www.geotron.de/software-ultraschall-messtechnik.html#umpc Accessed on: 05.04.2020
27. Menschik F (2015) Analysis of performance and wear of electrical rock hammer drills. Dissertation, Chair of Engineering Geology, Technical Unversity of Munich, Munich, pp 139
28. DIN EN 14146 (2004–06) Prüfverfahren für Naturstein - Bestimmung des dynamischenElastizitätsmoduls (durch Messung der Resonanzfrequenz der Grundschwingung) (Deutsche Fassung EN 14146:2004). Beuth Verlag GmbH, Berlin, 2004–06
29. www-09: https://www.geotron.de/software-ultraschall-messtechnik.html#lhdw Accessed on: 05.04.2020
30. www-10: https://www.geotron.de/schallgeschwindigkeitsmessung-p-und-s-wellen.html Accessed on: 05.04.2020

31. Beattie AG (1983) Acoustic emission, principles and instrumentation. J Acoustic Emission 2:95–128
32. Beattie AG (2013) Acoustic emission non-destructive testing of structures using source location techniques. Albuquerque, NM (Sandia National Laboratories), pp 128
33. Odegaard E, Avseth PA (2004) Well log and seismic data analysis using rock physics templates. First Break 22(10):37–43
34. Hossain Z, MacGregor L (2014) Advanced rock-physics diagnostic analysis: a new method for cement quantification. The Leading Edge, Special Section: Rock Phys 310–316
35. Gegenhuber, N. & Pupos, J. (2015): Rock physics template from laboratory data for carbonates. J Appl Geophys 114:12–18
36. Müller B, Pippig U, Sebastian U (2019) *Geotechnische Klassifikationen von Festgesteinen und Festgebirgen.* Berlin, Springer Spektrum, pp 197
37. Mutschler T (2004) Neufassung der Empfehlung Nr. 1 des Arbeitskreises "Versuchstechnik Fels" der Deutschen Gesellschaft für Geotechnik e. V.: Einaxiale Druckversuche an zylindrischen Gesteinsprüfkörpern. Bautechnik 81(10):825–834
38. DIN 51220 (2003–08) Werkstoffprüfmaschinen - Allgemeines zu Anforderungen an Werkstoffprüfmaschinen und zu deren Prüfung und Kalibrierung. Beuth Verlag GmbH, Berlin, 2003–08
39. ISRM (1999) Draft ISRM suggested method for the complete stress-strain curve for intact rock in uniaxial compression. Int J Rock Mech Mining Sci 36:279–289
40. Lepique M (2008) Empfehlung Nr. 10 des Arbeitskreises 3.3 "Versuchstechnik Fels" der Deutschen Gesellschaft für Geotechnik e. V.: Indirekter Zugversuch an Gesteinsproben - Spaltzugversuch. Bautechnik 85(9):623–627
41. Zoback MD, Barton CA, Brudy M, Castillo DA, Finkbeiner T, Grollimund BR, Moos DB, Peska P, Ward CD, Wiprut DJ (2003) Determination of stress orientation and magnitude in deep wells. Int J Rock Mech Mining Sci 40(7–8):1049–1076
42. Goldstein H (1980) Classical mechanics. Reading (Addison-Wesley Pub. Co.), pp 672
43. Stockinger G, Käsling H, Menschik F, Thuro K (2019) 3D rotation applied to in situ stress fields for 2D numerical modelling, borehole stability and drill core recovery in deep geothermal wells. In: Fontoura SD, Rocca RJ, Mendoza JP (eds) Rock mechanics for natural resources and infrastructure development–full papers. London, CRC Press, pp 3144–3151
44. Heidbach O, Rajabi M, Cui X, Fuchs K, Müller B, Reinecker J, Reiter K, Tingay M, Wenzel F, Xie F, Ziegler MO, Zoback M-L, Zoback M (2018) The World stress map database release 2016: crustal stress pattern across scales. Tectonophysics 744:484–498
45. Harris CR, Millman KJ, Van Der Walt SJ, Gommers R, Virtanen P, Cournapeau D, Wieser E, Taylor J, Berg S, Smith NJ, Kern R, Picus M, Hoyer S, Van Kerkwijk MH, Brett M, Haldane A, Del Río JF, Wiebe M, Peterson P, Gérard-Marchant P, Sheppard K, Reddy T, Weckesser W, Abbasi H, Gohlke C, Oliphant TE (2020) Array programming with NumPy. Nature 585(7825):357–362
46. Qian W, Pedersen LB (1991) Inversion of borehole breakout orientation data. J Geophys Res: Solid Earth 96(B12):20093–20107
47. Lisjak A, Tatone B, Mahabadi OK (2019) Irazu 2D verification manual. Toronto (Geomechanica), pp 39
48. Byerlee J (1978) Friction of rocks. Pure Appl Geophys 116:615–626
49. Mahabadi OK (2020) Personal communication by email: artificially increasing tangential contact penaltiy due to the small mesh size. Received by: Stockinger G on 22 June 2020
50. Jaeger JC, Cook NG, Zimmerman R (2007) In: Fundamentals of rock mechanics. 4th ed. Blackwell Publishing, pp 475
51. Whittaker BN, Singh RN, Sun G (1992) In: Rock fracture mechanics: principles, design, and applications. Amsterdam (Elsevier BV), pp 591
52. Lisjak A, Tatone B, Mahabadi OK, Kaifosh P (2019b) Irazu 2D theory manual. Toronto (Geomechanica), pp 70
53. Schulz R, Thomas R (2012) Geothermische Charakterisierung von karstig-klüftigen Aquiferen im Großraum München—Endbericht. Hannover (LIAG), pp 98

54. Strasser A (1988) Shallowing-upward sequences in Purbeckian peritidal carbonates (lowermost Cretaceous, Swiss and French Jura Mountains). Sedimentology 35:369–383
55. Koschel G (1991) Geologischer Überblick - Hydrogeologische Randbedingungen. In: Bayerisches Landesamt für Wasserwirtschaft and Geologisches Landesamt Baden-Württemberg (eds): *Hydrogeothermische Energiebilanz und Grundwasserhaushalt des Malmkarstes im Süddeutschen Molassebecken (Schlussbericht - Forschungsvorhaben 03 E 6240 A/B)* , München, Freiburg, pp 140 S
56. Detournay E, Chang AH-D (1993) Fundamentals of poroelasticity. In: Fairhurst C (ed) Chapter 5 in comprehensive rock engineering: principles, practice and projects, Vol. II, analysis and design method, Pergamon Press, pp 113–171
57. Drews MC, Bauer W, Caracciolo L, Stollhofen H (2018) Disequilibrium compaction overpressure in shales of the Bavarian Foreland Molasse basin: results and geographical distribution from velocity-based analyses. Marine Petrol Geol 92:37–50
58. Drews MC, Bauer W, Caracciolo L, Stollhofen H (2020) Corrigendum to "Disequilibrium compaction overpressure in shales of the Bavarian Foreland Molasse Basin: results and geographical distribution from velocity-based analyses [Mar. Pet. Geol. 92 (2018) 37–50]. Marine Petrol Geol 112:104102
59. Drews MC, Hofstetter P, Zosseder K, Straubinger R, Gahr A, Stollhofen H (2019a) Predictability and controlling factors of overpressure in the North Alpine Foreland Basin, SE Germany: an interdisciplinary post-drill analysis of the Geretsried GEN-1 deep geothermal well. Geothermal Energy 7(1):24
60. Drews MC, Hofstetter P, Zosseder K, Straubinger R, Gahr A, Stollhofen H (2020b) Retraction note to: Predictability and controlling factors of overpressure in the North Alpine Foreland Basin, SE Germany: an interdisciplinary post-drill analysis of the Geretsried GEN-1 deep geothermal well. Geothermal Energy 8(1):8
61. GeoService GmbH (2017) Geothermiebohrung GEN-1ST-A1 Abschlussbericht. (unpublished report), pp 20
62. Lemcke K (1976) Übertiefe Grundwässer im süddeutschen Alpenvorland. Bulletin der Vereinigung Schweizer Petroleum-Geologen und Ingenieure 42:9–18
63. Ellis DV, Singer JM (2008) In: Well logging for earth scientists. 2nd ed., Dordrecht, Springer, pp 692
64. www-11: https://www.glossary.oilfield.slb.com/en/Terms/m/multifinger_caliper.aspx Accessed on: 12.12.2020
65. Kahnt R, Gabriel A, Schlegel M (2019) Verbundprojekt Dolomitkluft: Erschließung, Test und Analyse des ersten kluftdominierten Dolomitaquifers im tiefen Malm des Molassebeckens, Teilprojekt C. Halsbrücke (G.E.O.S. Ingenieurgesellschaft mbH;), pp 47

Chapter 5
Results

5.1 Rock Mechanical Properties of the Analog Rocks

Thuro et al. ([1]: 81ff) published the rock mechanical parameters for all analog rocks from Fig. 4.2 (Sect. 4.1.1). This dissertation reconsiders these parameters uniquely and comprehensively, and cross-correlates them by statistical analyses with the Python© package Seaborn (https://seaborn.pydata.org). The cross-correlation forms the basis for choosing input parameters for numerical models and constitutes their choice. Further, this Chapter introduces results from the acoustic impedance (comp. Sect. 4.2.2.5), which allows to derive σ_u and τ_S. Appendix B lists the parameters of each analog rock by count, minimum, mean, median, maximum, and standard deviation.

5.1.1 Dynamic Rock Properties of the Analog Rocks

The dynamic rock properties—also elastic properties—were measured with the ultrasonic laboratory method (II), as explained in Sect. 4.2.2.4. They include the dynamic Young's modulus and the dynamic Poisson's ratio calculated from both P- and S-wave velocities of the analog rocks. The following Chapters present the elastic parameters. The elastic wave velocities are not listed individually, but they find themself in the cross-correlation and the acoustic impedance.

5.1.1.1 Dynamic Young's Modulus E_{dyn}

From the eleven analog samples, Fig. 5.1 shows the median values and the statistical dispersion of the dynamic Young's modulus of each rock. The two limestones BK and OKL, have the lowest median values for E_{dyn} with 40.3 GPa and 43.3 GPa, respectively, with only a narrow spread. The values of the de-dolomitic rock WD

G. M. Stockinger, *Fracturing in Deep Boreholes*, Springer Theses,
https://doi.org/10.1007/978-3-030-94569-5_5

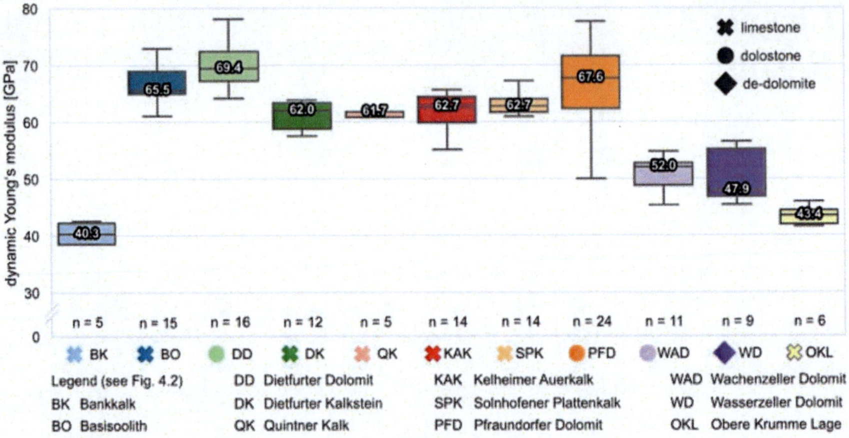

Fig. 5.1 Box plot of the dynamic Young's modulus E_{dyn} for all analog rocks, uniquely colored and classified after their carbonate type. The legend below repeats the corresponding rocks from Fig. 4.2

and the dolomitic rock WAD spread across almost the same range. However, WAD has a narrower interquartile range (IQR) and a slightly higher value of 52.0 GPa compared to WD with 47.9 GPa. The limestones DK, QK, KAK, and SPK values resemble strongly with values of 62.0, 61.7, 62.7, and 62.7 GPa. KAK spreads in total over 10 GPa and shares a slightly larger IQR of 5 GPa with DK, while QK and SPK have an IQR of 3 GPa at max.

DD and PFD, two of the three remaining dolostones, share the highest median of all samples with 69.4 GPa and 67.6 GPa, respectively. However, the IQR of 5 and the whisker-range with a minimum of 65 GPa, and a maximum of 78 GPa of DD is lower than values from PFD with an IQR of 10 and a whisker range with 50 GPa at its lowest and 78 GPa at its highest. Hence, values from PFD scatter more. BO has the highest values of all limestones, with a median of 65.5 GPa.

5.1.1.2 Dynamic Poisson's Ratio ν_{dyn}

Figure 5.2 shows the dynamic Poisson's ratio of all eleven analog rocks with their median and statistical dispersion. The values for each rock scatter only minorly and range between a value of 0.27–0.38. Comparably large IQRs and larger whisker ranges of 0.05–0.08 show DD, KAK, PFD, and WD. This scattering is also observed with the dynamic Young's modulus.

5.1.1.3 Cross-Correlation of all dynamic Rock Parameters

Figure 5.3 illustrates each specimen's dynamic parameter separately in a so-called pair plot. The colors correspond to the legend, with the analog rocks after Fig. 5.1. Specimens with an "x" mark limestone, an "o" dolostone, and a diamond the

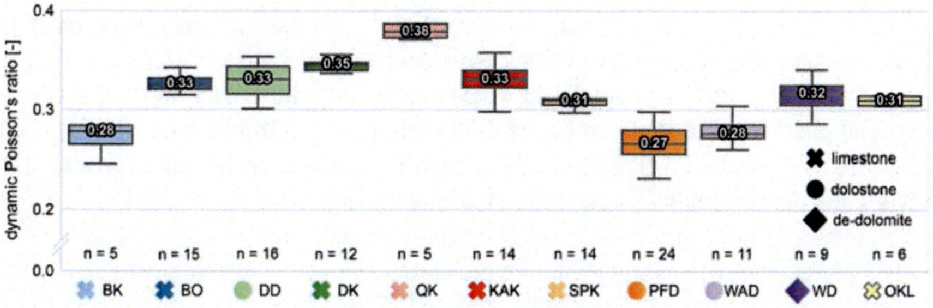

Fig. 5.2 Box plot of the dynamic Poisson's ratio ν_{dyn} for all analog rocks, uniquely colored and classified after their carbonate type, with the legend after Fig. 5.1

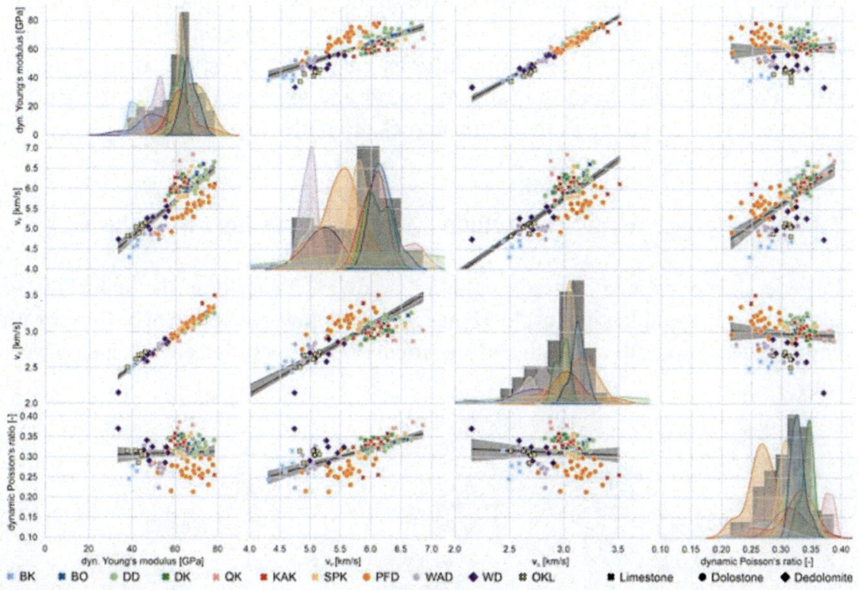

Fig. 5.3 Cross-correlation of the dynamic elastic rock parameters, the dynamic Young's modulus E_{dyn}, the Poisson's ratio ν_{dyn}, and their elastic wave velocities, v_P and v_S for all analog rocks, uniquely colored and classified after their carbonate type, with the legend after Fig. 5.1

dedolomitic WAD. Along the diagonal, a density plot shows the rocks' distribution individually, with a collective histogram in the background. A tall thin distribution implies a concentration around a specific value, while a lower but broader distribution indicates a larger range of values.

E_{dyn} ranges between 33 to 78 GPa and is mainly distributed around 60 GPa with a smaller peak at 40 GPa. Compared to v_P and v_S, a linear relationship is visible. Contrarily, ν_{dyn} is cloudily distributed and does not show any correlation with the E_{dyn}. The v_P ranges from 4300 to 6800 m/s, with a dense accumulation around 6000 m/s and a smaller second accretion around 5000 m/s. Although some rocks, like WD,

OKL, BK, and KAK, do not fit the linear relation with the ν_{dyn}, the others match the linear regression well. S-wave velocities range between 3500 and 2150 m/s. Like the maxima of E_{dyn} and v_P, a small peak develops at 2600 m/s and a larger accretion at around 3000 m/s. There is no correlation with the ν_{dyn}. Single values for ν_{dyn} vary between 0.22 and 0.38. The PFD, WAD, and BK build a small peak at around 0.27, while the other samples locate majorly between 0.30 and 0.33.

5.1.2 *Static Rock Properties of the Analog Rocks*

The static elastic parameters of the analog rocks presented in this Chapter were measured during the uniaxial compressive strength test, as explained in Sect. 2.1.3. This Chapter contains the results from the uniaxial compression test after [2], where the average and static Youngs' modulus is calculated from the axial strain, the static Poisson's ratio from the ratio of lateral and axial, and finally, the Uniaxial Compressive Strength σ_u as the ultimate rock strength.

5.1.2.1 Average Young's Modulus V_{stat} and static Young's Modulus E_{stat}

In general, the average Young's modulus is always lower than the static Young's modulus from the unloading cycle. Hence, Fig. 5.4 shows a box plot of both properties combined and notes the median and statistical dispersion for all eleven analog rocks.

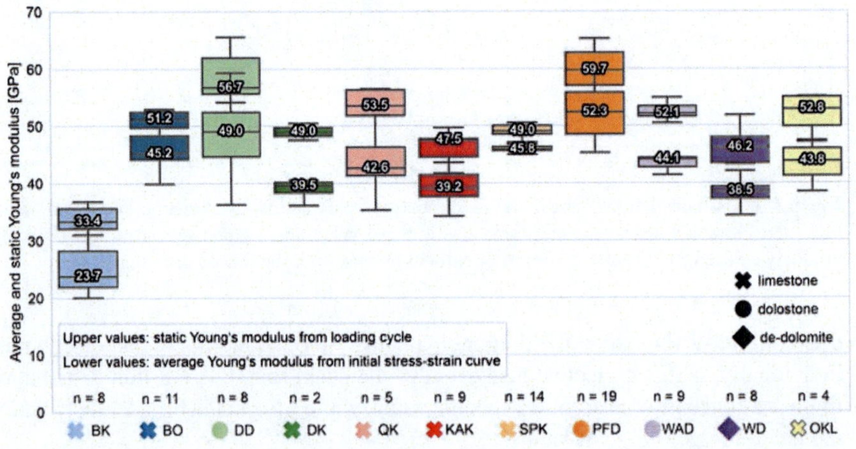

Fig. 5.4 Box plot of the average V_{stat} (lower boxes) and static Young's modulus E_{stat} (upper boxes) of all analog rocks, uniquely colored and classified after their carbonate type, with the legend after Fig. 5.1

BK has the lowest values of all rocks, with a median for E_{stat} of 33.4 GPa. The average Young's modulus' median is 23.7 GPa and can become as low as 20 GPa. Together with WAD and WD, the limestones' values are uniform with medians for E_{stat} between 53.5 GPa for QK and 46.2 GPa for WD. The median of the average Young's modulus for the same rocks lies between 45.8 GPa for SPK and 38.5 GPa for WD with an IQR of 5 GPa. Noticeably, SPK and WAD have IQR that are as low as 2 GPa, and so are the minimum and maximum whiskers (DD is excluded due to the low number of tests). The dolostones DD and PFD show the highest medians from all tested rocks. DD has an E_{stat} of 56.7 GPa and an average Young's modulus of 49.0 GPa. PFD has a median of 59.7 GPa for E_{stat} and a value of 52.3 GPa for the average modulus, which even exceeds E_{stat} of the majority of tested rocks. For both DD and PFD, the fluctuation is high. The IQR is slightly below 10 GPa for E_{stat} and around 10 GPa for the average Young's modulus. DD shows a minimum whisker limit at 37 GPa.

5.1.2.2 Static Poisson's Ratio ν_{stat}

Figure 5.5 shows the median and statistical dispersion of the static Poisson's ratio for all eleven analog rocks and a total of 120 single measurements. Although results showed larger than 0.5 for values of the Poisson's ratio, the illustration is limited to a range between 0 and 0.5 since 0.5 is the theoretical upper limit for rocks as published by [3]. Results above 0.5 may result from axial spalling during testing and are kept in the statistical evaluation since these tests are valid after all. Since the mechanical coupling of the lateral strain gauge to the specimen is difficult and prone

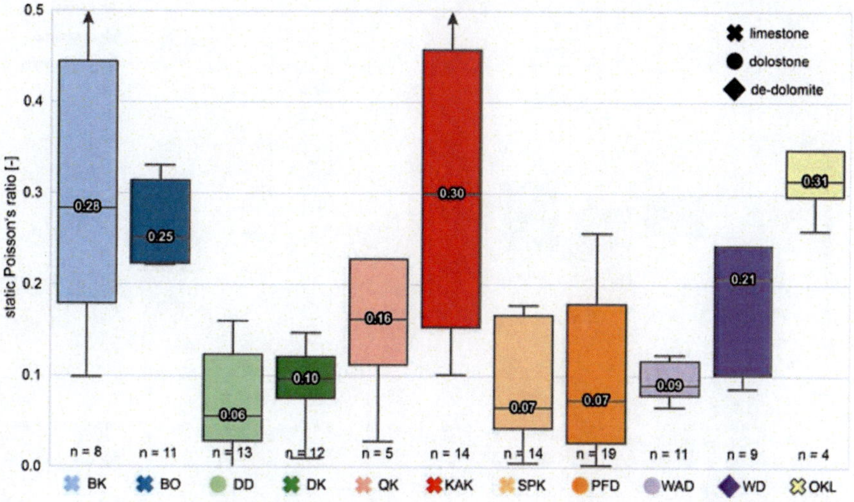

Fig. 5.5 Box plot of the static Poisson's ratio ν_{stat} of all analog rocks, uniquely colored and classified after their carbonate type, with the legend after Fig. 5.1

to major errors (e.g., unilateral loading of the LVDT mechanism, jammed or dirty ball bearings that lead to a delayed or gradual reaction of the chain, or axial spalling of the specimen during testing), this Chapter only presents the median values. From five recommended classifications after [3], the tested analog samples can be grouped into three major divisions:

$\nu_{stat} \geq 0.20$: Five rocks - BK, BO, KAK, WD, and OK—have median values for the static Poisson's ratio of 0.28, 0.25, 0.30, 0.21 and 0.31, respectively. According to [3], this falls into the classification of "medium".

$0.10 \geq \nu_{stat} > 0.20$:DK and QK have a median value of 0.10 and 0.16, resp., for the static Poisson's ratio and classifies as "low" after [3].

$\nu_{stat} < 0.10$:The values of the remaining four rocks—DD, SPK, PFD, and WAD—with a median of 0.06, 0.07, 0.07, and 0.09, respectively, fall below 0.1. After [3], these values must be classified as "very low".

Noticeably, the rocks, which are classified in the "low" to "very low" category, also show lateral strain ratios of almost zero, while the "medium" category has a lower limit of 0.1.

5.1.2.3 Uniaxial Compressive Strength σ_u

The analog rocks' peak strength, the so-called uniaxial compressive strength σ_u, was determined on a total of 125 specimens. Figure 5.6 shows the median and statistical dispersion of the uniaxial compressive strength σ_u for each of the eleven analog rocks.

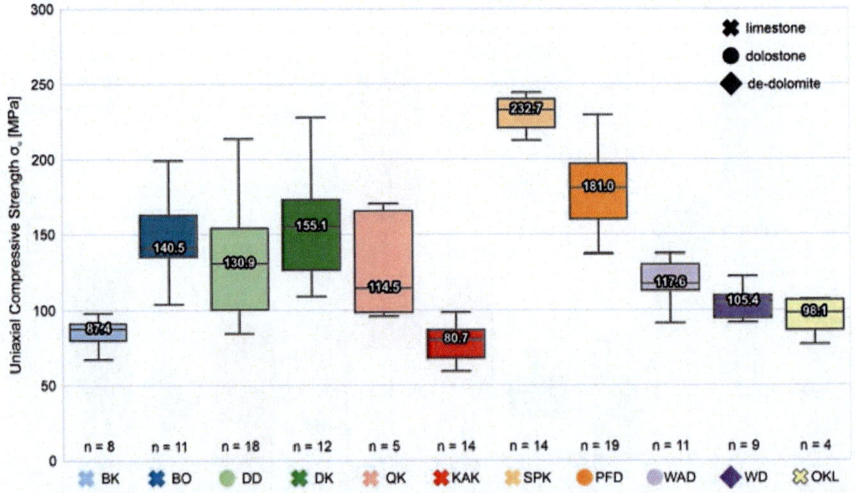

Fig. 5.6 Box plot of the uniaxial compressive strength σ_u of all analog rocks, uniquely colored and classified after their carbonate type, with the legend after Fig. 5.1

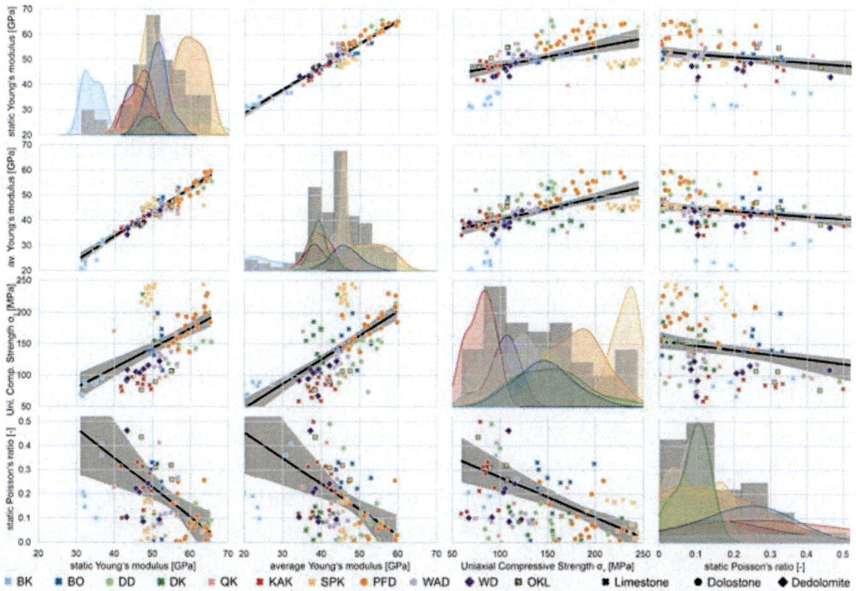

Fig. 5.7 Cross-correlation of the static rock parameters, the average and static Young's modulus and the static Poisson's ratio, with their uniaxial compressive strength σ_u of all analog rocks, uniquely colored and classified after their carbonate type, with the legend after Fig. 5.1

The limestones BK, KAK, and OKL, have the lowest strength of all tested rocks with median values for σ_u with 87.4, 80.7, and 98.1 MPa. While the IQR of BK is only 10 MPa, KAK and OKL vary by more than 25 MPa. According to the German standard [27], their strength between 50 to 100 MPa is classified as "high". All other rocks are above 100 MPa and therefore fall in the [27] category "very high". The strength values of WD, WAD, and QK move on the boundary between "high" and "very high" with median values of 105.4, 117.6, and 114.5 MPa, respectively. The medians of BO, DD, DK place at 140.5, 130.9, and 155.1 MPa, respectively. Together with QK's, Their IQRs vary by more than 50 MPa, and their whisker range is almost at 100 MPa (BO) and above (DD and DK). PFD is the dolostone with the highest strength at a median of 181.0 MPa. Like the rocks before, PFD fluctuated significantly around the median with an IQR of 40 MPa and a whisker range of 100 MPa. In contrast to these highly fluctuating strength values, SPK is the rock sample with the highest median strength of 232.7 MPa and the third-lowest IQR of 30 MPa of all rocks.

5.1.2.4 Cross-Correlation of all static Rock Parameters

Figure 5.7 illustrates each specimen's static parameter separately in a so-called pair plot. The colors correspond to the legend, with the analog rocks after Fig. 5.1. Specimens with an "x" mark limestone, an "o" dolostone, and a diamond the dedolomitic WAD. Along the diagonal, a density plot shows the rocks' distribution individually, with a collective histogram in the background. A tall thin distribution implies a concentration around a specific value, while a lower but broader distribution indicates a larger range of values.

The distribution of the static Young's modulus E_{stat} from all analog samples, in the top left corner of Fig. 5.7, allows a division of three major peaks. From left to right, the peak with the lowest values, at 30 - 40 GPa attributes to BK. The second and highest peak combines BO, DK, QK, KAK, SPK, WAD, WD, and OKL and reaches a value of 45–55 MPa. The highest values for E_{stat} have PFD, DD, and a small SPK share. E_{stat} correlates well, within a narrow 95% confidence interval, with the V_{stat}, showing that the unloading and reloading curve stayed within the rocks' elastic limits. The E_{stat} versus σ_u shows a positive trend, indicating that E_{stat} increases with σ_u and vice versa. However, due to the accumulation of the two distinct rocks BK- and SPK-data at least 10 GPa below the regression line also increases the 95% confidence interval. Hence, a clear correlation between E_{stat} and σ_u is not suggestive. The regression line from E_{stat} vs. the Poisson's ratio ν_{stat} falls slightly. However, a horizontal 95% confidence interval indicates no coherence at all, confirming the impression of the randomly scattered values.

In total, the average Young's modulus V_{stat} shows four peaks (Fig. 5.7, row 2, line 2). From left to right, BK between 20 and 30 GPa, followed by a peak with DK, KAK, and WAD just below 40 GPa. The third and highest accumulation show BK, SPK, and WD between 40 and 45 GPa. DD and PFD form the last peak at 50 GPa. When V_{stat} is the dependent variable, V_{stat} vs. E_{stat} shows an equally small 95% confidence interval as plotted vice versa. The trend when V_{stat} plots against σ_u is positive. However, besides BK and PFD, DK drifts far away from the regression line and the 95% confidence interval. The regression line indicates no correlation with ν_{stat}.

In contrast to E_{stat} and V_{stat}, the uniaxial compressive strength σ_u distributes more evenly (Fig. 5.7, row 3, line 3). A first maximum manifests at around 100 MPa. The strength distribution stays constant up to 175 MPa until it drops. The remaining rocks with a higher σ_u are PFD and SPK, which causes a final peak at the maximum end of 225 MPa and above. As the dependent variable, σ_u shows a positive trend when plotted vs. E_{stat} and V_{stat}. Although the regression line includes BK, the 95% confidence interval excludes most other rocks like KAK, WD, WAD, OKL, or SPK. However, the static Poisson's ratio values scatter even further.

The distribution of values from the static Poisson's ratio accumulates between 0 and 0.15. Around one-third of the values ranges between 0.15 and 0.3. As a dependent variable, the data points scatter strongly when ν_{stat} plots versus E_{stat} and V_{stat}. Although the regression line indicates a negative trend throughout all cross-correlated parameters, the large 95% confidence interval does not allow a reliable statement bringing it in any relation to the other parameters.

5.1.3 *Indirect Tensile Strength σ_t of the Analog Rocks*

Although the indirect tensile strength also counts to the destructive measured strength parameters, the results are separately presented from Sect. 5.1.2. While tests from the previous Chapter are perfectly comparable as they are performed on the same specimen, tensile strength testing was conducted on separately prepared specimens. With the Brazilian test method from Sect. 4.2.3.2, the indirect tensile strength of 146 rock specimens was measured for the eleven analog rocks. Figure 5.8 shows the median and statistical dispersion of the indirect tensile strength σ_t for each of the eleven analog rocks.

The lowest values show QK, BK, WD, and KAK with median values of 3.8, 5.5, 5.8, and 6.5 MPa. The median of WAD, OKL, BO, and DK with 7.3, 8.4, 8.7, and 9.7 MPa show intermediate values from all samples. For the dolostones, DD and PFD have the highest values with 10.2 and 10.9 MPa, respectively. SPK has the highest tensile strength from the limestones, with a median of 14.4 MPa, surpassing almost every other rock's whisker maximum. The highest value for SPK was measured at 18.3 MPa. The IQR for all rocks—with some exceptions—is around 2.5 MPa. For BK, PFD, and WD, the IQR is fairly lower in comparison. WAD shows an increased IQR of almost 4 MPa. The minimum and maximum whiskers are at an extend of 2 MPa for BO, DD, DK, QK, KAK, and PFD, showing a wide fluctuation of values.

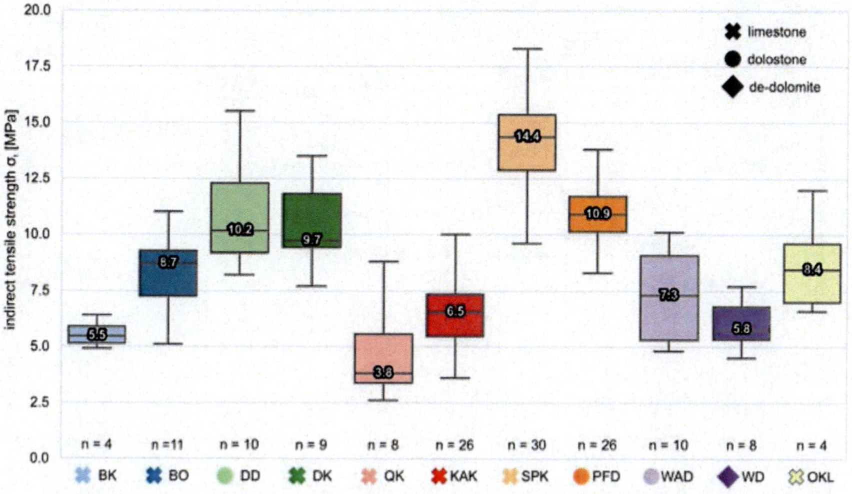

Fig. 5.8 Box plot of the indirect tensile strength σ_t of all analog rocks, uniquely colored and classified after their carbonate type, with the rock legend after Fig. 5.1

Especially SPK shows even larger whisker ranges of over 2.5 MPa, indicating even a higher spread of values.

5.1.4 Calculated and Derived Parameters from Analog Rocks

The previous Sects. (5.1.1–5.1.3) present properties from primary test results—i.e., parameters that are directly and purposely measured by a specific method. These parameters, e.g., the porosity, elastic wave velocity, or strength parameters, can produce new parameters by multiplying (e.g., the acoustic impedance) and can afterward be put in an empirical relationship with other parameters. This Chapter presents these newly calculated parameters and applies them to empirical models to derive supplemental results.

5.1.4.1 Acoustic Impedance Z_P and Z_S of the Analog Rocks

Figure 5.9 shows the acoustic impedance results as a box plot diagram with the median and statistical dispersion for the acoustic impedance of the P-wave Z_P (upper box plots) and the S-wave Z_S (lower box plots) for a total of 124 specimens. Z_P results from the P-wave velocity multiplied with the rock's bulk density, hence Z_S from the S-wave correspondingly, as explained in Sect. 4.2.2.5. Thuro et al. [1] published the values for the applied porosity. The unit generally used for Z is (km/s)*(g/cm^3).

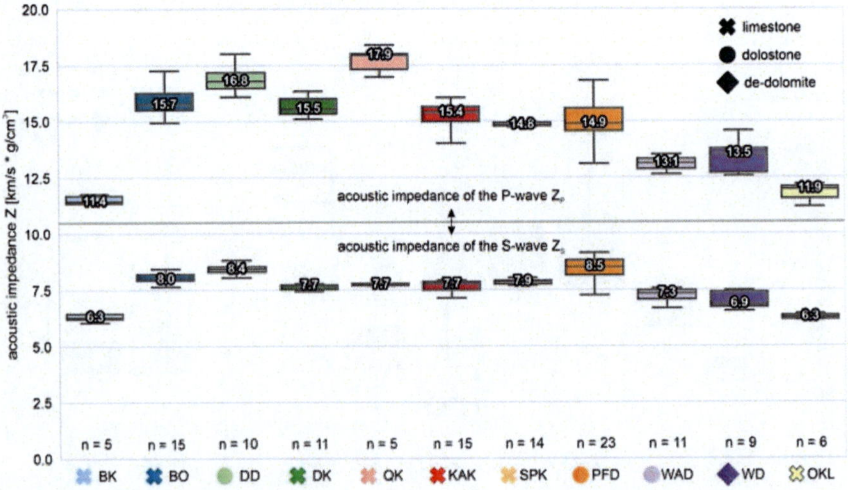

Fig. 5.9 Box plot of the acoustic impedance from the P-wave Z_P (upper boxes) and the S-wave Z_S (lower boxes) of all analog rocks, uniquely colored and classified after their carbonate type, with the rock legend after Fig. 5.1

The lowest values have the limestones BK and OKL with a Z_P of 11.4 and 11.9 and a Z_S of 6.3 for both. Followed by the dolostone WAD and the de-dolomitic WD, the median of Z_P is 13.1 and 13.5, with a median for Z_S of 7.3 and 6.9, respectively. Noticeably for both rocks, the larger Z_P has a smaller value for Z_S and vice versa. Subsequently, in increasing values, follow values of Z_S between 7.7 and 8.0 for all remaining limestones. While the values for Z_S are fairly constant at a maximum difference of 0.3, the medians for BO with 15.7, for DK with 15.5, for QK 17.9, for KAK with 15.4, and finally for SPK with 14.8 vary strongly by 3.1. The dolostones DD and PFD have a consistent median at 8.4 and 8.5 for Z_S but spread further for Z_P with 16.8 and 14.9. The statistical dispersion Z_S only shows slightly extended IQRs for PFD and WD with ranges in between one. Whiskers are negatively elevated for KAK, PFD, and WAD, while there is a slightly increased positive whisker for PFD. Z_P, in comparison, shows higher statistical dispersions. While the IQR for all rocks, except BK, SPK, WAD, and OKL, is at least one, PFD and WD are at their highest with 1.5. Whiskers are noticeably high for BK, DD, KAK, PFD, and WD. PFD remarkably shows a spread of 4 (km/s)*(g/cm^3).

5.1.4.2 Z_P and Correlating Parameters

Section 4.2.2.5 shows the P- and S-wave ratio (v_P/v_S) plotted against the acoustic impedance Z_P and therefore presented a way to classify carbonates in either Gas or Brine Carbonates after [4].

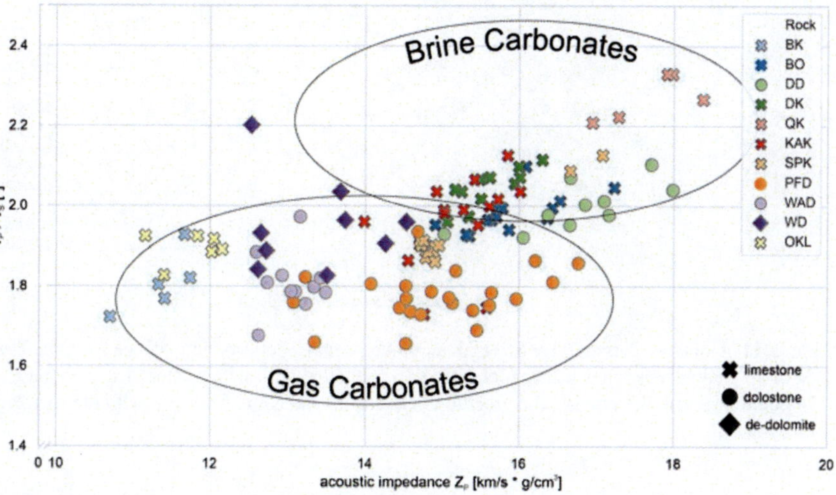

Fig. 5.10 v_P/v_S-ratio plotted against acoustic impedance from the P-wave Z_P of all analog rocks in a scatter plot, uniquely colored and classified after their carbonate type. The margins for the Gas and Brine Carbonates are after [4], with the analog rock legend after Fig. 5.1

Figure 5.10 acquires an excerpt from the diagram from Sect. 4.2.2.5, Fig. 4.49, and inserts 123 values for the eleven analog samples. Although some points fall in both categories and one outlier drops out of both, a clear subdivision is possible.

Hence, BK, OKL, WD (except one outlier), WAD, and PFD cluster into the "Gas Carbonates" ellipse. Also, a majority of SPK accumulates in there but has two outliers that locate clearly in the "Brine Carbonates". BO, KAK, DK, and DD are located in both ellipses, although most of these samples take their place in the "Brine Carbonates". QK plots in the "Brine Carbonates" sector only.

Remarkably, [4] created the two ellipses from tests on dry and saturated samples of three different carbonate rocks from alpine Triassic carbonates. These carbonates plot, when dry, in the "Gas Carbonates" section and when saturated in the "Brine Carbonates". However, the analog samples were tested after drying for at least 24 h at 105 °C. Hence, there still must be other criteria that influence V_P/V_S and Z_P.

Müller and Pipping [5] present a way to use Z_P to derive the Uniaxial compressive strength from rocks. Their empirical correlation suggests that the uniaxial compressive strength calculates from the equation attached to the blue line in Fig. 5.11, where 123 pairs of values from σ_u plot against Z_P. Only three single values from PFD, DK, and SPK meet the prediction from [5], while the majority of tested sets submerge that line. However, SPK shows higher strength than predicted by [5].

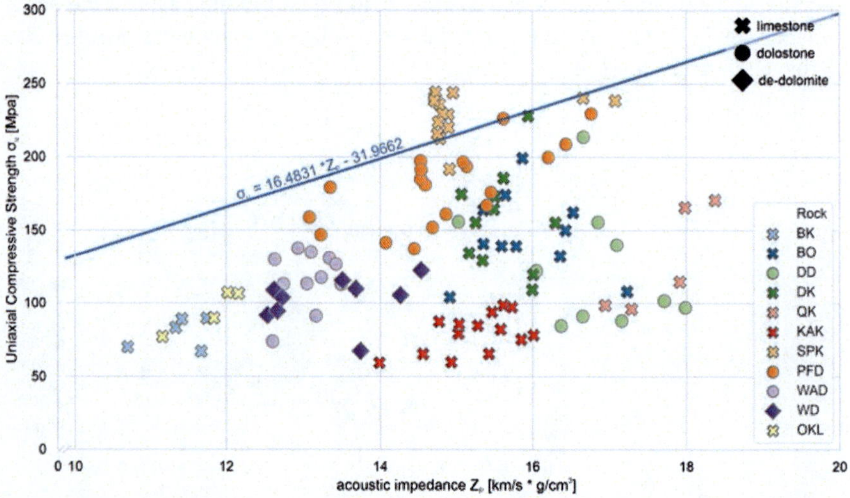

Fig. 5.11 Uniaxial Compressive Strength σ_u plotted against acoustic impedance from the P-wave Z_P of all analog rocks in a scatter plot, uniquely colored and classified after their carbonate type. The blue line with the empirical correlation between σ_u and Z_P is from [5], with the rock legend after Fig. 5.1

5.1.4.3 Z_s and Correlating Parameters

Besides calculating σ_u from Z_P, [6] suggest using the product of the S-wave velocity and the bulk density to estimate the shear strength τ_s of rock. Although no shear strength tests are available for the analog rocks, Z_S can be used with the empirical slope in Fig. 5.12 (blue line) to determine τ_s after [6]. Figure 5.12 plots the medians of Z_S of all analog rocks along the x-axis.

From the left to the right, BK and OKL plot first and show a shear strength of 27.5 MPa. Followed by the de-dolomitic WD with a τ_s of 31.2 MPa. The dolostone WAD has a calculated τ_s of 33.7 MPa. For the three limestones KAK, QK, and DK, which share the same median for Z_S, a τ_s of 36.1 MPa derives. Followed by SPK and BO with a τ_s of 37.4 and 38.0, respectively. At last, DD and PFD have the highest calculated shear stresses with 40.4 and 41.0. The maximum whisker of PFD would equal a maximum shear strength of 45 MPa. Table 5.1 summarizes the absolute values of τ_s for the equivalent analog rocks after [6].

5.1.5 *Density, Isotropy, Porosity and Grainsize of the Analog Rocks*

The determination of density, porosity, and grain size was part of the work on the analog rocks' lithological properties, published by [7]. Thuro et al. [1] extended

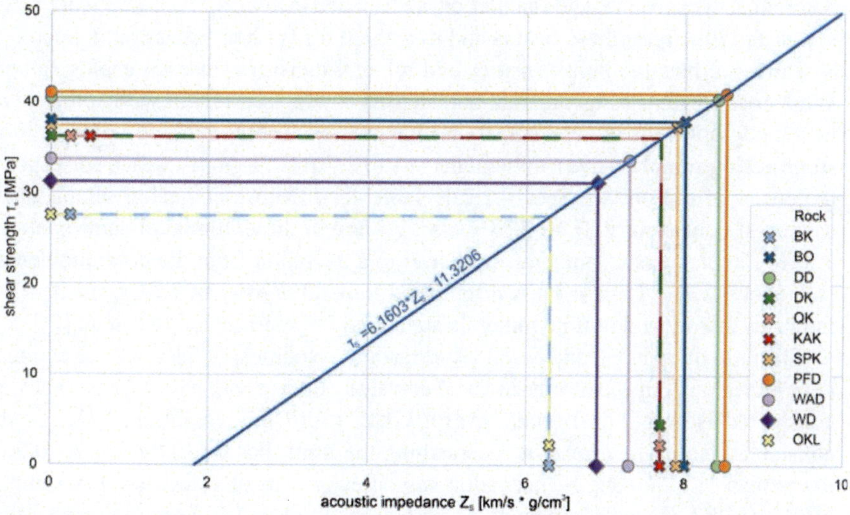

Fig. 5.12 The analog rocks in their unique color plot according to their median of Z_S along the x-axis and project vertically to the blue line, which correlates τ_s with Z_S, with the rock legend after Fig. 5.1

Table 5.1 Derived shear strength τ_s from values of the acoustic impedance Z_S after [6] for all analog rocks, uniquely colored and classified after their carbonate type, with the rock legend after Fig. 5.1

Analog rock	Lithology	Marker	Parameters	
			Z_S [(km/s)*(g/cm^3)]	Shear strength τ_s after [6] [MPa]
BK, OKL	Limestone		6.3	27.5
WD	De-dolomite		6.9	31.2
WAD	Dolostone		7.3	33.7
KAK, QK, DK	Limestone		7.7	36.1
SPK			7.9	37.4
BO			8.0	38.0
DD	Dolostone		8.4	40.4
PFD			8.5	41.0 (max: 45.0)

these studies and added the (An-)Isotropy-index. The development of this index, its theoretical background, and calculation is explained in Sect. 4.2.2.3. This work newly seizes and illustrates these results and uses them for further discussion in Sect. 6.1.

Table 5.2 lists the bulk densities and mineral densities for each analog sample. While rocks differ strongly in their bulk density—which correlates with porosity (see next paragraph), the mineral density is characteristic for the lithology. Hence, limestone analog samples have a mineral density of 2.70 - 2.71 g/cm^3, which corresponds to density of calcite (see Sect. 4.2.2.1, Table 4.1). Dolostone analog samples have a mineral density of 2.81 to 2.83 g/cm^3. Although the dolomite content is almost 100%, some accessory minerals may lead to a deviation from the dolomite density (see Sect 4.2.2.1, Table 4.1). A diminished mineral density of 2.81 g/cm^3 for WD indicates that most dolomite minerals are already transformed into calcite [7].

The ratio of axial v_{Pa} and radial v_{Pr} ultrasonic velocities, as Sect. 4.2.2.3 explains, determines an (An-) Isotropy-index. For values $0.95 < v_{Pa}/\ v_{Pr} < 1.05$, rocks are considered isotropic. All rocks, except OKL, fulfill this condition. WAD is still considered isotropic, even if it is scratching the limit. For OKL, the v_{Pa}/v_{Pr} ratio is lower than 1, indicating higher radial velocities than axial velocities. The sampled blocks of OKL show bedding with a small layer thickness [8]. Specimen preparation happened perpendicular to the bedding, indicating that the layering influences the rock's mechanical behavior.

Table 5.2 Density and Anisotropy for all analog rocks, uniquely colored and classified after their carbonate type, with the rock legend after Fig. 5.1

Analog rock	Lithology	Marker	Median density (g/cm^3)				Isotropy	
			n	Bulk density	n	Mineral density	Value	Label
BK	Limestone	✖	8	2.49	5	2.71	1.02	Isotropic
BO		✖	8	2.56	5	2.72	1.00	Isotropic
DD	Dolostone	●	10	2.67	10	2.81	1.01	Isotropic
DK	Limestone	✖	11	2.56	11	2.71	1.00	Isotropic
QK		✖	5	2.68	5	2.70	1.00	Isotropic
KAK		✖	14	2.53	14	2.71	1.00	Isotropic
SPK		✖	16	2.60	3	2.70	1.00	Isotropic
PFD	Dolostone	●	12	2.71	12	2.82	1.03	Isotropic
WAD	Dolostone	●	11	2.62	11	2.83	1.04	(an)-isotropic
WD	De-dolomite	◆	9	2.60	9	2.81	1.00	Isotropic
OKL	Limestone	✖	6	2.35	6	2.71	0.92	Anisotropic

Derived from the bulk density and the mineral density, Thuro et al. ([1]: 39ff) determine porosity values by a helium pycnometer, kindly conducted and provided by Daniel Bohnsack from the Chair of Hydrogeology, TUM. Figure 5.13 shows the median and statistical distribution for the effective porosity of all analog rocks. All analog samples show a median for effective gas porosity between 4.3% of the PFD and 8.2% for BK. BO, DD, DK, KAK, PFD, and WAD vary by more than 2% in porosity. While BK and SPK only fluctuate marginally, WD varies by more than 6%. Exceptionally, QK has a gas porosity of 0.8% with no fluctuation. With a median of 13.4%, OKL has the highest porosity that may reach up to almost 15%.

Mraz et al. [7] identified different grain sizes and skeletal structures with thin section examinations on each of the eleven analog rocks. Figure 5.14 shows the range of grain size within the rocks and colors them according to their skeletal structure. Since the publication of [7] contains only the upper and lower limits for the respective grain sizes and no count of grains, this range is shown as a bar chart with uniform distribution. The classifications of the limestones, written in the bars, are according to the modified Dunham Classification System [9, 10].

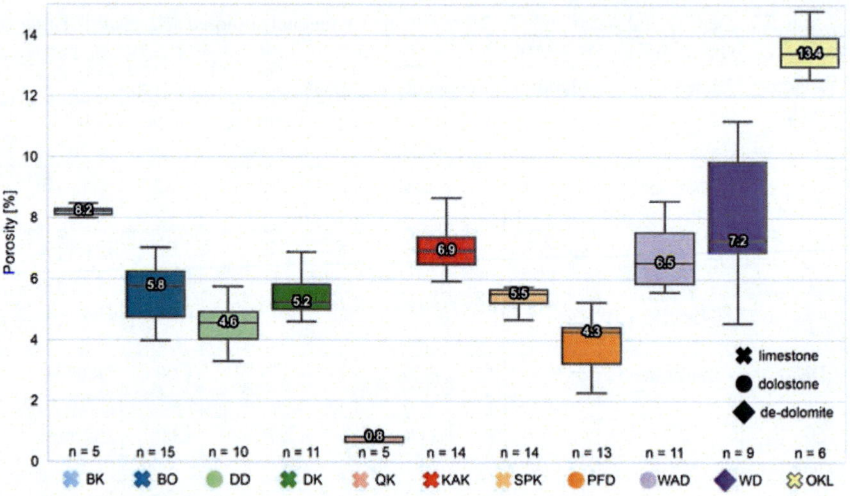

Fig. 5.13 Porosity values in percent for all analog rocks, uniquely colored and classified after their carbonate type, after Thuro et al. ([1]: Table 50), with the rock legend after Fig. 5.1

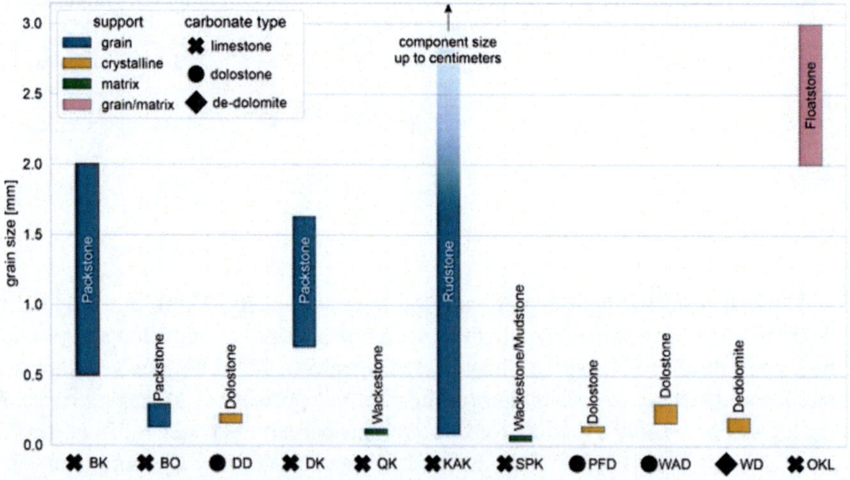

Fig. 5.14 Grain size for all analog rocks, colored according to their skeletal support and classified after their carbonate type, after [7], with the rock legend after Fig. 5.1

The dolostones and De-dolomite WD have a crystalline structure, built up by dolomite crystals with a grain size between 0.1 and 0.3 mm. PFD shows the lowest range with crystals between 0.10 mm and 0.14 mm, followed by DD between 0.16 and 0.23 mm, and WAD with the highest range of 0.16 to 0.30 mm, the maximum crystal size.

The Packstones BK, BO, and DK share a grain supported skeleton. This skeleton, however, is supported by grains of different sizes. BK shows grain sizes of 0.5 up to 2.0 mm. DK narrows this range down to 0.7 to 1.63 mm. BO ranges between 0.13 and 0.3 mm and hence is less than the minimum grain size of the other two Packstones. KAK is a grain-supported Rudstone that contains components up to cm. OKL is a partly grain- and matrix-supported Floatstone, where grain sizes in a range of 2–3 mm construct the skeleton. QK is a Wackestone and SPK a Wackestone/Mudstone. Both are matrix-supported, and their grain size is around 0.1 and 0.055 mm, respectively.

5.2 Drill Cores

Analog Samples give a good overview of the prevailing carbonate rocks representing the host rock. However, in situ rock samples reveal the reservoir's true rock properties and provide a real insight into the rock mass, including its integrity, exposure to stresses, its elastic and strength parameters, and the fracture network.

5.2.1 Geological and Rock Mechanical Description of the in situ Rock Cores

The description of the cores focuses on the engineering and mechanical properties according to application and nomenclature of the German standard [27] and the suggested methods from the international society for rock mechanics; specifically, [11], Suggested Methods for the Quantitative Description of Discontinuities in Rock Masses, and [8] with the classification for rock and rock masses. Thuro et al. [1] and its appended report from [12] provide information on the cores' lithology. The core runs, according to their depth, are listed in Sect. 3.3.2.1, Table 3.2.

Core Run 1 (4593 m TVD) consists of pure dark, black, micritic, and dense limestone. Single, macroscopically hardly recognizable joints have a bituminous coating. Visible joints are healed and did not break during drilling. Fragments of the core that did not abrade during the jamming fractured conchoidally. The Total Core Recovery (TCR) is 6.4% with a Rock Quality Designation Index (RQD) of less than 1% (see Fig. 5.15).

Core Run 2 (4599 m TVD) connects directly to the milled area of Core Run 1. The dense impression of the rock continues at the beginning of the CR until it is abruptly interrupted by a perceptibly less dense, grey, yet still micritic segment of around 45 cm. This segment delivered the only successful overcored 5 cm specimen (see Sect. 5.2.2, Fig. 5.16). A single joint, accompanied by a fine lamination, intersects the core with a dip of 76°. This joint also shows signs of stylolites. To the end of the CR 1, the micritic limestone becomes dense and darker again. Inactive joints are hardly visible in the original core but become evident in the 2 cm plugs (Fig. 5.18). The TCR of CR 2 is 40%, while the RQD is 15.4% (Fig. 5.15).

Core Run 3 (4652 m TVD) is 50 m deeper than the first two CRs. Still, dark, black, dense, and micritic limestone, fracturing splintery, and conchoidally, is characteristic. However, various joints match with joint sets, and newly—a joint system is indicative. One of these joint sets is dipping at around 83° with the core axis and recurs in extremely close to very close spacing distances (<20–60 mm) along the whole CR. These joints are partly open, hence limited mechanically effective. Their surface is—mirror-likely—polished, and after ISRM terms [11] smooth, planar. After [12], stylolites appear in sections on this surface. Perpendicular and steeply oblique to this joint set, several other joints emerge. Although visible, they differ from the previously described joint set as they are closed, hence mechanically effective. However, the cores' small diameter does not allow to properly measure these joints' orientation and group them in sets. Therefore, the determination of spacing was not possible. Despite a core recovery of 80% (TCR), the RQD is only 6.4% (Fig. 5.15).

The limestones from the consecutive **Core Run 4** (4652 m TVD) shares the properties of CR 3. The steeply inclining joint set has the same dip as in CR 3. While a part of the core run fragmented during the coring of CR 3 or CR 4, still 83% was recovered, from which 32% were longer than 100 mm (RQD). While conducting US long-term testing on these cores, they still were intact with a full lateral surface. However, after transporting the cores to Berlin and afterward to Munich, these cores

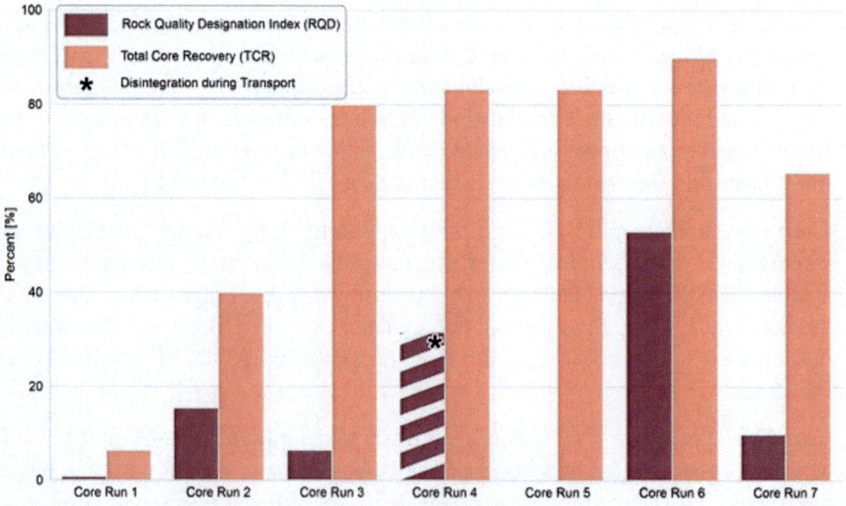

Fig. 5.15 Rock Quality Designation Index (RQD) and Total Core Recovery (TCR) of all seven core runs in percent. The striped RQD column of Core Run 4 shows the reduction of the RQD from 32% to zero as the core pieces disintegrated during transport (after [13])

disintegrated along with the steeply inclining joint sets, reducing the RQD of CR 4 to zero (Fig. 5.15).

Core Run 5 (4710 m TVD) started 50 m deeper than CR 4, maintains the attributes from the hanging dark, black and micritic limestones, and shares the bituminous coated joint surfaces with the stylolites. Unlike the previous runs, the core disintegrated axially in discs with thicknesses of 0.5–2 cm. This type of disintegration follows a clear pattern, while disintegration in the previous CRs seemed mostly random. Due to the fine discing, the RQD is 0%, despite a TCR of 83% (Fig. 5.15).

The first 40 cm of **Core Run 6** (4712.5 m TVD) start equally disintegrated as CR 5 but changes to longer consistent cores with a maximum length of 47 cm. A total core recovery of 90% (TCR) comes with an RQD of 53% (Fig. 5.15). Noticeably, bituminous coated joint surfaces decrease to the end, and the lithology varies around a Mud-, Wacke, and Packstone until it migrates to a strongly brecciated material at the end of the CR. Several calcite-healed joints with thicknesses up to several centimeters could be traced and measured alongside the cores. Although the core pieces are macroscopically in good condition, only slight torsion or minimal bending is sufficient to break the cores along their longitudinal axis. Core pieces that come from 5382.5 m MD and below show saddle-shaped surfaces, presumably from core disking (compare to Sect. 5.3.1.1). These saddle-shaped paths can also be traced alongside entire core pieces.

Core Run 7 (4714 m TVD) concludes the coring operation, which recovered 65.5% of cores with an RQD of 10%. The rock is limestone breccia in the transition to dolostone. While missing in CR 5 and 6, the steeply inclining joint set reappears with a dip of 85° even steeper than in the CRs 1, 2, 3 & 4. Two of these joints appear in the core pieces and have a spacing of approx. 60 mm. However, the joint surface lacks stylolites and the polished feature as it was the case for the previous sets.

5.2.2 Overcored Specimens and their Rock Mechanical Parameters

Thuro et al. [1] published the results on the overcored specimens already in the report for the research project "Dolomitkluft". The first overcoring happened on the intact rock core from 5036.1 to 5036.4 m MD. A successful overcoring job resulted in a specimen with a diameter of 5 cm and a length of 13 cm, despite the intersecting joint (Fig. 5.16). The joint dips with an angle of 76°. Table 5.3 lists the results from non-destructive testing. A discontinuity/lamination pierces the specimen and forecloses the sample from destructive testing since it likely would falsify the true rock strength.

The results show a dyn. Young's modulus of 35.9 GPa, with a Poisson's ratio of 0.32. Ultrasonic tests show three different velocities according to the direction of elastic waves. In a lateral direction, the P-wave velocity is 4820 ± 270 m/s, when acting parallel with the joint. Perpendicular to the joint, the velocity is reduced by

Fig. 5.16 Successfully overcored specimen from Core Run 2 from 5036.20 to 5036.31 m MD, modified after ([1]: Abbildung 74)

Table 5.3 Dynamic rock properties, elastic wave velocities, density, and porosity from the overcored specimen from 5036.20 – 5036.31 m MD (from [1]: Tabelle 30))

Parameter	n	Value		
Marker				
Dyn. Young's modulus (GPa)	1	35.9		
Dyn. Poisson's ratio (-)	1	0.32		
P-wave velocity v_P (m/s)		Lateral II joint	Lateral ⊥ joint	Axial
Lateral: n = 4, axial: n = 2		4820 ± 270	3900 ± 160	4500
S-wave velocity v_S (m/s)	2	2320		
Bulk density (g/cm^3)	1	2.52		
Mineral density (g/cm^3)	1	2.73		
Porosity (%)	1	7.63		

almost 1000 m/s to 3900 ± 160 m/s. Axially, the v_P is 4500, and v_S is 2320 m/s, resulting in a v_P/v_S ratio of 1.94. Furthermore, the specimen has a bulk density of 2.52 g/cm^3, a mineral density of 2.73 g/cm^3, and an effective gas porosity of 7.63%.

After obtaining the first 5 cm specimen successfully, three further attempts on rock cores with a full lateral surface and a length of approximately 15 cm between 5384.00 and 5384.43 m MD failed. During overcoring, the rocks decomposed into discs with thicknesses of 5–30 mm (Fig. 5.17). Section 5.3.1.1 shows a statistical distribution of the disc thicknesses. Analyzations of overcoring results showed residues, cuttings, and disc segments thinner than 5 mm, implying that thinner discs may exist. The rotational strain from drilling, however, destroys them.

The red rectangle in Fig. 5.17 shows a specimen from 5384.20 to 5384.23 m MD with a diameter of 5 cm and a length of approximately 3 cm. For ultrasonic testing, the length of 3 cm is too short. However, the diametrical distance of 5 cm allows

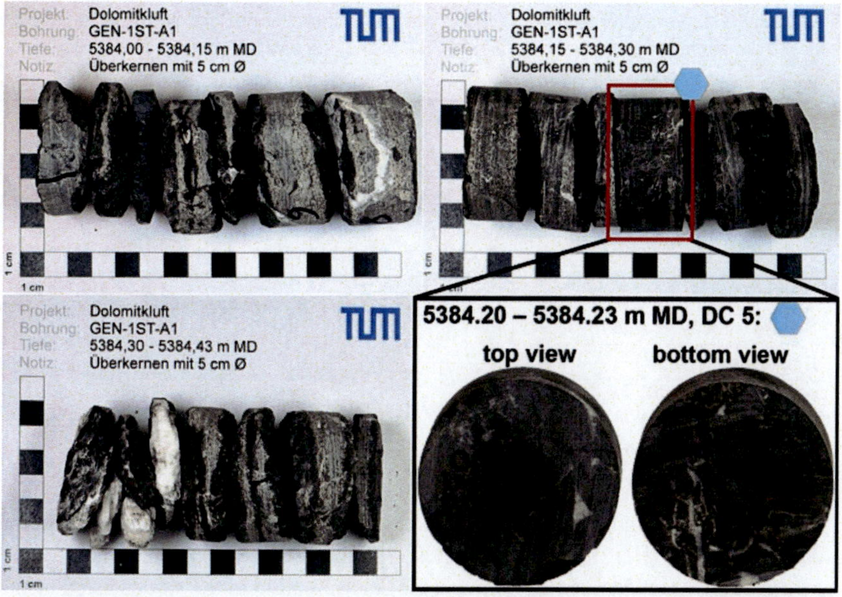

Fig. 5.17 Unsuccessful overcoring results on rock pieces from 5384.00 m MD to 5384.43 m MD. The intact cores disintegrated in discs during overcoring, modified after ([1]: Abbildung 75)

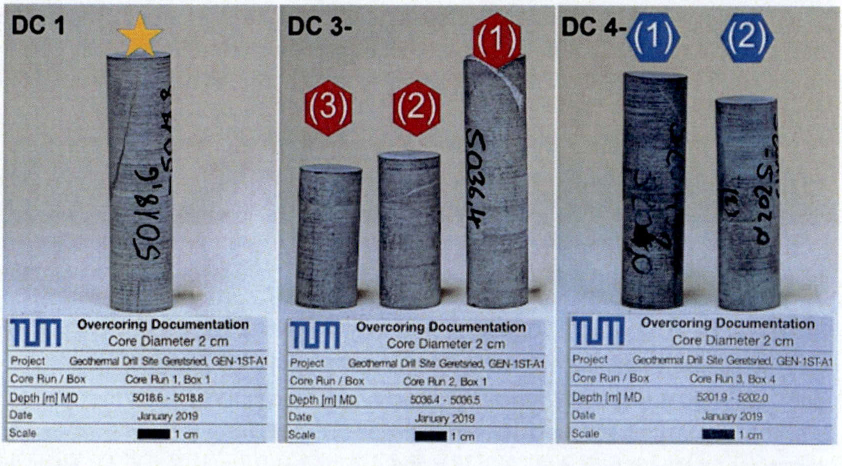

Fig. 5.18 Resulting specimens from overcoring with a 2 cm drill bit. DC 1) 7.8 cm long, from 5018.6 m MD, marked with a yellow star; DC 3) three specimens from 5036.4 m MD with 7.7 (1), 4.6 (2) and 4.2 (3) cm length, marked with red, numbered hexagons; DC 4) from 5201.9 m MD with 7.0 (1) and 6.3 (2) cm length, with blue numbered hexagons, modified after ([1]: Abbildung 76)

Table 5.4 Dynamic rock properties, elastic wave velocities, density, and porosity from the 3 cm long, 5 cm in diameter specimen from 5384.20 – 5384.23 m oMD, from ([1]: Tabelle 31)

Parameter	n	Value	
Marker			
Dyn. Young's modulus (GPa)	1	68.5	
Dyn. Poisson's ratio (-)		0.32	
P-wave velocity v_P (m/s)		5955	v_P/v_S-ratio
Lateral			1.93
S-wave velocity v_S (m/s)		3090	
Bulk density (g/cm^3)		2.69	
Particle density (g/cm^3)		2.71	
Porosity (%)		0.8	

measurements. Hence, the curved surface was milled flat, and planar S-wave transducers (method (III) from ultrasonic testing) allowed to obtain the rock's dynamic parameters. One measurement allowed the determination of the values listed in Table 5.4. This specimen has a dynamic Young's modulus of 68.5 GPa and a dyn. Poisson's ratio of 0.32. The ratio of v_P (5955 m/s) to v_S (3090) is 1.93. A bulk density of 2.69 g/cm^3 and a mineral density of 2.71 g/cm^3 assesses a 0.8% porosity.

Grinding curved lateral surfaces flat is difficult and time-consuming. Additionally, only a few of the discs fit for this procedure. Hence, the overcoring strategy changed from 5 to 2 cm drill bits. Thus, overcoring was then also possible for smaller and incomplete cores.

This adopted method resulted in six specimens from various depths with a diameter of 2 cm and different lengths (Fig. 5.18, DC 1 (yellow star), DC 3 (red hexagons), and DC 4 (blue hexagons)). All specimens fulfilled the requirements to obtain elastic properties with the resonance technique after method (II) from Sect. 4.2.2.4. Table 5.5 lists each specimen's mean values, calculated from 4 measurements (twice, rotated 90° around the axial length and again flipped upside down).

Figure 5.18, DC 1 shows a 7.8 cm long specimen from 5018.6 to 5018.8 m MD. Its dyn. Young's modulus is 75.3 GPa, and its Poisson's ratio is 0.35. The mean v_P and v_S is 6700 m/s and 3230 m/s, respectively, which results in a v_P/v_S ratio of 2.07.

From 5036.4 to 5036.5 m MD, three specimens could be overcored from **one** core piece with a length of 7.7 (1), 4.6 (2), and 4.2 cm (3) (Fig. 5.18, DC3). Their dyn. Young's modulus lies between 71.2 and 73.8 GPa, with a Poisson's ratio between 0.35 and 0.37. Their v_P ranges between 6543 m/s and 7030 m/s and their v_S between 3160 and 3213, resulting in a v_P/v_S ratio of 2.07, 2.06, 2.22, for (1), (2), and (3), respectively.

Figure 5.18, DC4 shows the deepest, successfully overcored plugs with lengths of 7 cm and 6.3 cm, also recovered from one piece of core. The longer specimen has a dyn. Young's modulus of 69.0 GPa with a Poisson's ratio of 0.38. Figure 5.18, DC4-2 has a Young's modulus of 80.9 GPa with a Poisson's ratio of 0.34. Their mean

Table 5.5 Dynamic rock properties, elastic wave velocities, density, and porosity from the overcored 2 cm specimens from 5018.6 to 5018.8 m MD, 5036.4 to 5036.5 m MD, and 5201.9 to 5020.0 m MD, assigned with their individual markers, modified after ([1]: Tabelle 32)

Parameter	Mean values (from four individual measurements)					
	5018.6–5018.8	5036.4–5036.5			5201.9–5020.0	
		(1)	(2)	(3)	(1)	(2)
Markers	★	1	2	3	1	2
Dyn. Young's modulus [GPa]	75.3	71.2	73.8	73.0	69.0	80.9
Dyn. Poisson's ratio [-]	0.35	0.35	0.35	0.37	0.38	0.34
P-wave velocity [m/s]	6700 ± 215	6543 ± 195	6615	7030 ± 280	7020 ± 264	6823 ± 110
S-wave velocity [m/s]	3230	3160	3213	3160	3050	3340
Bulk density [g/cm^3]	2.68	2.65	2.65	2.66	2.68	2.70
Particle density [g/cm^3]	2.70	2.68	2.69	2.71	2.71	2.73
Porosity [%]	1.52	1.59	1.24	2.13	1.03	1.05

v_P of 7020 and 6823 m/s, together with their mean v_S of 3050 and 3340 m/s, results in a v_P/v_S ratio of 2.30 and 2.04, respectively.

All plugs have a bulk density between 2.70 and 2.65 g/cm^3, with a mineral density between 2.68 and 2.73 g/cm^3. Helium pycnometer measurements show values of 1.03 to 2.13% porosity for all cores. Table 5.5 gives a holistic overview of dynamic properties, the density, and the porosity of all plugs.

5.2.3 Properties and Parameters derived from US Velocities

Similar to the analog samples, the drill cores' dynamic properties allow calculating the acoustic impedance Z_P and Z_S. Table 5.6 summarizes the for Z_P, and Z_S together with the v_P/v_S-ratio.

Section 5.1.4.1 shows that the analog samples classify into either a "Gas Carbonates" or "Brine Carbonates" category when the acoustic impedance Z_P is plotted against the v_P/v_S-ratio. Correspondingly, Figure 5.19 locates overcored specimen according to their parameter in the same plot as Fig. 5.10. In the background, the contours encompassing the distinct analog rocks' values, created by a bivariate kde-joint plot from Seaborn (Sect. 5.1, p. 82), underlie the in situ rock cores' markers transparently.

Table 5.6 v_P/v_S-ratio and acoustic impedance Z_P and Z_S for all overcored specimens, assigned with their individual markers.

Drill Core (DC) Depth [m MD] + (x)			Marker	v_P/v_S-ratio	Acoustic Impedance [(km/s)*(g/cm³)]	
					Z_P	Z_S
DC 1	5018.6–5018.8	-		2.07	17.9	8.6
DC 2	5036.2–5036.3	-		1.94	11.4	5.8
DC 3–1	5036.4–5036.5	(1)		2.07	17.3	8.4
DC 3–1		(2)		2.06	17.5	8.5
DC 3–1		(3)		2.22	18.7	8.4
DC 4–1	5201.9–5202.0	(1)		2.30	18.8	8.2
DC 4–2		(2)		2.04	18.4	9.0
DC 5	5384.15–5384.30	-		1.93	16.0	8.3

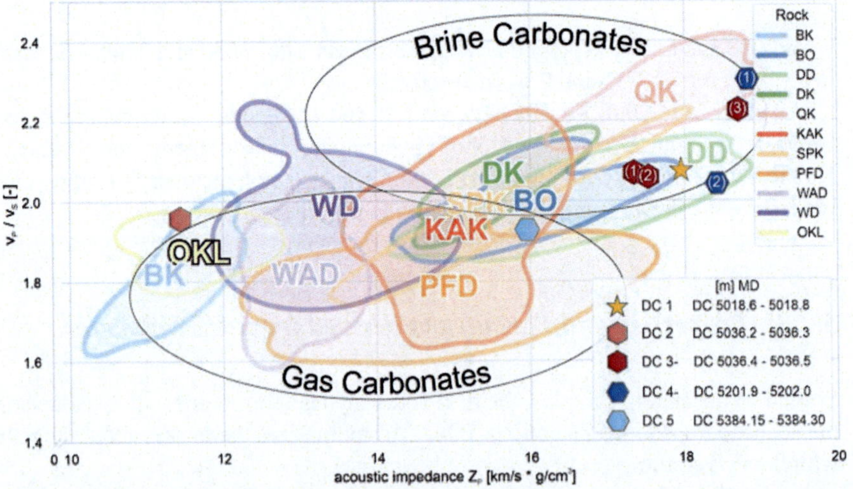

Fig. 5.19 v_P/v_S-ratio plotted vs. the acoustic impedance Z_P for all overcored specimens, assigned with their individual markers, with the analog rocks after Fig. 5.1

From 5018.6 to 5018.8 m MD, the shallowest specimen plots as a yellow star in Fig. 5.19 and finds itself directly in the contours of DD and at the border to BO. The only successfully overcored specimen with a diameter of 5 cm (5036.2 to 5036.3 m MD) has the lowest Z_P and plots as a pink hexagon on the left-hand side of Fig. 5.19, more or less at the intersection of the contours from BK and OKL. The specimens from 5036.4 to 5036.5 m MD, just a couple centimeters deeper than the previous sample, plot to Fig. 5.19' right-hand side in three, deep red hexagons. Number 1 and 2 of the red hexagons plot directly in the overlapping area of BO and DD. The third hexagon plots further to the right at the border of QK.

The 2 cm plugs from 5201.9 to 5202.0 m MD plot similar to Fig. 5.19's right-hand side in dark blue hexagons. Number 1 has the highest Z_P of all samples with 18.8 (km/s)*(g/cm^3). It plots directly in the contour of QK. Number 2 plots to the lower, right boundary of the DD contour.

At last, the deepest specimen plots as a light blue hexagon in Fig. 5.19. At this point, contours from KAK, BO, and DD overlap, and SPK and PFD margins are close. The contours of WAD and WD are not close to any of the overcored specimens.

As in Sect. 5.1.4.2, the acoustic impedance Z_P allows deriving the uniaxial compressive strength σ_u from the empirical equation after [5] also for the overcored specimens. Furthermore, [6] apply Z_S to derive the shear strength of the rock. Table 5.7 lists these values according to depth and marker.

According to the uniaxial compressive strength σ_u, the specimens DC 1, DC 3, and DC 4 exceed 250 MPa. DC 4-1 has the highest σ_u at 278.3 MPa, which is

Table 5.7 Uniaxial compressive strength σ_u and Shear strength τ_S, derived from Z_P and Z_S for all overcored specimens, assigned with their individual markers

Drill Core (DC) Depth [m MD] + (x)			Marker	Uniaxial compressive strength σ_u (MPa)	Shear strength τ_S (MPa)
DC 1	5018.6–5018.8	–		263.9	42.0
DC 2	5036.2–5036.3	–		156.6	24.7
DC 3–1	5036.4–5036.5	(1)		253.8	40.3
DC 3–1		(2)		256.9	41.1
DC 3–1		(3)		276.2	40.4
DC 4–1	5201.9–5202.0	(1)		278.3	39.1
DC 4–2		(2)		271.7	44.2
DC 5	5384.15–5384.30	–		231.1	39.7

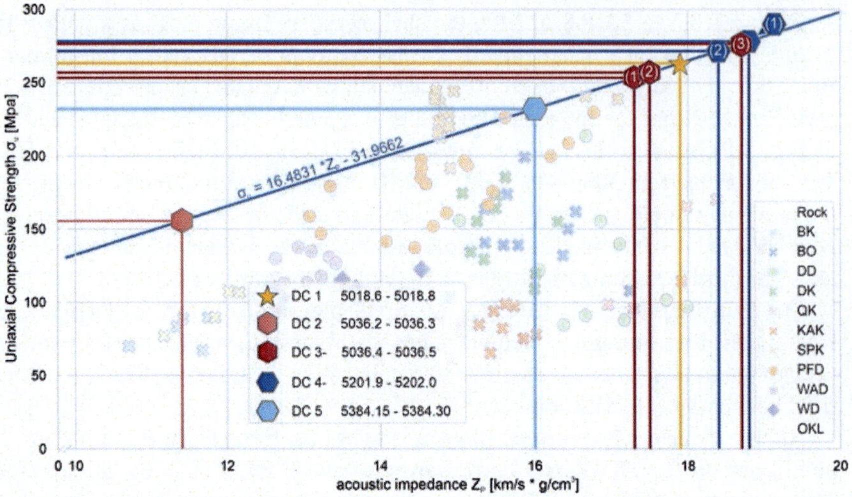

Fig. 5.20 Location of the overcored specimens, assigned with their individual markers, according to their acoustic impedance for Z_P along the x-axis. The blue line with the empirical correlation derives the uniaxial compressive strength for the overcored specimen after [5], with the analog rocks after Fig. 5.1 in the background

"exceptionally high" after [27]. These derived values exceed the measured parameters from the analog samples, as illustrated in Fig. 5.20. DC 5 has a derived σ_u of 231.1 MPa and lies congruently with one single value of DK (which hides behind the light blue hexagon, Fig. 5.20). This uniaxial compressive strength would also agree with results from SPK and single high values of PFD. Lastly, DC 2 has a calculated σ_u of 156.6 MPa, which is twice the strength of analog rocks with comparable Z_P.

Figure 5.21 locates Z_S of the drill cores plotted against the shear strength with the regression line after [6], equal to Fig. 5.12. DC 2, with a Z_S of 5.8, has the lowest calculated shear strength of 24.7 MPa. Its calculated shear strength is lower than τ_S of the analog samples BK and OKL. The samples from DC 4 frame the values of the remaining samples. DC 4-1 has a calculated shear strength of 39.1 MPa, while DC 4-2 has the highest shear strength of all samples with 44.2 MPa.

Ascending in between DC 4, DC 5 follows with a τ_S of 39.7 MPa. Consecutively, DC 3 ranges between 40.3 and 41.1 MPa. Finally, DC 1 has the second-highest calculated shear strength with 42.0 MPa. DC 1 and DC 3 to DC 4-1 are close to the values of Z_S of DD and PFD. DC 4-2 agrees with the maximum value of PFD at 45 MPa (compare Sect. 5.1.4.3).

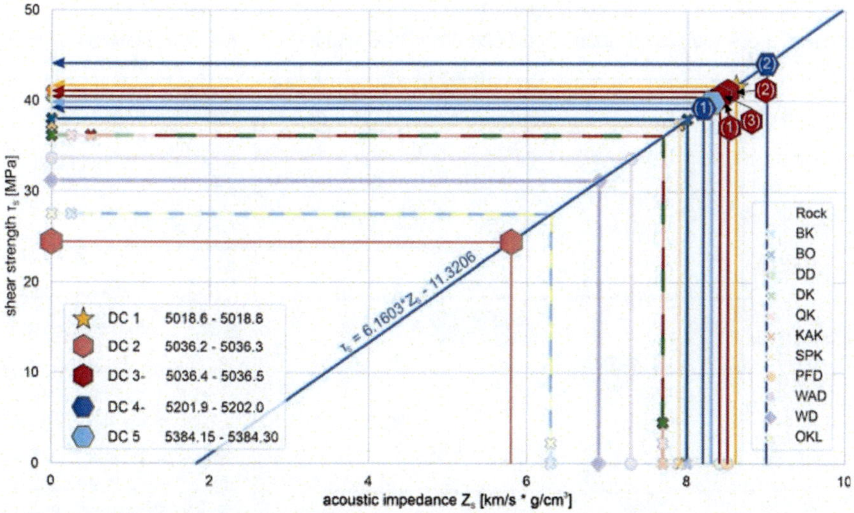

Fig. 5.21 Position of the overcored specimens, assigned with their individual markers, according to their acoustic impedance for Z_S along the x-axis. The blue line with the empirical correlation derives the shear strength for the overcored specimen after [6], with the analog rocks after Fig. 5.1 in the background

5.2.4 *US Measurements on the in situ Rock Cores*

While the previous Chapters deal with the visual, geological, and rock-mechanical description (Sect. 5.2.1), obtaining absolute dynamic parameters and looking for comparability and deriving strength parameter for these rocks, this Chapter focuses on the elastic wave velocities independently from the quantitative estimation of parameters. The observation of the ultrasonic velocities should bring new insights into carbonate rocks' mechanical behavior, scale-, stress-dependency, time-dependent development of US velocities, and hence, fracture propagation. Section 6.2 uses these results for further discussing this topic in depth.

5.2.4.1 US Velocities of the in situ Rock Cores

One of the latter defined conditions for successfully overcoring original rock cores with a 2 cm drill bit was an axial P-wave velocity larger than 4 km/s (Sect. 4.2.1.2). Therefore, the whole core run was examined with the US technique after Sect. 4.2.2.4 method (I) and, where flat surfaces were available or could be created, method (III). Hence, Section 3 lacks S-wave measurements. The overcored plugs are measured with the resonance method (II) (see Sect. 4.2.2.4). In total, 551 US measurements were conducted alongside the cores. Figure 5.22 shows the statistical distribution of the elastic wave velocities with mean and median, divided into three sections. Section 1 includes Core Runs 1 and 2, Section 2 includes Core Runs 3 and 4, and,

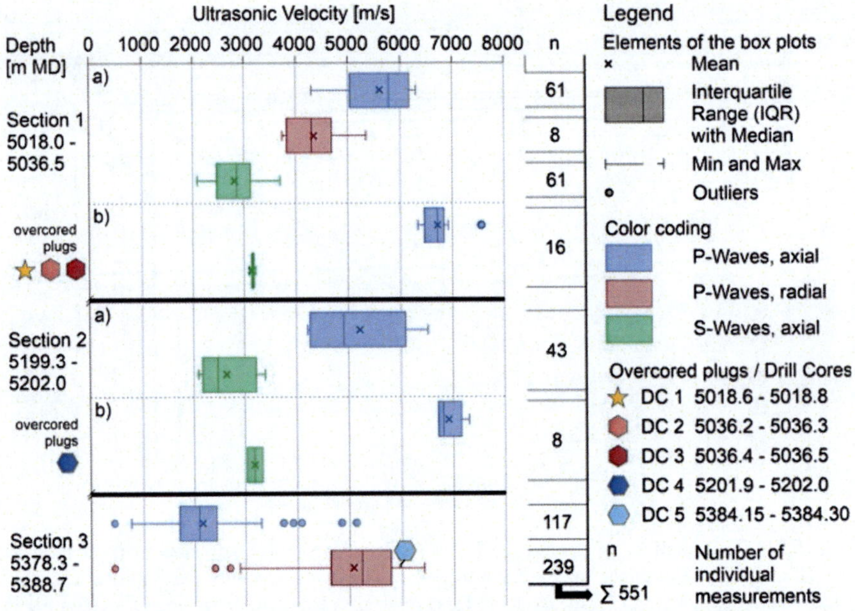

Fig. 5.22 Ultrasonic wave velocities of the original rock cores and their overcored specimens are summarized in three sections that correspond to the coherent core runs. Axial P-wave velocities are available for all sections (blue boxes). Depending on the cores' geometry, S-wave (green) and radial P-wave velocities (red boxes) were determined. The elastic waves' colors correspond to Fig. 5.27 and the long-term US velocities following in Sect. 5.2.4.2

finally, Section 3 combines all measurements from the core pieces of Core Runs 6 and 7. While Section 3 in Fig. 5.22 condenses all radial measurements in one boxplot, Sect. 5.2.4.3 looks more attentively at this section and plots the ultrasonic velocities versus the depth, subdividing minimum and maximum radial and axial velocities.

Section 1 opposes the US measurements from the original core pieces from 5018.0 to 5036.5 m MD (Fig. 5.22, Section 1a) against their overcored plugs (Fig. 5.22, Section 1b). From a total of 61 measurements with S-wave transducers, the axial v_P ranges between a minimum of 4300 m/s to a maximum of 6200 m/s. The median is at 5790 m/s, and the mean velocity at 5610 m/s. The IQR closely 1200 m/s and locates between 5000 m/s a little over 6000 m/s. Eight radial measurements on core sections with a full lateral surface or at least a distance of a full diameter range between a minimum of 3750 m/s to a maximum of 5380 m/s. Data shows that the median and mean lie close together with velocities of 4300 and 4360 m/s, respectively. The IQR with a range of almost 900 m/s starts at 3800 m/s and ends at 4700 m/s. S-wave velocities show the minimum and maximum velocities of 2090 and 3730 m/s. Median and mean lie together pretty evenly with 2880 and 2840 m/s. The IQR of 600 m/s lies majorly below 3000 m/s.

The overcored plugs (DC 1, DC 2 & DC 3) from Fig. 5.22, Section 1b show a median and mean for v_P of 6740 m/s from 14 measurements on 4 samples (Fig. 5.22a, b). The whiskers range between a minimum v_P of 6350 m/s and a maximum v_P of 6950 m/s. The IQR extends from 6480 m/s to 6850 m/s, covering a velocity range of 370 m/s. One outlier at 7590 m/s is excluded from the statistical evaluation. The median and mean of the shear wave velocity v_S specimens from Section 1 is 3180 m/s. The maximum extent of the whiskers is 100 m/s.

Section 2 opposes the US measurements from the original core pieces from 5199.3 m MD to 5202.0 m MD (Fig. 5.22, Section 2a) against their overcored plugs (Fig. 5.22, Section 2b). A lack of complete core diameter did not allow the measurement of radial v_P. In total, 43 measurements were conducted to obtain P- and S-wave velocities. The P-wave velocities of all cores range between 4230 m/s to 6540 m/s. Median and mean differ by almost 300 m/s at 4940 m/s and 5230 m/s, respectively. The IQR extends over 1800 m/s between 4000 m/s to 6000 m/s. v_S starts at 2100 m/s and climaxes at 3400 m/s. Similar to v_P, the median is lower than the mean, with a v_S of 2480 m/s and 2680 m/s, respectively. The IQR starts at 2200 m/s and ends at 3240 m/s, and locates mostly below 3000 m/s.

From 8 measurements on two overcored plugs (DC 4), v_P has a median and mean of 6850 m/s and 6920 m/s, respectively, with a minimum value of 6710 m/s and a maximum value of 7330 m/s. The IQR starts at proximity to the minimum at 6720 m/s and ends at 7170 m/s. The S-wave velocity shares the median and the mean at 3200 m/s with an equally negative and positive distribution of 150 m/s.

Noticeably, all the v_P of the overcored specimens from Sections 1 and 2 are higher than those of the core pieces. The S-wave velocities of the overcored specimens concentrate at the top of the IQR of the core pieces.

In total, 356 v_P-measurements were performed along the Core Runs 5 to 7. 117 axial measurements come along with 239 radial measurements, performed in 90° steps around the core pieces.

Figure 5.22, Section 3, shows the statistical distribution of the axial and radial v_P. The axial velocity has a median and a mean value of 2120 m/s and 2170 m/s, framed by whiskers with the lowest value of 780 m/s and fastest velocities of 3300 m/s. The IQR starts at 1710 m/s and ends at 2460 m/s. Six outliers are not accounted for in the boxplot. One with a velocity of 450 m/s to the lower side and five with faster velocities of 3750 m/s to 5150 m/s. The radial velocity shows a median and a mean of 5250 m/s and 5090 m/s. The whiskers cover a velocity range of 3500 m/s, starting at 2930 m/s and ending at 6450 m/s. The IQR starts at 4650 m/s and ends at 5800 m/s. The boxplot neglects three outliers with slower velocities of 2730 m/s, 2430 m/s, and 450 m/s.

The overcored sample DC 5 is illustrated in Fig. 5.22 and falls with a P-wave velocity of 5955 m/s in the upper whisker of the measured radial velocities.

Remarkably, there is a change in the ratio of axial to radial velocity from Sections 1 to 3, or in other words, from shallow to deep. The rock pieces from Section 1 show higher axial velocities than radial velocity, where the median of both differ by 1400 m/s, and the IQRs do not overlap overall. Conversely, Section 3 shows larger radial velocities than axial velocities by a difference in median of more than 3100 m/s and, also, the IQRs do not overlap.

5.2.4.2 Evolution of P-wave Velocities in the Rock Cores over Time

Thuro et al. ([1]: 97ff) presented these results in the research report "Dolomitkluft". Guided by the idea that the measurements specify either the prevailing in situ stress field, the process of fracture initiation and propagation, or temporal growth, these results are newly prepared and presented in the following Chapter.

Once recovered from the reservoir and described, the following criteria applied for the selection of rock pieces for long-term ultrasonic measurements:

- Full lateral surface
- Length of at least 10 cm
- Macroscopically undamaged habitus

Hence, from seven core runs, 6 samples met these criteria and were tested consequently. Table 5.8 gives an overview of the samples, labeled with LT-US and the chronological number of selections. From these six long-term examinations, three test series finished successfully, and three failed. The unsuccessful LT-US 1 and LT-US 3 did not allow the US transducers to connected properly on the shattered radial surface. LT-US 6 did not conduct a sufficiently strong signal.

The test series LT-US-2, from CR 4 in 5204.2–5204.4 m MD, lasted 456 h. Two steep inclining joints pervade this rock sample. These joints served as an orientation for the radial US measurements. One direction was selected perpendicular to the joints (Fig. 5.23b), black markers), the other direction turned by 90°, parallel to the

Table 5.8 Summary of the long-term ultrasonic conducted on the rock cores with the start right after recovery from the core barrels

Test /Sample	Depth [m MD]	Duration [h]	# of measurements	P-wave		Remark
				Axial	Radial	
LT-US 1	5199.10–5199.30	aborted	12	✖		Shattered radial surface
LT-US 2	5204.20–5204.40	456	36	✖	✔	Dissected by joints, measurement parallel and normal to the joints
LT-US 3	5374.20	aborted	12	✖		Shattered radial surface
LT-US 4	5379.55–5379.80	226	40	✖	✔	No arrival of the axial P-wave signal
LT-US 5	5381.00–5381.10	226	18	✔	✔	–
LT-US 6	5389.20–5389.50	aborted	14	✖		No arrival of the axial and radial P-wave signal

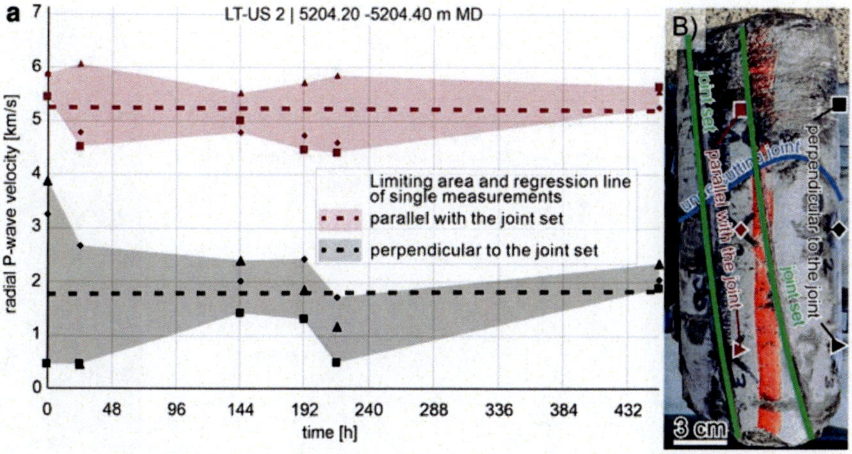

Fig. 5.23 **a** development of the radial P-wave velocity over time in a 20 cm long rock core from 5204.20 m MD to 5204.40 m MD; the direction of the black velocity measurements is perpendicular to a steeply inclining joint set (green lines, **b**)), while the red is parallel to it. **b** measurement points along the rock core with the markers corresponding to the markers in the diagram

joints (Fig. 5.23b), red markers). A marker assigns the individual location for the P-wave transducer on the core piece. From top to bottom, a square comes first, followed by a diamond, and finishes with a triangle. Figure 5.23a adopts these markers.

The temporal development of the radial v_P shows Fig. 5.23a. The P-wave velocity perpendicular to the joints starts diversely at 0 h between 0.5 to almost 4 km/s. With advancing time, v_P of all approaches at 192 h. After this timespan, v_P declines uniformly. However, v_P of all three points rises again, and the velocities align around 2 km/s. The black, dashed regression line is horizontal. Parallel to the joints, the velocities are higher by a factor of at least 2.5. The velocities start at 0 h between 5.5 to 5.8 km/s. Subsequently, the velocities start spreading (24 h), converge (144 h), and spread again (192 h and 216 h) until they finally approach a value between 5.2 to 5.6 km/s. Again, the red, dashed regression line is horizontal.

Conspicuously, the velocities marked with triangles are usually the highest, diamonds the second-highest and squares the lowest. A look at Fig. 5.23b shows that the squares locate in a highly disturbed area. Furthermore, a (secondary?) joint undercuts the diamond points, which is not noticeable for the triangles. The core piece disintegrated along the joints during the transport.

The second successful long-term US time series was recorded on LT-US 4, a 25 cm long core piece from Core Run 6 out of 5379.55–5379.80 m MD (Fig. 5.24b). In contrast to LT-US 2, there is no visible anisotropy. Hence, a first US measurement was placed randomly at the lateral surface, while the second oriented perpendicular to the first. A line of circles and crosses marks the measuring position every 5 cm along the core piece. The test series went for 226 h, with 6 test intervals.

Fig. 5.24a shows that the circular measuring points' velocities cover a huge spectrum. At 0 h, the velocity varies below 3 km/s and up to 6.7 km/s. While the lower

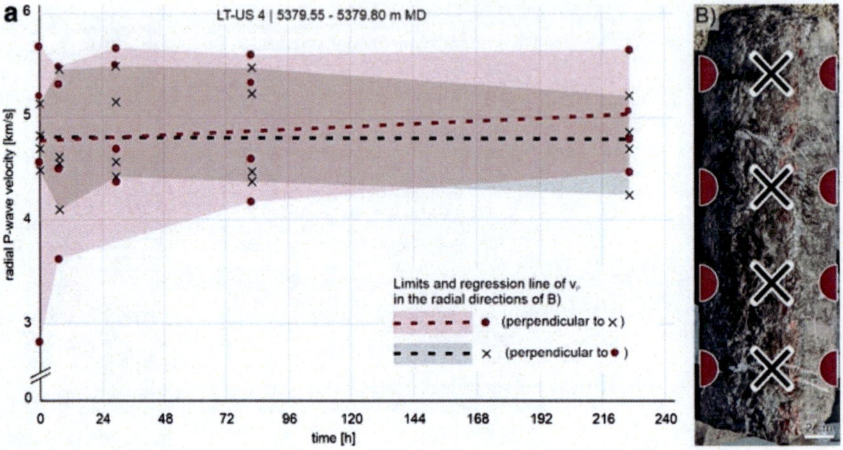

Fig. 5.24 **a** development of the radial P-wave velocity over time in a 30 cm long rock core from 5379.55 m MD to 5379.80 m MD from orthogonal measuring direction, indicated by red circles and black crosses on **b**

limit increases significantly up to 4.5 km/s after 226 h, the upper limit stays persistent with slight fluctuations at 6 h and 80 h. The cross-wise direction, marked with black crosses (Fig. 5.24a), starts in a narrow range of 0.7 km/s with velocities between 4.5 km/s and 5.2 km/s. Only after 6 h, velocities spread to velocities as small as 4.2 km/s and as big as 5.5 km/s. The lower velocity increases but falls back to 4.3 km/s, while the upper velocity steadily decreases to 5.3 km/s.

The black regression line of the cross-wise measurements is horizontal and does not show any changes. The red dashed regression line of the circles shows a slight increase in velocity.

LT-US 5 follows 1.2 m after LT-US 4 from Core Run 6 out of 5381.0–5381.1 m MD. While the results of LT-US 4 show a blurry distribution of v_P, the radial velocities of LT-US 5 can be distinguished. Furthermore, this core piece also allows measuring axial P-wave velocities over time.

Figure 5.25 shows the development of the two radial and the one axial v_P over time. One of the radial P-wave velocities starts at 5.1 km/s (red dashed line). The first three measurements show no change in velocity. However, the last two measurements show an increase to 5.5 km/s. The second radial velocity is around 1 km/s slower and starts at 4.2 km/s. For the first four measurements, velocity decreases by 0.5 km/s until it changes back to its initial velocity at 4.2 km/s after 226 h. The red dashed regression line for the higher radial velocity shares the slope with the black dashed regression line of the lower radial velocity; both increase slightly. The axial direction shows the lowest velocities. At 0 h, v_P is at 2.6 km/s and decreases during the first 80 h to 2.1 km/s. After 226 h, the velocity slightly rises to 2.3 km/s. Overall, the slope of the axial P-wave velocity regression line decreases.

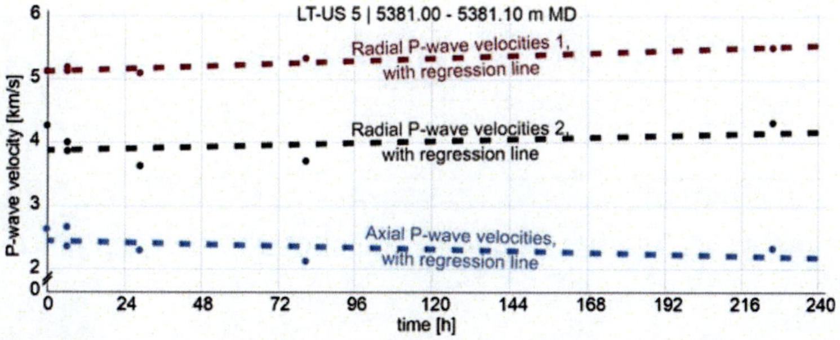

Fig. 5.25 Development of the P-wave velocity over time in a 10 cm long rock core from 5381.0 m MD to 5381.1 m MD. The red and black lines show two different radial v_P, the blue line shows the axial v_P

5.2.4.3 Differential Phenomena Measured by Ultrasonic Velocities in the Rock Cores

Section 5.2.4.1 divides the seven core runs into three sections and presents each section's US velocities as a summarized box-plot diagram. Section 3, consisting of Core Run 6 and Core Run 7, includes 356 axial and radial P-wave velocities on 40 core pieces over a length of 10.4 m coring distance that scatter over a wide range (see Fig. 5.22). For a better resolution of the v_P distribution, Fig. 5.26 plots the depth against the corresponding P-wave velocities. The colors of the P-wave velocities, axial in blue, maximal radial in purple and minimal radial in black, are according to Fig. 5.27. The axial P-wave velocity is blue, the maximum radial P-wave velocity is red, and the minimum radial v_P is black. In some segments, core disking with saddle-shaped curvatures could be identified (Sects. 2.4.2.3 and 5.2.1). In that case, the measurement of the radial v_P happened parallel and perpendicular to the saddle-axis. Each data point belongs to a single intact core piece and results from one or multiple v_P measurements. Radial v_P measurements were conducted every 5 cm along cores longer than 10 cm, and these compound measurements plot as mean values, framed by their standard deviation. The maximum and minimum deviations for each subsequent point are connected and form an area around the course of the means (transparent blue, black, and red areas in Fig. 5.26). The vertical lines show the mean velocity of all axial, minimum, and maximum radial velocities with their standard deviation as dashed lines to the left and right.

The mean axial v_P traverses Fig. 5.26 as a blue vertical line at 2120 m/s. Two dashed blue lines adjacent the mean velocity with a negative and positive amount of the standard deviation (SD) at 770 m/s. Six data points plot below the negative limit of the SD. Right at the beginning of the Core Run 6 between 5378 and 5379 m MD, three velocities range between 1120 m/s and 810 m/s. Below 5382 m MD, two pieces have velocities of 1070 and 1350 m/s. Finally, at the end of CR 7, the velocity decreases to 450 m/s. In contrast, three velocities exceed the positive limit of SD. One data point

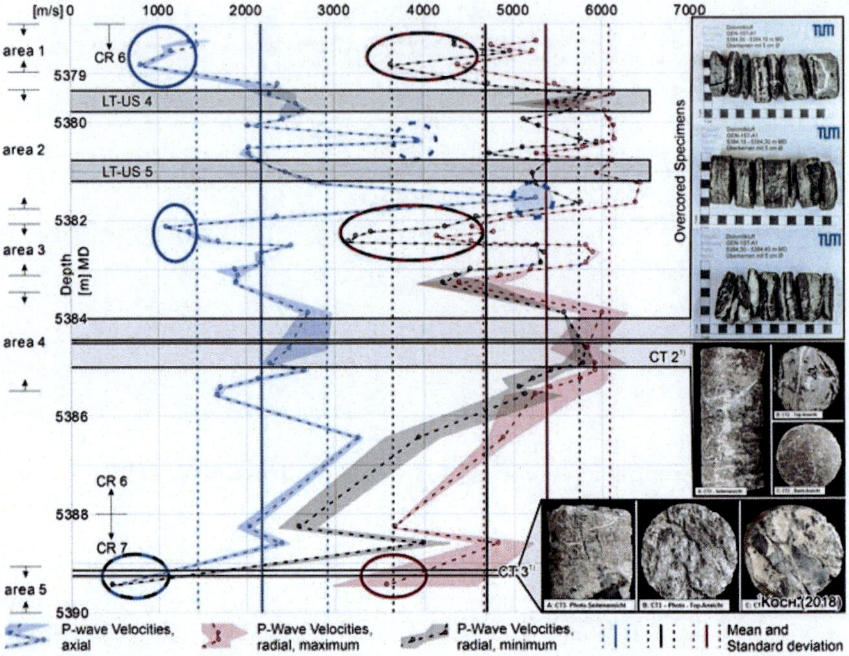

Fig. 5.26 P-wave velocities along Core Run 6 & 7 (Section 3 in Sect. 5.2.4.1) in axial (blue) and radial velocities (black = minimum, red = maximum), with mean and standard deviation. Five areas (left) distinguish the cores' velocities according to the circled peculiarities and the marked area of previous long-term results, overcored specimens from Sect. 5.2.2, and CT scans (CT2 & CT3) from Sect. 3.3.2.2

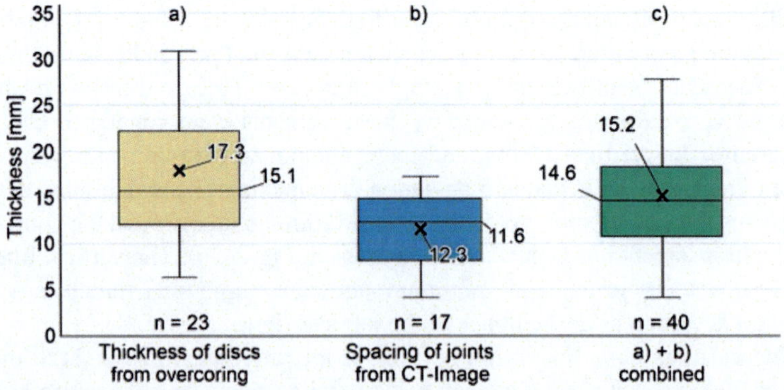

Fig. 5.27 Thickness and spacings of interpreted core disking; **a** thickness of discs from overcoring, **b** spacing of joints in CT scan CT2 (Fig. 5.29), **c** previous methods (**a**) and (**b**) combined

plots with a velocity of 3930 m/s in 5379.6 m MD, the second velocity directly at 5381 m MD with 4858 m/s, and the third velocity at 5386.5 m MD with 3260 m/s.

The minimal radial v_P intersects Fig. 5.26 as a black vertical line at a mean velocity of 4690 m/s. Its standard deviation (SD) is 1050 m/s, illustrated in Fig. 5.26 as black, dashed lines at a distance to the mean velocity. The mean and SD of the maximum radial v_P plots likewise as a red vertical and two red dashed lines, with velocities of 5370 m/s and 710 m/s, respectively. For minimal and maximal radial v_P, 3 and 6 velocities exceed, and 6 and 7 velocities fall below the SD-lines. A first set of the slow outliers is – likewise to axial v_P—at the beginning of CR 6 (Fig. 5.26, circle). A second wider set, only slightly above the upper limit, is between 5379.5 and 5381.7 m MD, where notably, the peaks from axial v_P coincide with a slight decrease in radial velocity. The most obvious negative deviation from the lower limit is between 5382.0 to 5382.5 m MD, where the minimal and maximal radial v_P becomes as low as 3150 m/s and 4130 m/s, respectively. The last outliers locate at the end of CR 7 with velocities of 450 m/s and 3580 m/s.

A look at the distribution and development of US velocities along the core axis allows interpreting five major areas along the core run. Figure 5.26 illustrated the extent of the areas on the left side.

Area 1 is the beginning of CR 6. The declining US velocities, both radial and axial, indicate an increased shattering of the rock mass. Considering the completely disintegrated CR 5, the beginning of CR 6 may be in parts disturbed rock from the previous core run. External mechanical damage might falsify further interpretation.

The two peaks of axial v_P distinguish area 2. While the rest of the axial velocities is close to the mean velocity, the radial velocities are significantly above the average. The associated measurement of the axial v_P shows a slight decrease in both radial velocities.

Area 3 shows a sudden decrease of all v_P, although the velocities come from a completely intact rock core with a length of 50 cm, while the consecutive rock core from the same core box shows the mean velocities for the axial v_P and above mean velocities for the radial velocities. Remarkably, the saddle-shaped curvatures appeared at 5382.5 m MD the first time.

In area 4, the results from overcoring and CT-scanning supplement the US velocity measurements. The US velocities are slightly higher than the means. For the axial velocities, the velocities range between 2800 m/s and 2300 m/s, for the radial velocities between 5500 m/s and 5900 m/s.

Area 5 is at the end of CR 7, where the velocities drop to their slowest. CT-Scan "CT 3" from [14] reveals a highly disturbed and fractured material. Although this may indicate a highly faulted rock mass, it may also result from core jamming and unintended milling during the drill bit. Particularly, jamming may be likely since the steeply inclining joint set from CR 3 and CR 4 reappears.

5.3 Stress- and Structurally-Controlled Phenomena

Evaluating the origin of the described phenomena as partly stress- and structurally-controlled intends already an interpretation to some degree. However, this Chapter presents the results, which provide the basis for interpretation and discussion in Chap. 6.

5.3.1 Indicators and Quantifiers for in situ Stresses on Rock Cores

5.3.1.1 Core Disking

Two separate methods revealed the phenomenon of Core Disking. Firstly, during overcoring with the 5 cm drill bit, a successful recovery of specimens for rock-mechanical experiments (presented in Sect. 5.2.2) was not successful; instead, the samples disintegrated in discs of various thicknesses (Fig. 5.17). Secondly, as described in Sect. 5.2.1 and detected in a CT-Image (Sect. 3.3.2.2), saddle-shaped patterns occur with various spacing distances. The measuring of twenty-three discs resulted in a mean thickness of 17.3 mm and a median of 15.1 mm, where discs were as thin as 6 mm and as thick as 31 mm (Fig. 5.27a). The 17 spacings, over a length of 18 cm, between the saddle-shaped pattern in specimen CT2 (Fig. 5.28), determined a mean of 12.3 mm, a median of 11.6 mm, a minimum of 4 mm, and a maximum

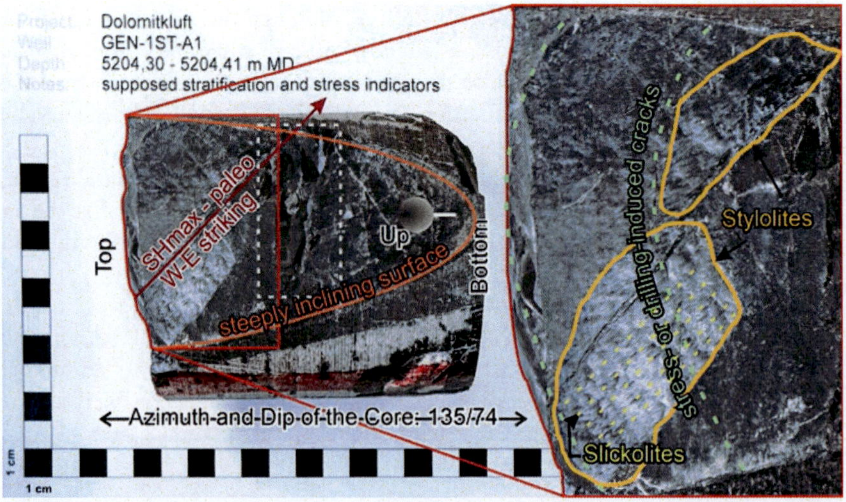

Fig. 5.28 Stylolites (orange) protrude from a set of steeply inclining joints, directed by slickolithes (yellow) in an EW-direction. Newly emerged curved drilling-induced fractures (green) show on these surfaces

of 17 mm (Fig. 5.27b). In accumulation, this results in a mean of 15.2 mm and a median of 14.6 mm. Figure 5.27c shows the distance distribution in both methods (a) and (b) and combined (c).

The thickness/spacing differs between both methods and decreases by 5 mm significantly from overcoring to CT-Image method. On the one hand, the overcoring may overestimate the total thickness since thinner discs may disintegrate during the destructive drilling process. Hence, thinner discs fall out of consideration. The difference of 2.2 mm between the mean and median supports this argument. On the other hand, the spacing in CT-Images may misjudge the distance and may recognize traces that are not mechanically relevant. The minor difference between the mean and median indicates that this effect is less significant. Combining both methods converges the mean and the median and implies a reasonable estimation of real core disking distance. An incline of the axial P-wave velocity of 300 m/s from 5384 to 5385 m MD, as Fig. 5.26 shows, may also explain the decrease in disc thickness. The decline, however, is only half the standard deviation and close to the measuring error. As an assumption, all drill cores within the standard deviation share an intermediate spacing distance of **15.2 ± 7.3 mm**, tending to a closer spacing for lower axial velocities and vice versa.

5.3.1.2 Geological Stress Indicators

The geological stress indicators, as presented in Sect. 2.4.2.3, appear on the steep inclining joint set through the core runs CR 2 to CR 4 and in CR 7. They indicate the probable direction of both paleo and present stresses and enable realigning the drill cores to their original position. Figure 5.28 shows a drill core from CR 4, from 5204.3 to 5204.4 m MD. Its axis strikes with 135° azimuthal direction and inclines with an angle of 74°. The enlarged area to the right of Fig. 5.28 (right) shows three different signs of possible stress patterns: Firstly, Slickolithes run along this surface with an angle of 45° counterclockwise to the drill core axis. Secondly, stylolites protrude perpendicular to this area and align with the slickolithes. Thirdly, stress- or drilling-induced cracks indicate that the least principal stress is acting perpendicular to the major principal stress parallel with this surface.

Since the red and black lines alongside the core can distinguish the top and bottom of the core piece, these patterns allow the interpretation of likely in situ stress fields.

The slickolithes strike 45° counterclockwise to the core's axis in a WE-direction (as shown in Fig. 5.28). Hence, they present the paleo stress regime that prevailed during sedimentation and subsidence. The stylolites align parallel to S_v as the major principal stress, which agrees with ([15]: 393), who identified a paleo normal faulting stress regime. Under these conditions, the surface could either point upwards or downwards. If the surface points upwards with the core's original position, this steeply inclining surface dips with an angle of 10° southwards. However, following this approach, stress- or drilling-induced fractures indicate a change of the stress field from a normal faulting regime during deposition and early alpine orogeny to a strike-slip regime. The curvature of these stress-induced cracks coincides with the spatial orientation of the saddle-shaped core disking patterns, mapped together with the fracture traces on the 360° core images from Sect. 5.4.3, Fig. 5.37.

5.3.2 *Stress- and Structurally-Controlled Phenomena in HMI Logs*

The results in this Chapter originate from the HMI logs presented in Sect. 3.3.4. The methodology from Sect. 4.1.2.1 applies.

5.3.2.1 Borehole Breakouts (BBOs) and Unilateral Patterns (UPs)

The rose plots, which illustrate the strike and direction of Borehole Breakouts and Unilateral Patterns, point in the direction of the progressing borehole (up to bottom), where 0° (green H for High-Side) is top, clockwise, right (90°), down (180°) and left (270°).

The 1500 m long HMI log from well GEN-1 reveals 76 borehole breakouts. The left rose plot in Fig. 5.29 illustrates the orientation of the BBOs. A majority of 35 borehole breakouts strike between 80° and 100°, hence from left to right. 24 BBOs strike between 100° and 120°. A marginal share appears at the abutments and in the vertical direction and strikes with an azimuthal direction between 60° and 80°. BBOs that strike between 80° to 120° in the borehole strike N-S in a geographical coordinate system.

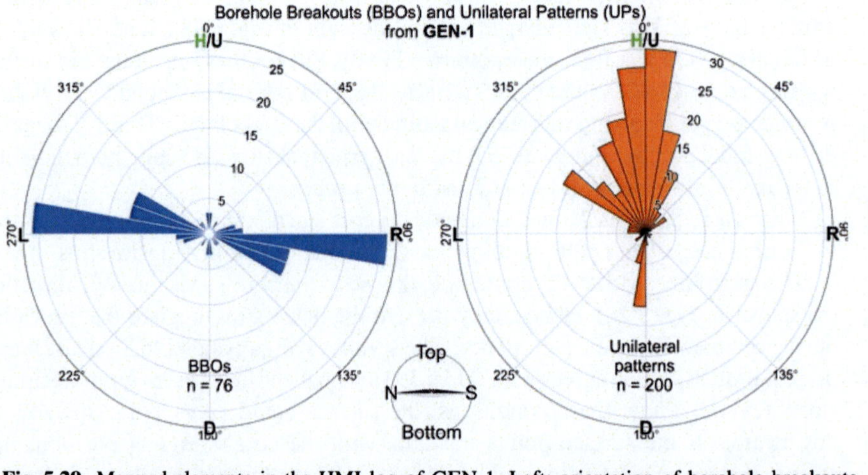

Fig. 5.29 Mapped elements in the HMI log of GEN-1. Left: orientation of borehole breakouts (blue), emerging at symmetrically at the left and right wall, striking NS in a geographic coordinate system. Right: orientation of Unilateral Patterns emerging majorly at the top of the well and at the top left abutment

The right-hand side of Fig. 5.29 shows the distribution of the Unilateral Patterns (UPs) along the borehole circumference. From 200 recorded patterns, a vast majority of 118 UPs develops at the top of the borehole between 330° and 30°. A minor share of 30 UPs appears in the upper left abutment and another at the borehole floor with 17 UPs. There are no UPs to the right and hardly any to the left side.

The 400 m long HMI log of GEN-1ST-A1 only shows two borehole breakouts, picked at the borehole's top/floor (Fig. 5.30, left). Whereas 72 Unilateral Patterns (UPs) were mapped that show a widespread distribution (Fig. 5.30, right). The majority of 34 UPs locates in the left top of the borehole and its abutment. 28 UPs accumulate at the floor of the borehole and its right abutment. Minor quantities of UPs appear at the top and its right abutment. However, there are no UPs to the right side and hardly any to the left side. Besides the direction of previous features, the HMI log also provides information on their total width, which is a fraction of the borehole's circumference. Figure 5.31 shows plots the widths of each mapped features as a histogram with their counts, and at the same time, shows the statistical distribution over all widths. Due to their low quantity, Fig. 5.31 does not consider the BBOs from GEN-1ST-A1 in the chart.

The breakouts in GEN-1ST-A1 show two peaks in the probability density curve. The most probable breakout width is slightly below 90°, the other one is half a borehole perimeter at 180°, which accounts for 13 counts. The width with the most count extends around and over 90°. Minorly, some range between 50° and 80° and a few slightly above 110° (Fig. 5.31, blue bins). The Unilateral Patterns of GEN-1 show a larger width, with a maximum at 110° to 135° (over 75 counts) and a second peak at 150° to 180° (68 counts). Subordinately, the rest distributes to their mid and side, and at 225° (Fig. 5.31, orange bins). The Unilateral Patterns of GEN-1ST-A1

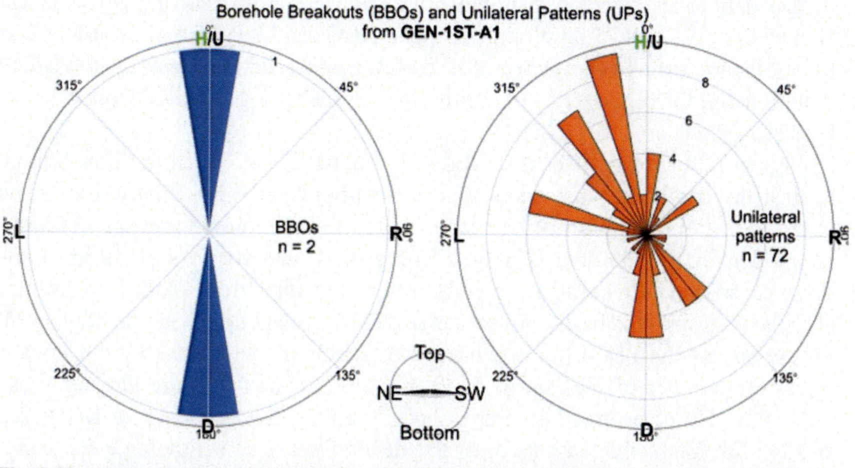

Fig. 5.30 Mapped elements in the HMI log of GEN-1ST-A1. Left: orientation of borehole breakouts (blue), emerging at symmetrically at the top and bottom. Right: orientation of Unilateral Patterns emerging majorly at the top left abutment, at the bottom right and at the bottom

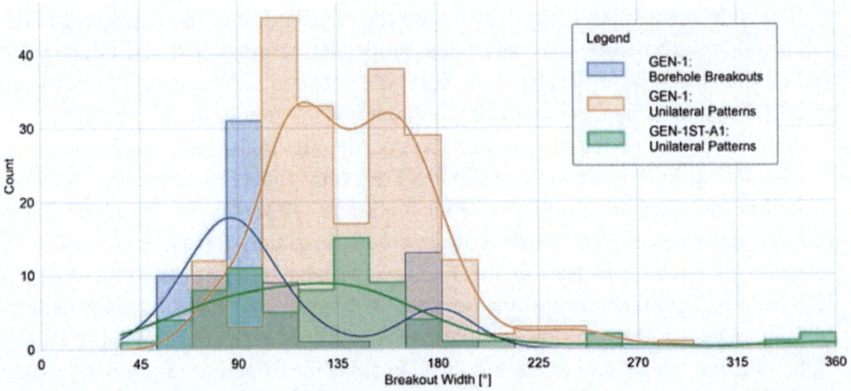

Fig. 5.31 Histogram (bins) with a density distribution (lines) of the widths of the borehole breakouts from GEN-1 (blue) and Unilateral Patterns of both wells (orange & green) in degrees of the borehole perimeter

only have one peak at a width of 130°, with a maximum count of 15 slightly over 135°. Some also extend over the whole diameter (Fig. 5.31, green bins).

5.3.2.2 Occurrence of Borehole Breakouts (BBOs) and Unilateral Patterns (UPs) along the Borehole

The previous Chapter concludes the distribution of BBOs and UPs, summarized for the whole HMI distance. However, the spatial distribution of BBOs and UPs could further help to analyze and distinguish the connection of occurring BBOs or UPs. Hence, Figs. 5.32 and 5.34 plot the mapped BBOs and UPs as a qualitative density plot (kde plot, from Seaborn, Sect. 5.1, p. 82)) against each geothermal well's depth. Additionally, the plot overlays the lithology and HMI log, adapted from Sect. 3.3, Fig. 3.2.

Figure 5.32 shows the estimated density of BBO and UP from GEN-1 plotted against the depth, starting in the platform carbonates (PCs). Firstly, UPs appear in the transition to platform-carbonates, where light limestone prevails. According to the density distribution, UPs dominate at first, and BBOs surpass for a short distance before UPs dominate clearly at the transition from platform-carbonates (PCs) to platform- and basin-carbonates (PBCs) with dark limestones in 4800 m MD. To the end of the PBCs, the occurrence of BBOs increases steeply but does not outdo the number of UPs up to 5150 m MD, although PCs are already present for 100 m. However, from the transition to the PCs, the frequency of BBOs stays steady with slight fluctuations until the density peaks at a lithological transition from dark limestone to dolostone in 5350 m MD. Controversially, the frequency of UPs decreases strongly in this area but simultaneously rises with the BBOs from 5300 m MD. At 5450 m MD, the frequency of UPs supersedes the density of BBOs.

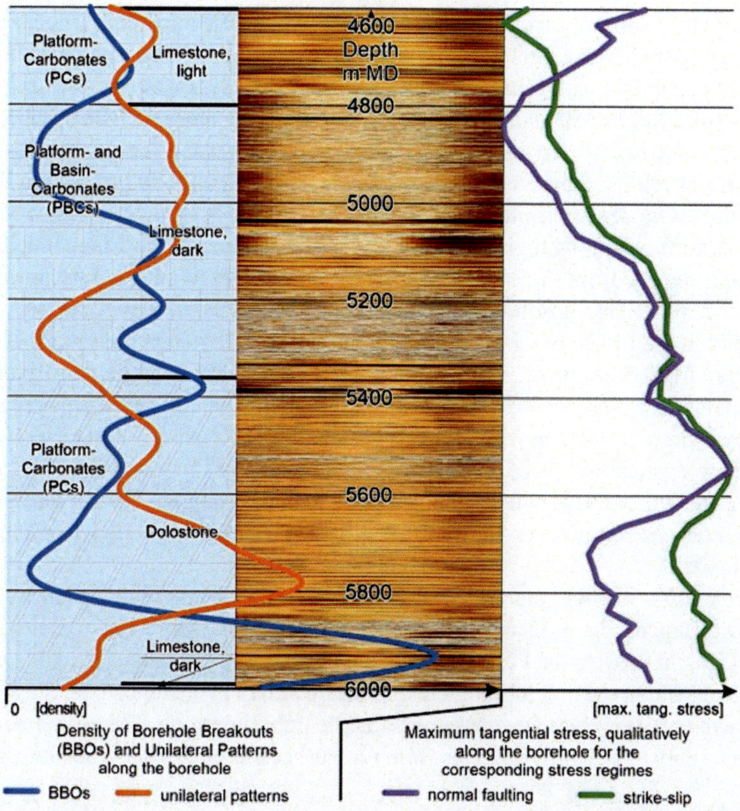

Fig. 5.32 Density of borehole breakouts (BBOs) as blue line and Unilateral Patterns (UPs) as orange line qualitatively along the borehole of GEN-1 with the facies, lithology and HMI log from Sect. 3.3, Fig. 3.2 in the background. Right: the unsteady development of maximum tangential stresses, as a purple line for a normal faulting and in green for a strike-slip stress regime decrease to the right and increase to the left, qualitatively adopted from Sect. 5.5.2

Remarkably, a peak of PU coincides with a trough in BBOs and vice versa. 70 m in front, another lithology change from dolostone to dark limestone, the frequency of BBOs climaxes, while the density of UPs continuously declines.

In an elastic, homogeneous, and isotropic rock mass, borehole breakouts correlate with the maximum tangential stress at the borehole wall as long as the rock strength stays unaltered. Hence, this correlation should show an increase in the density of BBOs with increasing maximum tangential stress and vice versa (compare to Sects. 2.3.2.2 and 2.4.2.3). For this reason, Fig. 5.32 compares the maximum tangential stress along the borehole for a normal faulting regime (Fig. 5.32, purple line) and a strike-slip regime (Fig. 5.32, green line) with the density of BBOs (Fig. 5.32, blue line). The values for the maximum tangential stress anticipate the results of the stress rotation in Sect. 5.5. However, the course of the line is purely qualitative and does not show absolute stress values.

In a strike-slip stress regime (Fig. 5.32, green line), the maximum tangential stress increases steadily in a shallow slope with depth. A minor decrease below 4800 m MD goes along with a strong and major drop in BBO density. However, a second decline ahead of 5400 m MD is accompanied by a strong increase of BBOs, whereas the third decline in between 5600 m MD and 5800 m MD also corresponds with a lower density of BBOs. However, the stresses stay steady from 5800 m MD to the final depth, BBOs' density triples.

In a normal faulting regime (Fig. 5.32, blue line), the maximum tangential stresses decline steeper from the start of the open hole section to an absolute minimum at 4850 m MD. This absolute minimum precedes an absolute minimum in BBOs' density in the PBCs. A subsequent incline is followed by an increase in BBOs up to 5100 m MD. From there, stresses increase even further while the density of BBOs oscillates in a non-correlating pattern. Consistently to the strike-slip regime, the stresses drop at 5400 m MD, but the BBOs' density inclines. After a consecutive increase of stresses, where the density of BBOs declines for another 150 m, stresses decrease again, and the concentration of BBOs keeps dropping. Finally, the maximum tangential stresses increase in a shallow slope to the end of the well, where the BBOs' density triples.

For GEN-1ST-A1, there is insufficient information to create a density plot of BBOs. Hence, Fig. 5.33 marks the location of BBOs with a horizontal blue line. For UPs, that covers the Purbeck-facies up to platform-carbonates in a final depth of 4448 m MD. In the first lithological unit of the Purbeck, the dolostone, the density is low and only increases when approaching the light limestone. The estimated density rises steeply to a change in facies from the Purbeck to platform-carbonates. Directly

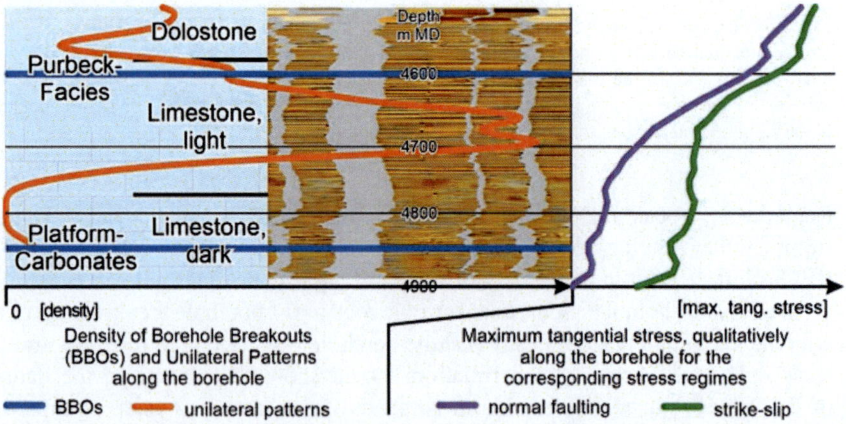

Fig. 5.33 Position of borehole breakouts (BBOs) as blue lines absolutely and the density distribution of the Unilateral Patterns (UPs) as orange line qualitatively along the borehole of GEN-1ST-A1 with the facies, lithology and HMI log from Sect. 3.3, Fig. 3.2 in the background. Right: the unsteady development of maximum tangential stresses, as a purple line for a normal faulting and in green for a strike-slip stress regime decrease to the right and increase to the left, qualitatively adopted from Sect. 5.5.2

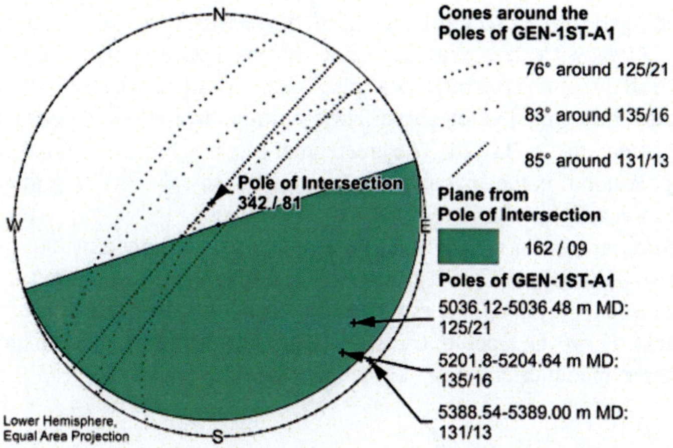

Fig. 5.34 Reconstruction of the original stratification of the steep inclining joint sets in from CR2, CR4, and CR7. Their individual inclination in the rock cores is plotted as cones around vector of the drill path (Poles of GEN-1ST-A1), resulting one possible plane for the steeply inclining joint set: 162/09

at the change in facies, the density drops slightly but resumes rising until 4700 m MD. From this point, the density falls steeply to zero shortly above 4800 m MD.

Calculations from the possible occurring stress regimes show that for both strike-slip and normal faulting regime, the maximum tangential stresses decrease continuously with depth for GEN-1ST-A1. Neither the BBOs nor the Unilateral Patterns (UPs) indicate a link with these stresses, although UPs density climax, where the steepest gradient stops and flattens just above 4700 m MD (Fig. 5.33, purple and green line).

5.4 Fracture Networks from Core Runs, HMI Logs and 360° Core Images

5.4.1 Joint and Fracture Traces recorded on Cores in Core Boxes

Section 5.2.1 recorded and described discontinuities and spacing distances for all core runs. In the Core Runs 2, 4, and 7, steeply inclining, open fractures dominate. These planes share similarities, e.g., stylolites, dark coating, equal spacing distance, or polished surfaces. However, all of these fracture sets increase slightly in dip with each ascending core run. In Core Run 2, this joint set inclines with 76° to the core's axis, in Core Run 4, at 83°, and in Core Run 7 even steeper with 85°. For each coring section, the drill trajectory is different, too. For CR 2, the azimuthal direction and dip are 125/21, for CR 4 135/16, and for CR 7, it is 131/13, respectively. Figure 5.34 shows a stereographic, equal-area projection, in which each pole of the various drill directions is plotted as a "plus"-sign. Around each pole, a cone plots with the corresponding inclination angle. The resulting three cones intersect in one point with the orientation of 342/81. The green great-circle in Fig. 5.34, which corresponds to this pole point, is the spatial orientation of the steeply inclining joint set with a dip direction and dip of 162/09.

Fracture orientations from other joints were only assumed or interpreted in the core boxes. Although different joints were recorded in core runs, the small diameter of the cores and the joints' low extension did not allow a determination of their spatial orientation. However, fracture tracing in HMI logs and 360° core images enable to measure the spatial orientation of other joints, too.

5.4.2 *Fracture Traces recorded on the HMI Log of GEN-1ST-A1*

Besides stress-related (BBOs) and lithological-related (UPs) features, the HMI log from GEN-1ST-A1 allows for picking 575 fracture traces. Macroscopically, the HMI log transitions from a very fine laminated, diverse pattern to thicker, consistent traces at 4720 m MD. Hence, the fractures' spatial distributions from 4487 m MD to 4720 m MD, with a length of 233 m and from 4720 m MD to 4888 m MD, with 168 m, are analyzed separately.

The first 233 m of the HMI log contain 204 picked planes. The rose plot in Fig. 5.35a shows that most fractures strike between 170° and 80°. A lesser part strikes between 80° to 170°. From the density distributions of the pole points, four

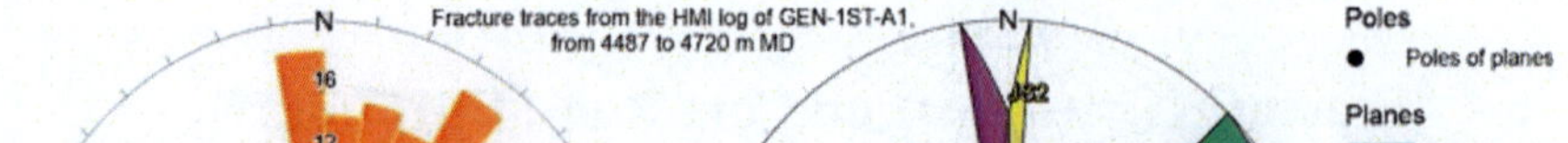

Fig. 5.35 Results from the discontinuity mapping in the HMI log of GEN-1ST-A1 from 4487 to 4720 m MD. **a** A rose plot illustrates the strike of all 204 mapped discontinuity traces. **b** A stereographic projection shows four resulting planes JS1 (green), JS2 (yellow), JS4 (purple) and JS5 (blue) from 201 poles that became visible in a density plot weighted after [16]

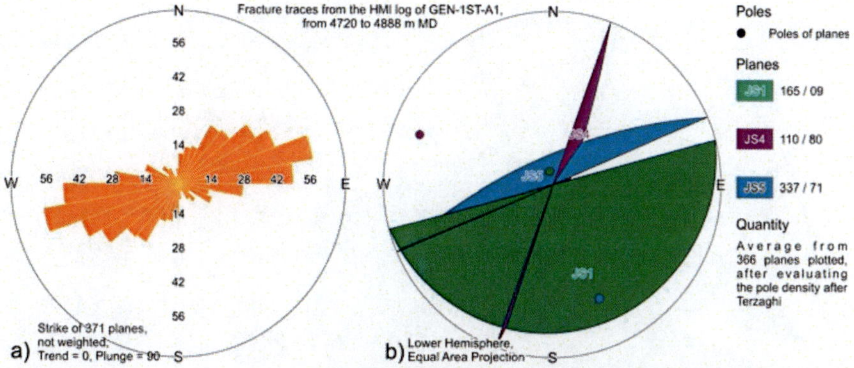

Fig. 5.36 Results from the discontinuity mapping in the HMI log of GEN-1ST-A1 from 4720 to 4888 m MD. **a** A rose plot illustrates the strike of all 371 mapped discontinuity traces. **b** A stereographic projection shows three resulting planes JS1 (green), JS4 (purple) and JS5 (blue) from 366 poles that became visible in a density plot weighted after [16]

major planes, or joint sets (JS), form. Descending, the green, shallow dipping great circle JS1 in Fig. 5.35b with the dip direction and dip of 135/12 has the highest density concentration, followed by the purple great circle JS4, which strikes N-S and dips steeply to the east (81/75). The two lowest density concentrations show the yellow great circle JS2, which also strikes N-S but steeply dips to the west (274/83), and the blue great circle JS5 that strike ENE-WSW and dips with an angle of 67° to the north (337/67). While the rose plot in Fig. 5.35a contains all 204 fracture traces, the great circles from Fig. 5.35b contain 201 planes. Hence, 3 traces fall out of statistical consideration. The mean spacing distance of JS1 is 0.97 m, with a minimum of 0.03 m and a maximum of 10.8 m. For JS2, JS4 and JS5 deviate the mean spacing distance strongly and could not be practically calculated.

The second section, which starts at 4720 m MD and runs for 168 m, reveals 371 fracture traces. The strike of these traces distributes mainly between 50° and 90° with a maximum between 70° and 80° (Fig. 5.36a). Subordinately, striking directions between 0° and 50° follow.

From these 371 fracture traces, the mean of 353 traces forms the green, shallow, southwards dipping JS1 in Fig. 5.36b, with a dip direction and dip of 165/09. Five traces form the purple JS4 that steeps deeply to the SE (110/80). Finally, eight traces define the blue JS5, which dips with a slope of 71° in the NW direction (337/71). Five traces do not allocate to any of these joint sets. The mean spacing distance of JS1 is 0.46 m, with a minimum of 0.04 m and a maximum of 4.76 m. Again, a determination of the spacing distance from JS4 and JS5 is not possible. Worth mentioning, the pole points that belong to the JS4 and JS5 disappear below 4740 m MD. Although the HMI log macroscopically changes, a transitional zone still contains some of the governing fracture traces that completely fade out, while JS1 remains.

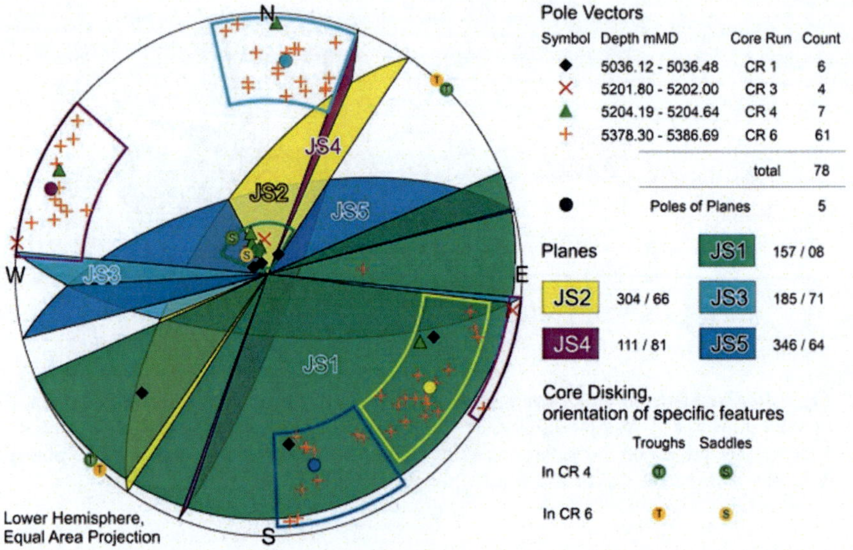

Fig. 5.37 Results of discontinuity mapping on the 360° core images as poles from CR1 (black diamonds), CR3 (red crosses), CR4 (green triangles), and CR6 (orange plus). Five discontinuity sets show: JS1 (green), JS2 (yellow), JS3 (turquoise), JS4 (purple), and JS5 (blue). Core disking features, such as the horizontal troughs (T) and subvertical saddles (S) show as green and orange poles

5.4.3 360° Core Images

Alongside the manual mapping of the rock cores from Sect. 5.2.1 and the fracture picking from HMI logs in the previous Chapter, 360° core images facilitate a much more detailed recording of the fracture traces. The methodology follows the HMI logs' evaluation from Sect. 4.1.2.1 using the 360° core images instead of the image log. Similar to the HMI log, ".las"-files provide the spatial data for the rock cores. However, the only orientation known about the 360° core images is the top and bottom. Hence, fracture tracing only returns the apparent dip and apparent azimuth of these sinusoidal line features, where the apparent dip equals the true dip in a vertical borehole.

In total, 78 fracture traces were recorded along with the 360° core images. 360° core images are available for rock pieces in 5036.12–5036.48 m MD, 5201.80–5202.0 m MD, 5204.19–5204.64 m MD and cover almost the entire CR 6. The legend in Fig. 5.37 lists the amount of fracture traces for each section.

The rock pieces from 5036.12–5036.48 m MD, 5201.80–5202.0 m MD, 5204.19–5204.64 m MD share the steeply inclining joints in the rock cores. Section 5.4.1 proves this joint set's real spatial orientation, undoubtedly due to the three intersecting cones. Therefore, each set of traces was in a first step tilted by the inclination and in a second step rotated by the azimuthal direction of the drill path so that the steeply inclining joint sets dips shallowly to the south. Figure 5.37 illustrates the results after tilting and rotating the poles from the steeply inclining planes by their drill path trajectory. The black diamonds, red crosses, and green triangles plot in the green quarter circle area with the central green pole point. These pole points create the green great circle JS1 in Fig. 5.37 with the dip direction and dip of 157/08. Besides their centrally oriented pole points, each of these three rock pieces has additional pole points scattered over the stereographic projection. For instance, JS2 contains a black diamond and a green triangle. JS4 contains a red cross and a green triangle, and JS3 and JS5 have poles from traces in 5204.19–5204.64 m MD (green triangle) and 5036.12–5036.48 m MD (black diamond), respectively.

However, none of the 360° core images from Core Run 6 contained this steeply inclining marker joint set. Therefore, all spatial information from 78 traces was merged and rotated to a congruent overlap. All sets were rotated combined, knowing the apparent and original location of one of the three first sets from the previous rotations. Figure 5.37 illustrates the final results. The orange plus signs are the relocated pole points from the fracture traces of CR 6. From this stage, four additional joint sets identify. JS2, with a dip direction and dip of 304/66, contains 18 pole points. The turquoise great circle JS3, which steeply dips to the south with 71° (185/71), contains 17 pole points. The purple plane JS4, derived from 14 pole points, is almost vertical with and dips with 81° eastwards (111/81). Finally, JS5 frames 11 pole points and dips 64° to the NW (346/64). JS4 and JS5 are almost identical to JS4 and JS5 from Sect. 5.4.2. Two poles, one black diamond, and an orange plus find no entry in any joint set.

Besides the fracture traces, Fig. 5.37 also illustrates the mapped core disking features. The stereographic projection plots the linear, intersecting the troughs and the saddles as a pole vector (Sect 2.4.2.3). Core disking only appeared in CR 4 and CR 6. In CR 4, the saddle (green point with an S) is very steep and dips with an angle of 73° towards NW (azimuthal direction: 315°). The linear of the trough in CR 4 is horizontal and strikes 45° NE-SW. The saddles' linear in CR 6 is even steeper and dips with 81° towards NW (azimuthal direction: 312°), while the troughs strike horizontally with 42° in NE-SW. In a local borehole coordinate system, the saddles would be at the top and bottom, and the troughs at the walls for all core disking phenomena in CR 4 and CR 6.

5.5 Stresses acting along the Drill Path

The newly developed stress rotation method illustrates perfectly the effective occurring stresses along the boreholes for any stress field with arbitrary stress gradients. When applied for the geothermal wells GEN-1 and GEN-1ST-A1, certain sections with specific characteristics emerge presented in the subsequent Sect. 5.5.1. A NS striking strike-slip stress regime applies for both wells, with stress ratios after [17]. Section 5.5.2 specifically deals with the development of stresses acting on the drill paths for a normal faulting (NF) to strike-slip (SS) stress regime as promoted by [18]

(Sect. 3.3.5, Table 3.3) and two pure Strike-slip regimes after [17, 17] (Sect. 2.4.2.2, Table 2.1. The maximum horizontal stress strikes NS; the least horizontal stress strikes EW. The calculation of the effective stresses considers only the formation pressure from the carbonate reservoir. In reality, the formation pressure in the Tertiary strata is higher than in the reservoir [19]. However, for simplicity, better comparability, and the fact that a casing seals these sections, formation pressure is persistent.

5.5.1 *Sections of resulting Stresses along different Well Trajectories*

GEN-1 and GEN-1ST-A1 share the same borehole up to 4313 m MD (4200 m TVD). As such, there is no distinction between the stresses along this common trajectory. However, from the point where the boreholes split, the stresses develop differently. When the stress rotation method applies to both GEN-1 and GEN-1ST-A1, the results allow the partitioning of four major sections presented in the successive Chapters.

5.5.1.1 GEN-1

Figure 5.38 illustrates the course of the drill trajectory from the original well GEN-1, which starts vertically up to 2900 m TVD. This large area covers Section 1, outlined with a circled 1 in Fig. 5.38. Based on Sect. 4.3's definition, a deviation of < 3° from the vertical does not alter the initial stresses. Hence, the stress state is constant, and σ_{xx} and σ_{yy} act perpendicular, and σ_{zz} acts parallel to the borehole. Shear stress occurs minorly in the yz/zy-direction, indicating a beginning tilting of the well.

From this level, the trajectory's inclination exceeds 3° and moves to an azimuthal direction of around 60° and marks the beginning of Sect. 2 (Fig. 5.38, circled 2). However, before the trajectory reaches its temporary azimuthal direction of 60°, it strongly fluctuates between 22° to 114° and back to 60°. These fluctuations appear illustratively in Fig. 5.38 between 2950 m TVD and 3200 m TVD, where σ_{yy} drops and σ_{xx} has a sudden peak. The inclination increases from 12° to 34° up to 3700 m TVD, which entails a constant decline of σ_{zz}. The azimuth in this range is consistently 60°, hence σ_{xx} decreases to a constant value of around 0.0150 MPa/m, while σ_{yy} increases by the amount of reduced σ_{xx} and σ_{zz}. The inclination becomes steeper from 3700 m TVD, and consequently, σ_{zz} recovers for a few stress gradient points, whereas σ_{yy} weakens. The azimuthal direction, which defines the horizontal rotation, influences the horizontal stresses strongly. As a consequence, shear stresses in xy-yx-plane emerge up to 0.005 MPa/m. Shear stresses in xz/zx and yz/zy-plane develop due to the increasing inclination. However, since the inclination is lower than the azimuthal rotation, they do not exceed 0.0027 MPa/m in Section 2. Up to this point, the development of stresses along to trajectory is likewise to GEN-1ST-A1.

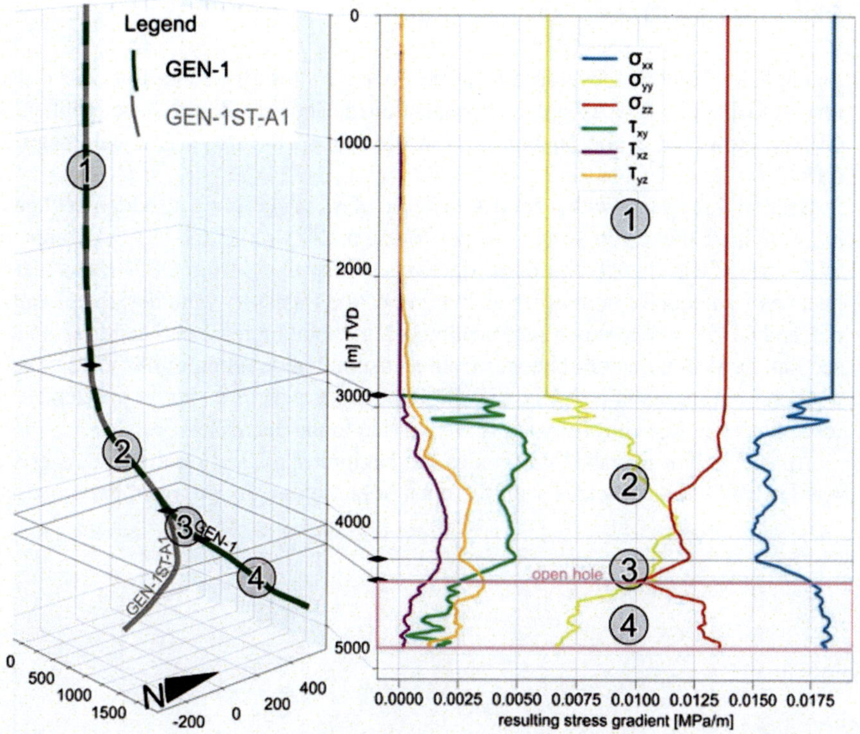

Fig. 5.38 Development of principal stresses and emerging of shear stresses along the well GEN-1, which starts vertically (1), inclines slightly (2), exceeds 45° inclination (3), and approaches a sub-horizontal course (4). Due to the inclination and deviation of the orthogonal stress field, shear stresses develop, while principal stresses approach or depart (2), (3), (4)

The beginning of Section 3 marks the segregation of both wells. For GEN-1, the azimuthal direction and also the inclination increases from 60° to 70°, and from 30° to 45°, respectively. With this right-hand twist, the trajectory heads increasingly east and becomes flatter. Hence, σ_{xx} increases, while σ_{yy} and σ_{zz} decrease. Up to this point, the stresses acting effectively on the borehole are σ_{xx} and σ_{yy}. Also, with a diverging the horizontal stresses, the shear stresses on the xy/yx-plane decrease, while the shear stresses in the yz/zy-plane increase due to the shallower trajectory.

However, the stresses acting effectively on the borehole change as the inclination exceeds 45°, which marks the start of Section 4 and the borehole's open hole section. To the end of Section 4, the trajectory approaches an inclination of 80° and an azimuthal direction of 83°. Hence, with a decreasing slope of the trajectory to almost horizontal, σ_{zz} increases. Also, σ_{xx} increases, whereas σ_{yy} declines.

Since the drill trajectory is nearly horizontal and strikes almost EW, the stresses apply almost with their initial gradient. Consequently, all shear stresses diminish in the open hole section, compared to the previous Sections 2 and 3. However, shear stresses develop differently in GEN-1ST-A1.

5.5.1.2 GEN-1ST-A1

Figure 5.39 shows the development of the effective stress gradients along the trajectory of GEN-1ST-A1. As previously mentioned, Sections 2 and 3 are identical to GEN-1. However, at the beginning of section three, the stress gradients begin to differ.

Section three starts with a sudden increase of σ_{xx} and σ_{zz} and a sudden decline of σ_{yy}, which are the results from a steeper course of 24° from GEN-1ST-A1 compared to 35° from GEN-1. Afterward, σ_{zz} develops similar to σ_{zz} from GEN-1 as a consequence of an equally increasing inclination of the trajectory. The horizontal stress, σ_{xx}, and σ_{yy}, however, experience an almost horizontal increase and decrease. In this section, the drill trajectory changes from an azimuthal direction of 68° to 95°. Since the borehole strikes in a purely EW direction, σ_{xx} resets to its initial gradient. This reset, however, does not fully apply for σ_{yy} due to the decrease of σ_{zz}.

An inclination over 45° designates the beginning of Section 4, which starts at 4390 m TVD and coincides with the open hole section. Again, from here, σ_{xx} and

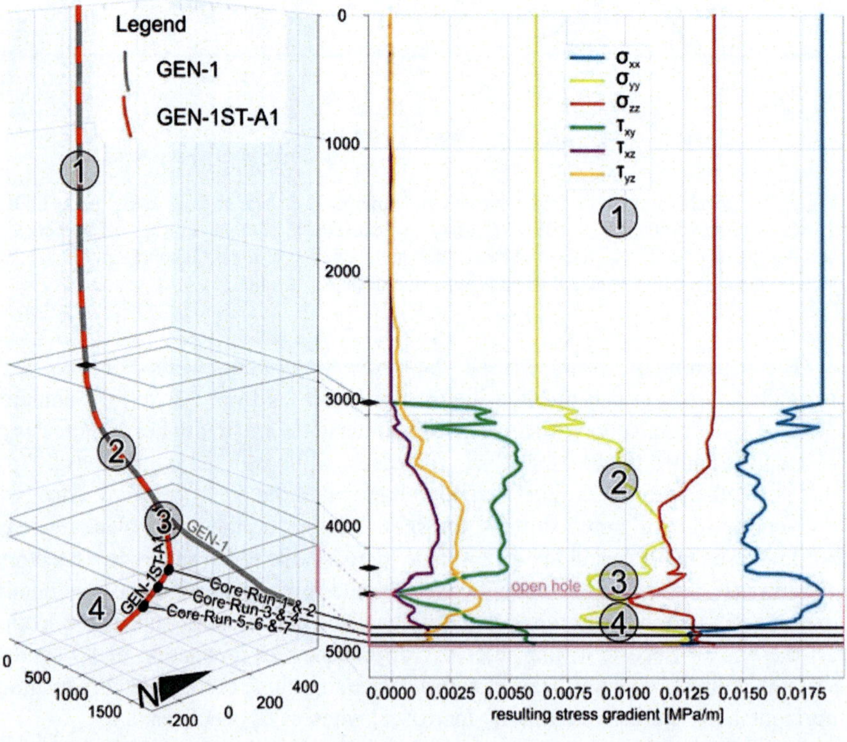

Fig. 5.39 Development of principal stresses and emerging of shear stresses along the well GEN-1ST-A1, which starts vertically (1), inclines slightly (2), exceeds 45° inclination (3), and approaches a sub-horizontal course (4). Due to the inclination and deviation of the orthogonal stress field, shear stresses develop, while principal stresses approach or depart (2), (3), (4)

σ_{zz} effectively act on the borehole. Both inclination and azimuthal direction steadily increase until values of 85° and 135°, striking exactly SE. At first, σ_{xx} and σ_{yy} steeply decrease, while σ_{zz} increases with a rising inclination up to 4500 m TVD. At this point, the inclination becomes constant at 70°, and σ_{zz} becomes steady, while the azimuthal direction continuously rises from 110° to 135°. As a consequence, σ_{yy} rises, and σ_{xx} declines. When the trajectory reaches 135° azimuthal direction, both σ_{xx} and σ_{yy} are equal. As the drill path reached 135° azimuthal direction, the inclination continued rising from 74° to 85°, resulting in a growing σ_{zz}. Due to the principal stresses' horizontal rotation by 45°, maximum shear stresses in the xy/yx-plane emerge (green line in Fig. 5.39). The remaining shear stresses in the xz/zx- and yz/zy-plane also occur but annihilate while approaching horizontal bearing.

Figure 5.39 shows the 7 core runs from the three distinct sampling points along the drill path. Coring of CR 1 and 2 happened at an inclination and azimuthal direction of 70° and 124°. Hence, the horizontal stresses do not converge yet. The drill path at Core Run 3 and 4 inclines with 74° and strikes with 135° inclination. With a strike of 45° to the principal stresses, the horizontal stresses σ_{xx} and σ_{yy} become equal. The vertical stress σ_{zz} is larger than during the previous CRs due to the shallower dip. With the applied stress ratio, σ_{zz} exceeds both σ_{xx} and σ_{yy}. At the deepest three core runs CR 5 to 7, the borehole inclination increases from 76° to 85°, and the azimuthal direction is steady at 132°. Hence, the horizontal stresses σ_{xx} and σ_{yy} diverge again slightly, while σ_{zz} further increases.

5.5.2 *Resulting Stresses of different Stress Regimes on the Drill Trajectories*

While the previous Chapter explains the drill path in full detail and distinguishes four sections, this Chapter deals with the Sections 2, 3, and 4 under the consideration of different stress regimes, stress gradients and focus particularly on the absolute stresses. Stockinger et al. [13] published similar results regarding the stress gradients only. All stresses start at the surface at zero.

5.5.2.1 Strike-Slip Regimes

Seithel et al. [17] identifies a strike-slip stress regime for the carbonate rocks in the nearby geothermal well of Sauerlach. He defined a stress ratio of $S_{Hmax} = 28.0$, $S_v = 23.0$, $S_{hmin} = 15.5$ MPa/km, with a formation pressure p_f of $0.4{*}S_v$.

Figure 5.40 illustrates the absolute effective stresses acting on the borehole for both wells GEN-1 and GEN-1ST-A1. These plots are similar to the gradient plots from Fig. 5.38 (GEN-1) and Fig. 5.39 (GEN-1ST-A1), with the little difference that the stresses accumulate with depth. Therefore, the new plots allow determining the absolute stress directly with a look at the figures. In Fig. 5.40, σ_{xx}, σ_{yy}, and σ_{zz} enter

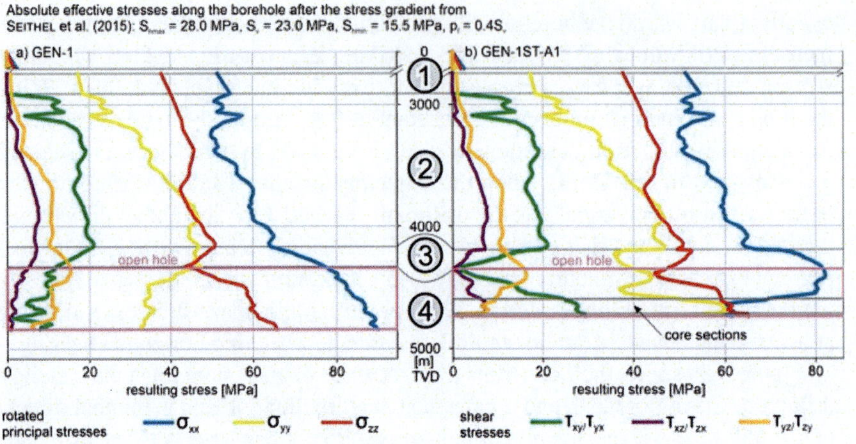

Fig. 5.40 Absolute effective stresses along the drill paths of the geothermal wells **a** GEN-1 and **b** GEN-1ST-A1, after a stress rotation in a strike-slip stress regime from [17]

Section 2 at 54, 18, and 40 MPa. In Section 2, σ_{yy} approaches σ_{zz} and outdoes it in 3790 m TVD for almost 200 m. At the transition to Section 3, σ_{xx}, σ_{yy}, and σ_{zz} have stress values of 63, 46, and 51 MPa. Also, first shear stresses emerge in Section 2 that act at the crossing to Section 3 on the xy/yx-plane with 21 MPa, on the xz/zx-plane with 7 MPa, and on the yz/zy-plane with 11 MPa.

Along GEN-1, as traversing Section 3, the stresses σ_{xx} and τ_{zy} increase to 76 and 16 MPa, σ_{zz}, τ_{xy} and τ_{zy} decrease to 45, 11, and 5 MPa, and σ_{yy} stays unchanged at 46 MPa. In the open hole section, σ_{yy} decreases to 32 MPa, and shear stresses attenuate to 8, 1, and 6 MPa. However, the stresses that effectively act on the borehole increase to $\sigma_{xx} = 88$ MPa and $\sigma_{zz} = 66$ MPa.

In Section 3 of GEN-1ST-A1 (Fig. 5.40b), the drill trajectory takes a dextral turn, and the inclination changes from below to above 45°. As a consequence, the shear stresses on the τ_{xy}- and τ_{xz}-plane disappear at the beginning of Sect. 4, while the principal stresses σ_{yy} and σ_{zz} take a sharp turn and become equal with 45 MPa, and σ_{xx} keeps rising to a little over 80 MPa. As the trajectory drifts further to the right and becomes flatter, the stresses approach along Section 4. Consecutively, τ_{xy} climaxes as the drill trajectory reaches a maximum angle of 45° to the initial in situ stress field. Since the drilling path becomes less than 135° again, the stresses drift slightly apart to a final stress state of $\sigma_{xx} = 60$, $\sigma_{yy} = 56$, $\sigma_{zz} = 65$, $\tau_{xy} = 29$, $\tau_{xz} = 2$, and $\tau_{yz} = 2$ MPa.

For the core runs obtained in section four, Table 5.9 lists the stresses after the stress gradient from [17], where σ_{xx} and σ_{zz} act perpendicular to the cores, and σ_{yy} acts axially.

For the stress gradient after [20] with $S_{Hmax} = 41.1$, $S_v = 24.1$, $S_{hmin} = 17.3$ MPa/km and a formation pressure $p_f = 0.39*S_v$, the development of resulting stresses is similar. However, the absolute effective stresses are different due to their basic gradient and their ratios. Hence, Section 2 in Fig. 5.40a, b shows a weakened

Table 5.9 Absolute effective stresses acting on the drill path at the position of the three stages of coring for a strike-slip stress regime after [17]

Core runs	Effective resulting stress values on the drill cores (MPa)					
	σ_{xx}	σ_{yy}	σ_{zz}	τ_{xy}	τ_{xz}	τ_{yz}
CR 1 and CR 2	68	49	59	24	6	9
CR 3 and CR 4	59	58	61	27	7	7
CR 5, CR 6 and CR 7	61	57	63	27	6	7

difference between σ_{yy} and τ_{xy}. With 43 MPa, τ_{xy} is double the amount compared to τ_{xy} from [17], whereas σ_{yy} is just 50% larger. Also, σ_{yy} already exceeds σ_{zz} in around 3300 m TVD.

The open hole section of GEN-1 starts with stresses $\sigma_{xx} = 129$, $\sigma_{yy} = 57$, $\sigma_{zz} = 50$, $\tau_{xy} = 27$, $\tau_{xz} = 5$, and $\tau_{yz} = 14$ MPa and approaches a finals state of resulting stresses of $\sigma_{xx} = 151$, $\sigma_{yy} = 41$, $\sigma_{zz} = 70$, $\tau_{xy} = 15$, $\tau_{xz} = 1$, and $\tau_{yz} = 6$ MPa.

The open hole section of GEN-1ST-A1 starts with a stress state of $\sigma_{xx} = 138$, $\sigma_{yy} = 50$, $\sigma_{zz} = 50$, $\tau_{xy} = 9$, $\tau_{xz} = 2$, and $\tau_{yz} = 15$ MPa. Subsequently, σ_{xx} and σ_{yy} approach again to 93 MPa, and σ_{zz} becomes 66 MPa in the coring section. Concurrently, shear stresses in the xy-plane emerge and reach up to 54 MPa, while the remaining shear stresses taper off. A final stress state at the borehole bottom for this stress state is $\sigma_{xx} = 98$, $\sigma_{yy} = 90$, $\sigma_{zz} = 69$, $\tau_{xy} = 56$, $\tau_{xz} = 2$, and $\tau_{yz} = 2$ MPa (Fig. 5.41).

For the core runs obtained in section four, Table 5.10 lists the rotated stresses after the stress gradient from [20], where σ_{xx} and σ_{zz} act perpendicular to the cores and σ_{yy} acts axially.

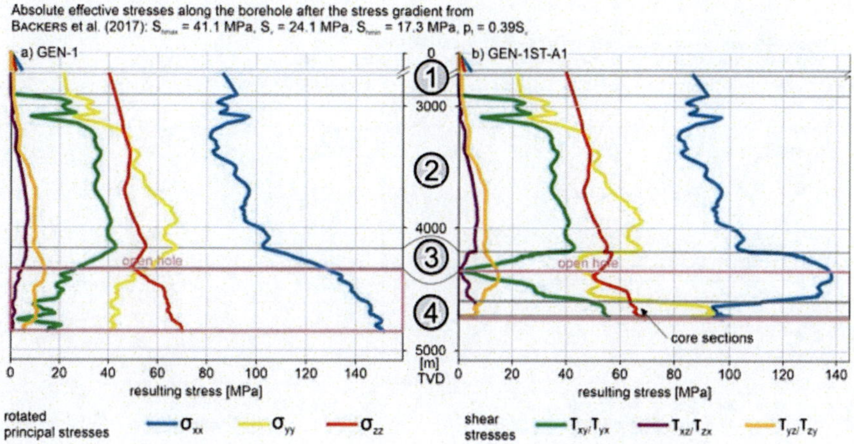

Fig. 5.41 Absolute effective stresses along the drill paths of the geothermal wells **a** GEN-1 and **b** GEN-1ST-A1, after a stress rotation in a strike-slip stress regime from [20]

Table 5.10 Absolute effective stresses acting on the drill path at the position of the three stages of coring for a strike-slip stress regime after [20]

Core runs	Effective resulting stress values on the drill cores (MPa)					
	σ_{xx}	σ_{yy}	σ_{zz}	τ_{xy}	τ_{xz}	τ_{yz}
CR 1 and CR 2	111	73	64	49	6	8
CR 3 and CR 4	93	93	66	54	6	6
CR 5, CR 6 and CR 7	98	90	67	55	5	6

5.5.2.2 Transition from a Strike-Slip Regime to a Normal Faulting Regime

Wolfgramm et al. [18] point out that a possible transition from a normal faulting stress regime to a strike-slip stress regime might be possible for the host rock of Geretsried with a stress gradient of $S_{Hmax} = S_v = 25.0$ and $S_{hmin} = 13$ MPa/km and a formation pressure of $p_f = 0.38\ S_V$. Figure 5.42 plots the resulting effective stresses for GEN-1 and GEN-1ST-A1.

Hence, S_{Hmax} and S_v have an equal slope when entering Section 2 with an absolute value of 45 MPa. S_{hmin} arrives in Section 2 with 10 MPa. Consecutively, during deflection of the borehole, S_{Hmax} and S_{hmin} approach up to a point where τ_{xy} almost overlaps with S_{hmin}. At the transition to Section 3, the absolute stresses on the borehole are $\sigma_{xx} = 55$, $\sigma_{yy} = 38$, $\sigma_{zz} = 52$, $\tau_{xy} = 17$, $\tau_{xz} = 11$, and $\tau_{yz} = 19$ MPa.

Along the trajectory of GEN-1, the stresses develop similar to the previous examples, except that τ_{yz} peaks at the beginning of Section 4 with 25 MPa. All other stresses are $\sigma_{xx} = 64$, $\sigma_{yy} = 42$, $\sigma_{zz} = 43$, $\tau_{xy} = 8$, and $\tau_{xz} = 8$ MPa. At the bottom

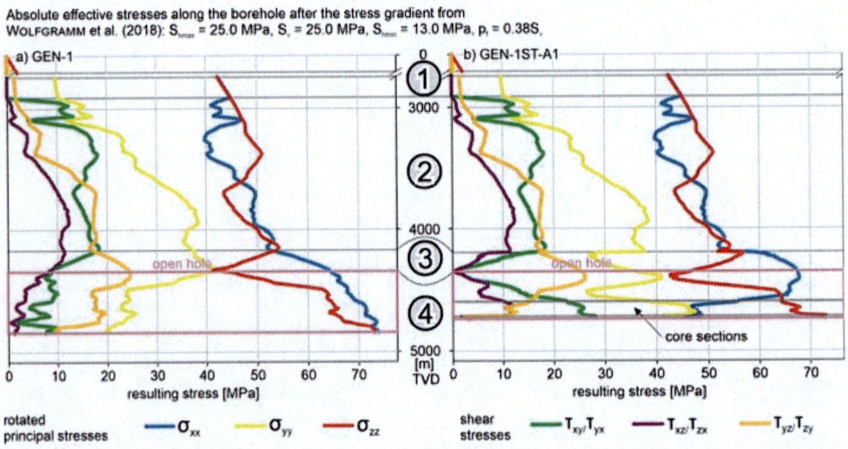

Fig. 5.42 Absolute effective stresses along the drill paths of the geothermal wells **a** GEN-1 and **b** GEN-1ST-A1, after a stress rotation in a transitional normal faulting to strike-slip stress regime from [18]

Table 5.11 Absolute effective stresses acting on the drill path at the position of the three stages of coring for a transitional normal faulting to strike-slip stress regime after [18]

Core runs	Effective resulting stress values on the drill cores (MPa)					
	σ_{xx}	σ_{yy}	σ_{zz}	τ_{xy}	τ_{xz}	τ_{yz}
CR 1 and CR 2	56	38	65	23	10	15
CR 3 and CR 4	46	46	68	26	10	10
CR 5, CR 6 and CR 7	49	45	69	26	10	10

of GEN-1, the resulting stresses are: $\sigma_{xx} = 74$, $\sigma_{yy} = 20$, $\sigma_{zz} = 73$, $\tau_{xy} = 7$, $\tau_{xz} = 1$, and $\tau_{yz} = 10$ MPa. Noticeably, the effective stresses σ_{xx} and σ_{zz} acting perpendicular to the borehole are equal.

The stresses develop parallelly, with the exceptional peak of τ_{yz} at 25 MPa, also applies for GEN-1ST-A1. The stresses, however, amount slightly differently at the transition to the open hole section with: $\sigma_{xx} = 68$, $\sigma_{yy} = 41$, $\sigma_{zz} = 43$, $\tau_{xy} = 2$, and $\tau_{xz} = 2$ MPa. Table 5.11. registers the resulting stresses on the core runs within Sect. 4. The final stress state at the bottom of GEN-1ST-A1 under the give stress regimes is $\sigma_{xx} = 47$, $\sigma_{yy} = 43$, $\sigma_{zz} = 73$, $\tau_{xy} = 28$, $\tau_{xz} = 3$, and $\tau_{yz} = 4$ MPa. Remarkably, σ_{zz} exceeds σ_{xx} by more than 25 MPa.

5.5.3 *Tangential (Hoop) and Radial Stresses at the Borehole Wall*

The previous Chapter's absolute resulting stress values serve as basic input parameters for the elastic derivation of the maximum and minimum tangential stresses, acting on the borehole wall. Section 2.3 presents the background to these calculations, basically performed after [21]. The following figures show the drill trajectories in a 3D coordinate system of the open hole section. From the resulting stresses, σ_{xx} is larger than $\sigma_{zz.}$ Hence, the maximum tangential stress is horizontal, while the minimum tangential stress is vertical.

5.5.3.1 Tangential Stresses at the Borehole Wall

In the sequence of previously presented absolute resulting stresses, Fig. 5.43 shows the development of maximum tangential stresses alongside the geothermal wells, displayed over a range from 100 to 380 MPa. Noticeably, for the two pure strike-slip stress regimes Fig. 5.43a, b, the upper limits of the maximum tangential stress appear in the well GEN-1, while the lower limits appear in the well GEN-1ST-A1. At the transition from a normal faulting to a strike-slip regime, the tangential stresses are larger in GEN-1ST-A1.a, Figure 5.43 with the stress regime after [17], shows that the effective max. Tangential stresses for GEN-1 start at 179 MPa and end at 193 MPa

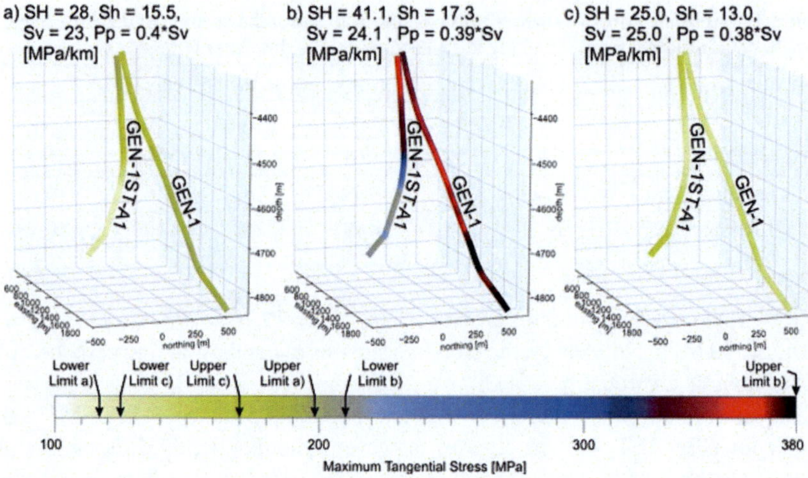

Fig. 5.43 Maximum tangential stresses along the drill paths from GEN-1 and GEN-1ST-A1 with different stress ratios **a** and **b** strike-slip (SS), and **c** transitional from normal faulting to SS regime. The color gradient indicates that the maximum tangential stresses are less in GEN-1ST-A1 than in GEN-1 for a SS regime

with an upper limit of 197 MPa. GEN-1ST-A1 starts at 195 MPa and decreases to 130 MPa, with a lower limit of 116 MPa. The maximum tangential stresses with the stress gradient after [20], illustrated in Fig. 5.43b, progress in the same manner. GEN-1 starts at a max. hoop stress of 332 MPa and ends at 380 MPa, which is its upper limit. This limit also occurs at 4750 m TVD. GEN-1ST-A1 starts at a max. tangential stress of 363 MPa and constantly decreases to 220 MPa, with a lower limit of 210 MPa at 4750 m TVD.

Figure. 5.43c illustrates the max. hoop stress of the stress gradient after [18]. GEN-1 starts at a max tang. stress of 146 MPa and also ends at 145 MPa. During the drill trajectory, the max. tang. stress does not fall below 135 MPa. For GEN-1ST-A1, the max. tangential stress is 162 MPa initially and returns to 168 MPa while reaching a minimum of 125 MPa during its dextral turn.

Figure 5.44 shows the minimum tangential stresses along the drill trajectories. In Fig. 5.44b, GEN-1 starts at a minimum tangential stress of 53 MPa and steadily increases to 104 MPa. GEN-1ST-A1 starts at a slightly lower min. hoop stress of 35 MPa and increases to 110 MPa, with a maximum 116 MPa at 4710 m TVD. In the stress state, after [20], as illustrated in Fig. 5.44, the min. hoop stress of GEN-1 starts at 18 MPa and increases to 55 MPa. The min. tang. stress for GEN-1ST-A1 is negative for the first 20 m before it becomes positive and accounts for a value of 106 MPa for $\sigma_{\theta,min}$ at the bottom and an upper limit of 110 MPa.

Figure 5.44c shows the development of minimal hoop stresses in the hybrid stress field. The min. hoop stress of GEN-1 starts at 62 MPa and culminates at 140 MPa. These are the highest values for the minimum tang. stress from all three considered stress gradients. Furthermore, $\sigma_{\theta,min}$ approaches $\sigma_{\theta,max}$ with increasing depth of the

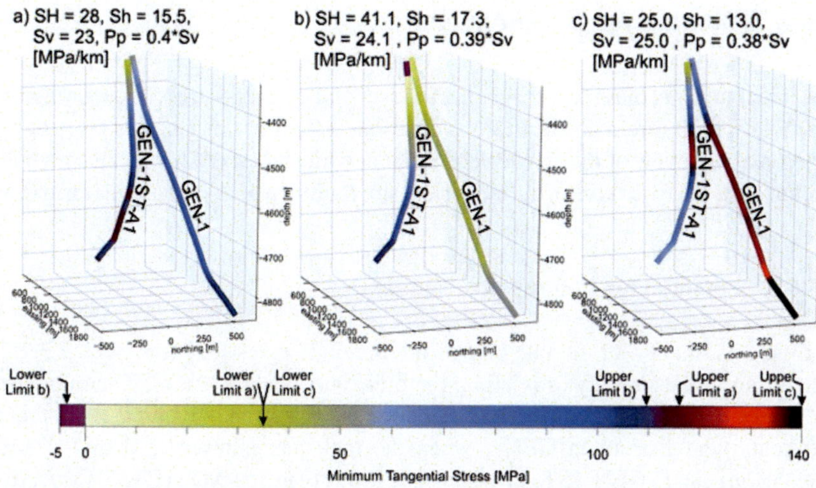

Fig. 5.44 Minimum tangential stresses along the drill paths from GEN-1 and GEN-1ST-A1 with different stress ratios **a** and **b** strike-slip (SS), and **c** transitional from normal faulting (NF) to SS regime. The color gradient indicates that the minimum tangential stresses are in general higher in GEN-1ST-A1 than in GEN-1 for a SS but lower in a NF regime

borehole. In GEN-1ST-A1, the min. hoop stress starts at 36 MPa, crowns at 123 MPa in 4540 m TVD and drops back to 64 MPa at the end.

5.5.3.2 Radial Stresses

The difference between formation pressure p_f and the mud weight p_w defines the radial stress acting supportive in the borehole against the effective stresses. Since these two kinds of pressure are hydrostatic, the course of the drill trajectory has no influence. The formation pressure presented in this Chapter is from [20], with a ratio of 0.39*24.1 MPa/km. The mud weight p_w is determined from the originally used mud's density of 1.05 g/cm^3, multiplied by the depth and the gravitational acceleration, and starts at the surface. At 4350 m TVD, where the open hole section of GEN-1ST-A1 starts, p_f is 41 MPa.

To the bottom of the borehole at 4735 m TVD, p_f increases to 44.5 MPa. The mud weight rises from 44.8 MPa to 48.8 MPa. Hence, the fluid pressure difference is 3.9 MPa at the beginning and 4.3 MPa at the borehole deepest. In GEN-1, the difference is 4.4 MPa at the bottom. Since the differences in fluid pressure are consistent over both wells, no image is provided.

5.5.4 Potential for Borehole Failure

The maximum and minimum tangential stresses and the radial stress allow estimating borehole stability from an elastic consideration. However, this Chapter gives only a superficial overview of the elastic deformation, while the next Chapter also includes fracturing of the rock mass in the numerical simulations and thus plastic deformation.

5.5.4.1 Potential for Compressive Failure

A simple estimation of the integrity of the borehole gives [22] (see Sect. 2.3.2.2). According to the stability estimation, the breakouts occur if the maximum tangential stress exceeds the rock's uniaxial compressive strength. Therefore, borehole breakouts would occur in GEN-1 when the rock has a lower σ_u than 197 MPa (Fig. 5.43a) and in GEN-1ST-A1 when σ_u is lower than 168 MPa (Fig. 5.43b). After [20], maximum stresses occur as high as 380 MPa for GEN-1. Hence, for lower rock strength, breakouts occur.

Zoback et al. [23] approaches the minimum required rock strength σ_u to avoid borehole failure by using an equation based on the Mohr–Coulomb fracture criterion, presented in Sect. 2.1.2.1. This equation uses the friction coefficient μ. Solving this equation with the max. tangential stresses from the gradient after [20], the radial stress from Sect. 5.5.3.2 and a friction coefficient of $\mu = 0.6$, the rock strength must be at least 107 MPa to prevent borehole failure in GEN-1. For GEN-1ST-A1, the required rock strength at the beginning of the open hole must be at least 102 MPa. However, due to the decreasing $\sigma_{\theta,max}$, a required rock strength of 62 MPa is sufficient to avoid BBOs at the borehole deepest.

5.5.4.2 Potential for Tensile Failure

Tensile failure depends on the minimum tangential stresses. Once the minimum tangential stress becomes zero or falls below zero, a crack opens in tension (mode 1). An opening would occur perpendicular to the smallest principal stress, which strikes EW for all three considered stress regimes. Hence, the cracks would form vertically in a NS-orientation independently from the trajectory of the borehole. Projected on GEN-1, these cracks would show as jagged lines—or slightly curved lines along the circumference, and for GEN-1ST-A1, the fractures would show as en-echelon DITFs on the roof and floor of the borehole.

Stress calculations show no hazard for tensile failure for both wells from stress regimes after [17, 17] (Fig. 5.44a, c). However, after the stress gradients from [20], the minimum hoop stresses from the beginning of the open hole section to 4365 m TVD (4515 m MD) in GEN-1ST-A1 is negative. The difference in fluid pressure further exceeds $\sigma_{\theta,min}$ to 4375 m TVD (4535 m MD). Under the condition that fractures were mapped in the corresponding direction (Fig. 5.35, JS2 and JS4), no tensile strength would apply, and fractures would open.

5.5.4.3 Depth of the Excavation Damaged Zone (EDZ)

Hoek [24] allow the empirical estimation of the depth of the excavation damaged zones (EDZ), as presented in Sect. 2.3.2.3. For the strike-slip regime after [20], which is also used in the numerical modeling, Fig. 5.45 shows the development of the depth of the EDZ along GEN-1 and GEN-1ST-A1. At the beginning of the open hole section, the depth of the EDZ is indistinguishable in both boreholes. However, with a progressing drill path, differences become perceptible. Three different rock strengths (σ_u) illustrate the development depth of the EDZ. Figure 5.45a shows the EDZ's depth with a σ_u of 80 MPa, which start at 38 cm for both wells and increases in GEN-1 to 42 cm, while it decreases in GEN-1ST-A1 to 22 cm. Likewise, yet at a smaller extent, the depths develop with a rock strength of 200 MPa and 250 MPa. In Fig. 5.45b, with a σ_u of 200 MPa, the depth starts at 12 cm, increases to 14 cm in GEN-1, and decreases to 6 cm in GEN-1ST-A1. For a σ_u of 250 MPa, the EDZ initially extends to 9 cm, and grows to 11 cm in GEN-1, while it shrinks to 4 cm in GEN-1ST-A1 (Fig. 5.45c).

Arrows in Fig. 5.45 mark the depth of the EDZ in 4600 m TVD, which is the according depth of the numerical models (Sect. 4.4.4). For an increasing σ_u, the depth decreases in GEN-1 from 40 cm (80 MPa) to 14 cm (200 MPa), and 10 cm

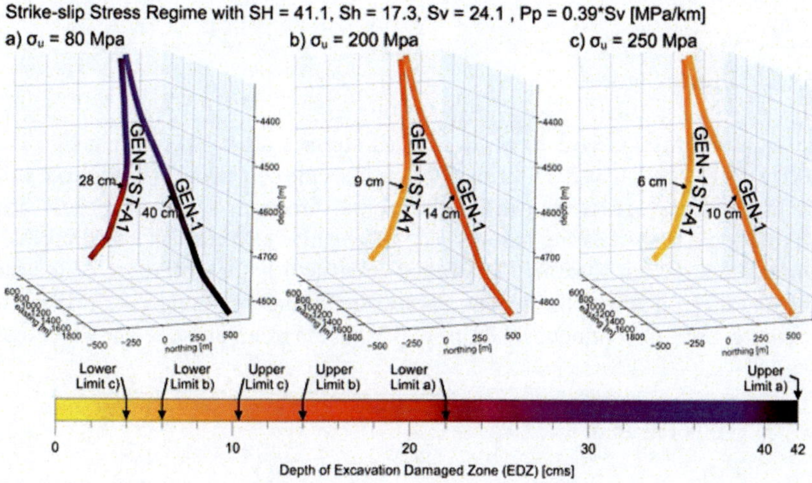

Fig. 5.45 Empirical estimated depth of the excavation damaged zone (EDZ) around the boreholes GEN-1 and GEN-1ST-A1, with the maximum tangential stress from a strike-slip regime after Backers et al. (2017) (Fig. 5.43b) and three different uniaxial compressive strengths σ_u: **a** 80 MPa, **b** 200 MPa, and **c** 250 MPa. The arrows in **a**, **b**, and **c** mark the depth of the EDZ in 4600 m TVD

(250 MPa). Likewise, in GEN-1ST-A1, the depth also declines from 28 cm (80 MPa) to 9 cm (200 MPa), and 6 cm (250 MPa).

5.6 Fracture Initiation and Propagation from Numerical Modeling

5.6.1 Results for the Model Setup

Each numerical model consists of a specific set of parameters for the material, the in situ stresses, and the DFN parameters. All models simulate an excavation of a circular borehole in 4600 m TVD.

5.6.1.1 Material Parameter

Section 4.4.1 presents the source and parameters assigned to the surfaces that embody the rock. Three different lithologies represent the host rock: "weak limestone", "limestone", and "dolostone". These lithologies share only the viscous damping factor and the friction coefficient but differ in all other parameters. Hence, Table 5.12 lists these parameters as they apply in the numerical simulation. Sections 6.1.2 and 6.5.1 copes with these parameters and discusses their suitability and reliability to represent each rock type.

5.6.1.2 In situ Stresses

The in situ stresses derive from the rotated stresses after [20] in Sect. 5.5.2 for 4600 m TVD for both wells. Section 6.4 discusses the choice of stress regime. Table 5.13 lists the stresses according to their horizontal direction (σ_{xx}), their vertical direction (σ_{yy}), and their shearing motion (τ_{xy}). A negative sign indicates a compressive regime and a dextral shearing. The radial stress (σ_r) is the difference of the mud weight (1.05 g/cm^3) with the formation pressure. Initially, starting at 4.5 MPa at the excavation time, σ_r continuously drops to 0 MPa, the reservoir's regular water table.

5.6.1.3 DFN Parameters

The DFN parameters consist of various properties. First of all, an adaption of the real fracture network, projected to the models' 2D cross-section, requires determining each joint set's apparent dip. For GEN-1 (5120 m MD), the borehole's azimuth and inclination are 84° and 70°. The cross-section perpendicular to this line has the dip

Table 5.12 Material properties as input parameter for the numerical models of the three different lithologies: "weak limestone", "limestone" and "dolostone"

Properties	Parameters			Lithologies		
				Weak limestone	Limestone	Dolostone
Elastic properties	Density	ρ	kg/m^3	2.5	2.65	2.75
	Young's modulus (static)	E_{stat}	GPa	35	50	60
	Poisson's ratio	ν	–	0.25	0.15	0.2
	Viscous damping	μ	kg/(m s)	1		
Strength properties	Tensile strength	σ_t	MPa	7	17	13
	Cohesion	c	MPa	12	50	40
	Mode I fracture energy	G_{IC}	J/m^2 N/m	23	86	47
	Mode II fracture energy	G_{IIC}		230	860	470
	Friction coefficient		–	0.6		
Penalties	Normal contact penalty	p_n	GPa m	350	500	600
	Tangential contact penalty	p_t	GPa/m	900	13,000	1600
	Fracture penalty	p_f	GPa	350	500	600

Table 5.13 Effective rotated stresses as input parameter for the numerical models of the geothermal wells GEN-1 and GEN-1ST-A1 (adopted from Table 5.10)

Wells	Absolute effective stress values (MPa)			
	σ_{xx}	σ_{yy}	τ_{xy}	σ_r
GEN-1	−145	−64	−1	Declining from 4.5 to 0
GEN-1ST-A1	−111	−64	6	

direction and dip of 264/70; it strikes more or less NS and dips to the west. GEN-1ST-A1 (5011 m MD) drill path points to a SE direction with an azimuth of 135° and an inclination of 74°. The resulting plane strikes NE-SW and dips with 74° to the NW. The resulting apparent dips from each joint set, identified on the rock cores in Sect. 5.4.3, are listed in Table 5.14 with their dip angle and dip direction.

Secondly, the five joint sets form clusters of three individual DFN sets. The reduction from 5 joint sets to 3 DFN sets eases the model setup and reduces possible meshing artifacts within the model. JS 1, which is discussed as bedding in Sect. 6.3.1, is always DFN-1, which dips with 8° to the south in GEN-1 and with 3° to the SW in GEN-1ST-A1.

Table 5.14 Apparent Dip of the joint sets from Sect. 5.4.3 (Fig. 5.37) on the 2D cross-sections of the numerical models from GEN-1 and GEN-1ST-A1

		Wells	
		GEN-1	GEN-1ST-A1
Dip direction and dip of the modelled cross-section		264/70	315/74
Strike and dip of the modelled cross-section		174/70 E	045/74 NW
Joint sets	True orientation	Apparent dip	
JS 1	157/08	8° S	3° SW
JS 2	304/66	66° N	45° SW
JS 3	185/71	41° S	53° SW
JS 4	111/81	65° S	43° NE
JS 5	346/64	60° N	61° NE

From this point, DFN-2 and -3 start to differ according to the well. GEN-1 integrates JS2 and JS5 into DFN-2, which dips between 60° and 66° in the north direction. DFN-3 merges JS3 and JS4 and dips between 41° and 65° to the south. Southward inclining joint sets have a negative sign. For GEN-1ST-A1, DFN-2 composes from JS2 and JS3 and dips between 45° and 53° towards SW. DFN-3 combined JS4 and JS5 and dips with 43° to 61° in the NE direction. Although the dip of DFN-1 is always constant, DFN-2 and DFN-3 distribute uniformly in the range of the dip in Table 5.15.

Thirdly, length, and spacing additionally define the DFN sets. DFN-1 is persistent over the whole 2D cross-section, with a spacing between 0.05 m and 0.6 m. DFN-2 and -3 have normally distributed discontinuity lengths (distance, where the DFN properties apply) and bridge length (distance, where the DFN properties do **not** apply) with 0.5 ± 0.05 m 0.2 ± 0.05 m. The spacing is also normally distributed with a span of 0.5 ± 0.05 m. Table 5.16 summarizes the length and spacing of all DFN sets.

Table 5.15 Distribution of the inclination values of each DFN set for the 2D cross-section of GEN-1 and GEN-1ST-A1

DFN set	Distribution	Dip (°)							
		GEN-1				GEN-1ST-A1			
		Joint sets	Range/value			Joint sets	Range/value		
DFN-1	Constant	JS 1	−8°			JS 1	3		
DFN-2	Uniform	JS 2 + JS 5	60	–	66	JS 2 + JS 3	45	–	53
DFN-3		JS 3 + JS 4	−65	–	−41	JS 3 + JS 4	−43	–	−61

Table 5.16 Length and Spacing of the DFN sets for the 2D cross-section of GEN-1 and GEN-1ST-A1

DFN set	GEN-1 & GEN-1ST-A1					
	Length (m)				Spacing (m)	
	Discontinuity length		Bridge length			
	Distribution	Range/single value	Distribution	Range/single value	Distribution	Range/single value
DFN-1	Constant (persistent over the whole cross section)				Uniform	0.05 – 0.6
DFN-2	Normal (stdv)	0.5 (0.05)	Normal (stdv)	0.2 (0.05)	Normal (stdv)	0.5 (0.05)
DFN-3						

Table 5.17 Properties of the defined DFN sets for all models with implemented DFNs

Properties	Parameters			DFN sets	
				DFN-1	DFN-2 & -3
Type				Broken	Cohesive
Strength properties	Tensile strength	σ_t	MPa	–	0.1
	Cohesion	c	MPa	–	0.5
	Mode I fracture energy	G_{IC}	J/m^2 N/m	–	1
	Mode II fracture energy	G_{IIC}		–	5
	Friction coefficient		–	0.3	0.4
Stiffness properties	Normal contact penalty	p_n	GPa m	50	50
	Tangential contact penalty	p_t	GPa/m	50	50
	Fracture penalty	p_f	GPa	–	50

Finally, properties need to be assigned to the newly created DFN sets. Due to the open joints, DFN-1 is of the type "broken" and cannot convey tensile forces within the rock mass. Furthermore, this shallow inclining DFN-1 has a very small friction coefficient of 0.3 (Table 5.17). DFN-2 and -3 are of a "cohesive" type and share the same parameters. They resist tensile stresses up to a σ_t of 0.1 MPa and a Mode I Fracture Energy of 1 and oppose any shear forces by a cohesion of 0.5 MPa and a Mode II Fracture Energy of 5. Although manually changed for the material properties, the penalties stay unaltered at a default value of 50 (Table 5.17).

5.6.2 Results from Numerical Modeling

The following Chapters present the results of the modeling scenarios. The first two Sects. 5.6.2.1 and 5.6.2.2 present the results from the models with only one "lithology" (Scenario 1, Sect. 4.4.4.1), excluding (Sect. 5.6.2.1) or including DFNs (Sect. 5.6.2.2); And Sects. 5.6.2.4 and 5.6.2.5 present the results from Scenarios 2-1 to 2-6, with the Scenarios including DFNs (SC2-3_DFN and SC2-5_DFN).

A comparative view of the models opposes GEN-1 to the left with GEN-1ST-A1 to the right. The models' excerpts show the full extent of fracturing, resulting in customized, varying scales for the comparisons. Initially, all Figures start at a time step of 200,000 (200 k, Time: 200), one time step before the actual excavation of the hole, where the Young's modulus lessens to a hundredth of its initial value (compare to the methodology in Sect. 4.4.3.2). The excerpts below this significant stage show the models at their ultimate stage, usually at 500 k time steps. Slight variations from the final time appear when models prematurely stop due to numerical instabilities, due to a lack of deformation/fracturing, or when computational time exceeds one week. All figures are synchronized to the same range of stresses between 0 MPa (dark blue) and 370 MPa (dark red) and share the same color code for tensile fractures (mode I, white) and shear fractures (mode II, green). The stress-free, blue area is the excavation damaged zone (EDZ), where failure leads to a significant opening and slip and causes highly and irreversible damage to the rock mass and stability issues for the cavity [25]. Beyond the EDZ, fracturing continues in the so-called fracture process zone (FPZ), where new fractures propagate and the rock mass softens, without necessarily causing stability issues ([26]: 497).

5.6.2.1 Scenario 1—Single Lithologies

Each model of Scenario 1 consists of one lithology only. Hence, the following paragraphs present the fracture initiation and propagation around the borehole of GEN-1 and GEN-1ST-A1 with the material parameters of "weak limestone," "limestone," and "dolostone." Scenario 1 with "weak limestone" is an exemplary case for a model with excessive computational time.

Weak Limestone

Fracturing in GEN-1 and GEN-1ST-A1 for the "weak limestone" extends over the vast extent of the whole refined mesh area in a 3 m square around the borehole (Fig. 5.46).Advanced fracturing in a mm-spacing range already after 76 k time steps, 124 k time steps before excavation of the borehole, led to the abortion of the run in GEN-1 (Fig. 5.46, left) for two reasons: an encountered of the material in that thickness in the reservoir is highly unlikely (see Discussion for model setup, Sect. 6.5.1.2) and due to the extensive computational time (>1 week). The same applies to GEN-1ST-A1, where less fracturing allowed running the model up to 303 k time steps (Fig. 5.46, right). However, mm-spaced fractures accumulate 1 m around the borehole and squeeze the borehole already by a few cm.

Limestone

GEN-1 at timestep 200 (Fig. 5.47, left, Time: 200) already shows an initiation and propagation of fractures even before the borehole's excavation. The stresses within the borehole strongly decline due to the decrease in Young's modulus. The stresses at the horizontal borehole walls compare well to the minimum tangential stresses in Fig. 5.44.

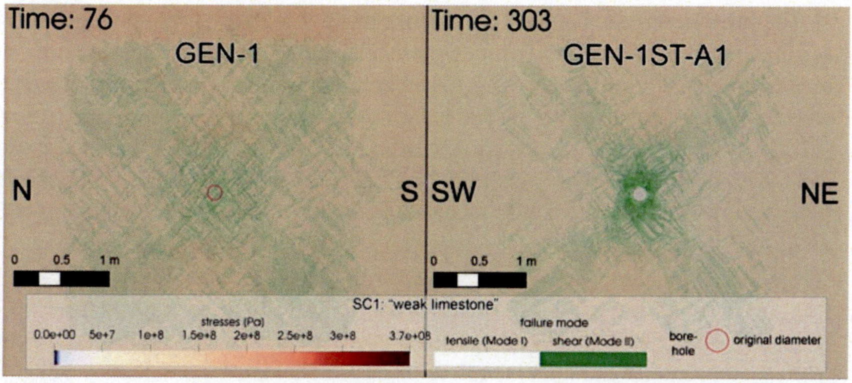

Fig. 5.46 Scenario 1, weak limestone: modelled fracture initiation and propagation around the borehole with the material parameters of "weak limestone". Left: GEN-1 at the time step 76 k, right: GEN-1ST-A1 at the time step 303 k

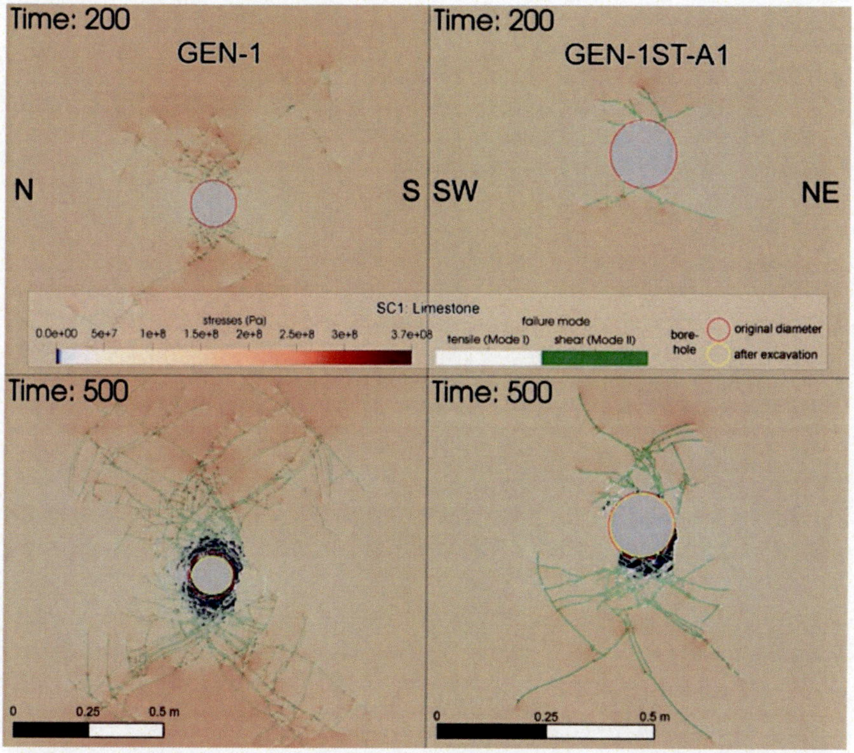

Fig. 5.47 Scenario 1, limestone: modelled fracture initiation and propagation around the borehole with the material parameters of "limestone". Left: GEN-1 at the time steps 200 k and 500 k, right: GEN-1ST-A1 at the time steps 200 k and 500 k

The stresses at the top and bottom decline due to fracturing, whereas high stresses accumulate at the crack tips. With the borehole excavation up to the final time step of 500 k (Fig. 5.47, left, Time: 500), the borehole supporting pressure decreases to 0, and fractures extend up to 1 m further in the rock. Also, an "excavation damage zone" (EDZ) develops in a vertical direction with an extension of around 0.1 m to each side, and the borehole tightens vertically, characterized by blue. The blue color generally indicates "stress-free" areas disconnected by fractures from the surrounding stress field.

GEN-1ST-A1 shows fewer and less wide progressed fractures before excavation (Fig. 5.47, right, Time: 200). After excavation, up to the final time step of 500 k (Fig. 5.47, right, Time: 500), the fractures grow and augment, forming intersecting fractures that lead to flaking at the floor of the borehole.

Dolostone

The fracture initiation in GEN-1 starts earlier, and fractures also spread to 2 m in the rock, forming a conjugated shear fracture network that interconnects well, even

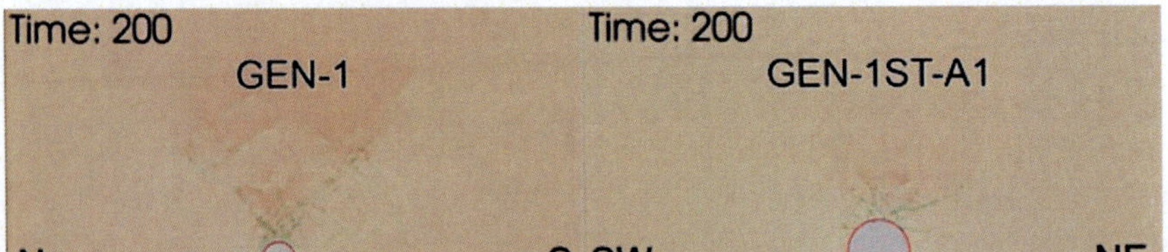

Fig. 5.48 Scenario 1, dolostone: modelled fracture initiation and propagation around the borehole of with the material parameters of "dolostone". Left: GEN-1 at the time steps 200 k and 500 k, right: GEN-1ST-A1 at the time steps 200 k and 500 k

before removing the rock (Fig. 5.48, left, Time: 200). Ultimately, after removing the rock and the support pressure, this network grows further, entangles the borehole, and an EDZ emerges that surrounds the borehole twice its diameter (0.3 m) (Fig. 5.48, left, Time: 500). The borehole diameter contracts by 2 cm.

For GEN-1ST-A1, the fractures' initiation starts at a later stage than for GEN-1 (Fig. 5.48, right, Time: 200). Also, fractures do not exceed a propagation of 0.3 m into the rock. For the final time step (Fig. 5.48, right, Time: 500), the FPZ spreads 1 m from the borehole. A surrounding blue ring of 5 to 10 cm marks the EDZ. The original diameter of the borehole is hardly altered.

5.6.2.2 Scenario 1 with Discrete Fracture Networks (DFNs)

This Chapter presents the same models from Sect. 5.6.2.1 but altered and supplemented with individual DFNs. While "weak limestone" and "limestone" share the same DFN of all three sets applied, "dolostone" only implements DFN-2 and -3 (compare Sect. 5.6.1.3).

Weak Limestone + DFN

Drilling into the "weak limestone," while considering a present DFN, still causes severe fracturing in the borehole and its proximity, already before the excavation. In GEN-1, dense fracturing extends around half a meter vertically around the borehole. Limited by a broken DFN (Fig. 5.49: Bedding), fracturing diminishes from a mm-spacing to a cm-spacing. GEN-1ST-A1 is also severely pierced by fractures.

However, these fractures restrict to a vertical extension of 0.5 m only, whereas remote fracturing breaks rock bridges from the DFNs. Again, the broken DFN limits the fracture propagation (Fig. 5.49, top row, Time: 200). After 428 k time steps, extensive computation time (>one week) forced a simulation stop.

Furthermore, the results show a strongly increasing fracturing that is not controlled by DFN properties anymore. The borehole diameter decreases continuously, and elements shove into the open hole area. The borehole begins to collapse (Fig. 5.49, left, Time: 428). In total, fractures extend over one square meter around the borehole, and even further lying rock bridges break and connect. The borehole begins to collapse. For GEN-1ST-A1, the influence of the DFN is still visible. Hence, a gradual decrease in the extent of fracturing shows along with the rock layers, disconnected by the broken DFN. Fractures extend twice as far vertically than horizontally, although the broken DFN damps the vertical extension (Fig. 5.49, right, Time: 500).

Limestone + DFN

At a time step of 200 k (Fig. 5.50, left, Time: 200), fracturing already started and orients on the pre-existing DFN-2 and -3. While reduction the Young's modulus links these single DFN sets, stress alteration in the proximity of one meter around the borehole also connects more distant rock bridges. Fractures in no relation with DFN sets' direction do not spread more than 10 cm away from the borehole. Also, the broken fracturing of the Bedding confines the growth.

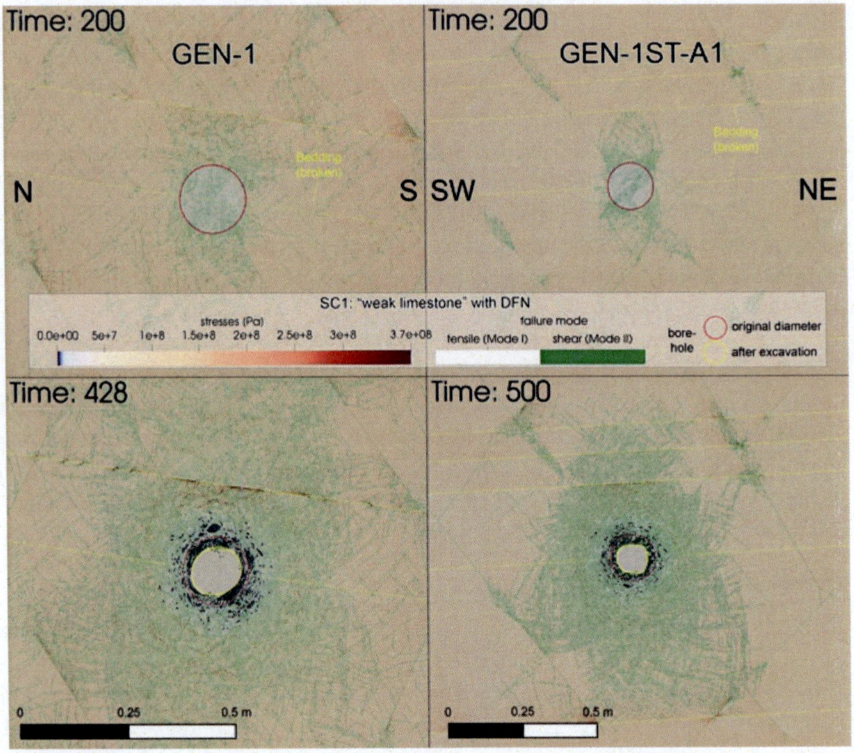

Fig. 5.49 Scenario 1, weak limestone + DFN: modelled fracture initiation and propagation around the borehole with the material parameters of "weak limestone" and all DFN sets. Left: GEN-1 at the time steps 200 k and 428 k, right: GEN-1ST-A1 at the time steps 200 k and 500 k

This constraining effect stays when removing the borehole and the support pressure. The rock disintegrates up to the smallest mesh element in the borehole roof until the broken DFN stops this process. From there upwards, fracturing continues and extends up to 1 m in distinct lines from points where multiple DFNs intersect. Under the borehole, no broken line confines the propagation, and fractures become more spaced and spread fan-like 0.5 m below the borehole floor (Fig. 5.50, left, Time: 450). While part of the strongly disintegrated rock in the roof is still under stress, the EDZ extends 10 cm in the roof, while it is only 5 cm at the walls and the floor. The severe fracturing leads to a detachment of single mesh elements from the wall and allocation into the borehole.

Again, fracturing in the pre-excavation phase is less in GEN-1ST-A1. Although one DFN set is close to the borehole, it does not connect at that stage. The extend of fracturing is reduced to 5 cm around the borehole (Fig. 5.50, right, Time: 200). Due to the randomness of the DFNs, they do not intersect with the borehole directly. Hence, initial fracturing starts at the nearest DFNs' tips and, after that, connects with the tips of the remaining discontinuities. Controlled by the DFNs, distinct fracture

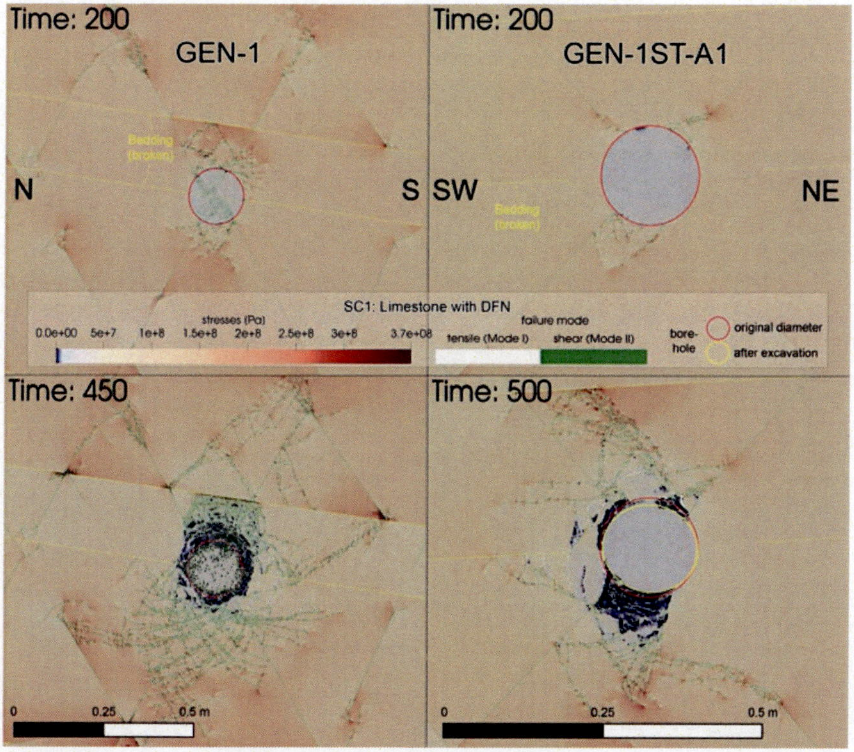

Fig. 5.50 Scenario 1, limestone + DFN: modelled fracture initiation and propagation around the borehole of with the material parameters of "limestone" and all DFN sets. Left: GEN-1 at the time steps 200 k and 450 k, right: GEN-1ST-A1 at the time steps 200 k and 500 k

branches align parallel to a given DFN set or aim to another discontinuity. Thus larger, continuous areas separate from the rock. These areas extend a few centimeters at the roof near the DFNs and up to 15 cm below the floor (Fig. 5.50, right, Time: 500). Minor spalling slightly reduces the borehole diameter.

Dolostone + DFN

For the "dolostone" material with DFNs, the borehole's rock also interfuses with fractures densely. The absence of the broken, slightly inclined Bedding discontinuity from Fig. 5.50 does not limit the fractures' extent. Hence, they develop in a rejuvenating cone-shaped form up to 20 cm above the borehole and in a fan-shaped form below the borehole (Fig. 5.51, left, Time: 200).

After an additional 290 k time steps (Fig. 5.51, left, Time: 490) and removing any supporting pressure, fractures propagate vertically by more than half a meter and horizontally by 0.25 m from the borehole. In the vertical direction, the rejuvenating cone-shape is visible in the dense fracture network. However, the EDZ develops in all directions uniformly and extends around 0.1 m from the borehole. In the EDZ,

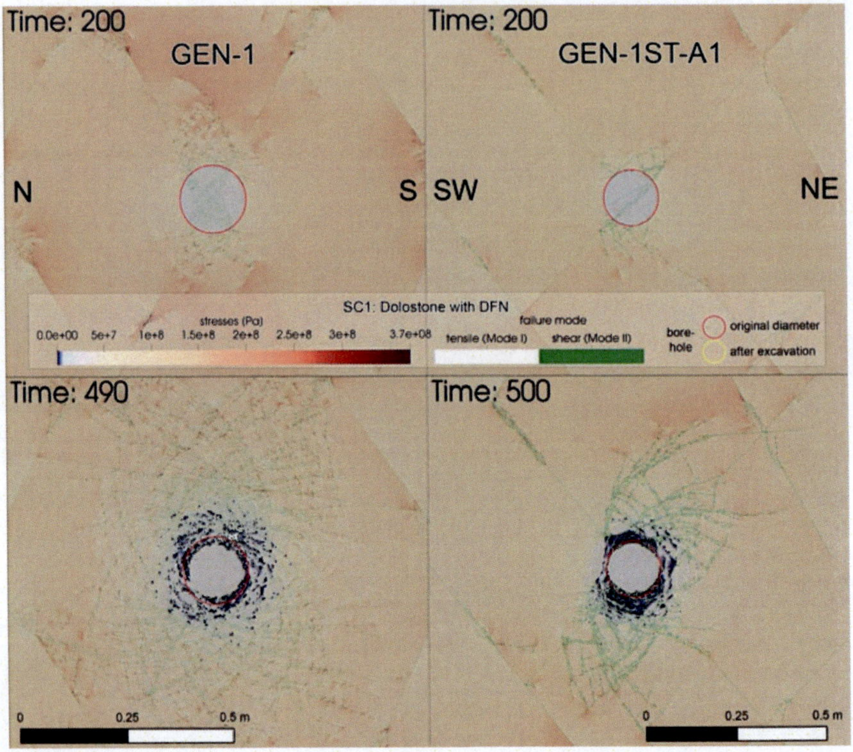

Fig. 5.51 Scenario 1, dolostone + DFN: modelled fracture initiation and propagation around the borehole with the material parameters of "limestone" and the DFN sets 2 and 3. Left: GEN-1 at the time steps 200 k and 490 k, right: GEN-1ST-A1 at the time steps 200 k and 500 k

some accumulated mesh elements are dark blue, some light blue, indicating that these pieces still interlock. The overall loosening of the rock mass leads to a shrinkage of the borehole to 10 cm.

In contrast to Fig. 5.50 (right, Time: 200), the reduction in Young's modulus in Fig. 5.51 (right, Time: 200) leads to the breakage of the rock bridge with the borehole. Furthermore, single fractures even propagate further and connect with the 0.25 m remote DFN. Fractures build a cone-like formation, slightly inclined to the SW dipping DFN set at the ultimate stage. In contrast to GEN-1 at this stage (Fig. 5.51, left, Time: 490), the EDZ consists of several larger, connected, dark blue elements that accumulate at the top and bottom of the borehole diameter. They extend up to 15 cm in the rock mass and squeeze the borehole diameter by a few centimeters vertically. (Fig. 5.51, right, Time: 500).

5.6.2.3 Fracture Opening and Slip in Scenario 1

The previous Chapter coped with the extent and type of fractures that emerge during the borehole excavation in the aquifer but does not give further information on the fractures' aperture or the amount of shearing. Shear-induced failure leads to borehole breakouts, and in the worst case, to a borehole collapse (2.3.2.2). Although stresses at the borehole margin give a good estimation of borehole failure, they do not reflect the total amount of slip at the shear planes. Hence, this Chapter presents the data on fracture opening, i.e., the thickness of the four-nodded elements that arise as soon as fractures occur, and on fracture slip, the shear offset of the triangular elements (Sect. 2.5.2).

Fracture Opening

Figure 5.52 shows the fractures' apertures from the final stages of all models of the GEN-1 scenarios 1. Taking the "weak limestone" aside due to the preliminary cancelation of the simulation, "limestone" and "dolostone" show openings based on the fracture pattern from Figs. 5.47 and 5.48. Although the actual fractures propagate further, two micrometer wide openings mark the minimum in a one meter square around the borehole. Towards the borehole, fractures open up to 20 mm. In the DFN models of GEN-1, DFN-1 limits open fractures' propagation and confine single open branches with a minor opening of 20–50 μm.

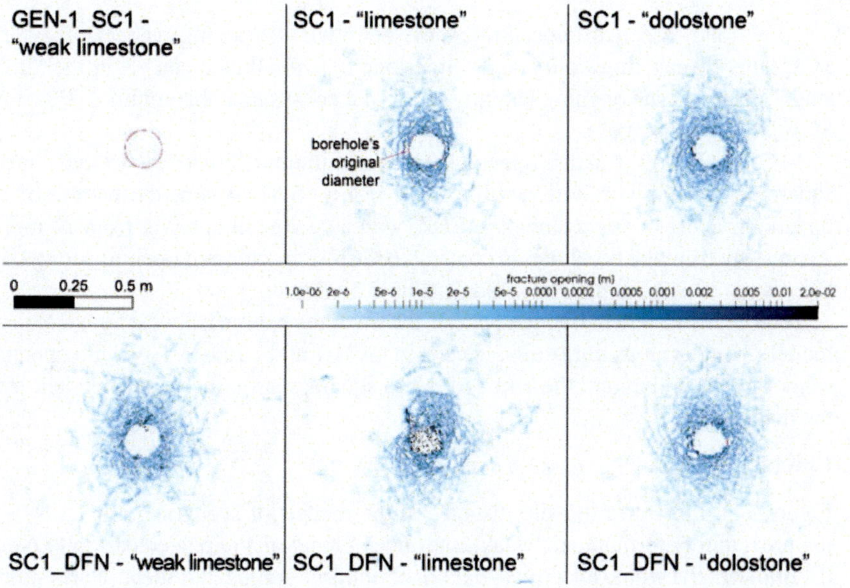

Fig. 5.52 Width of fracture opening [m] at the final time step of the models from GEN-1, scenario 1: "weak limestone," "limestone," and "dolostone" (top row) with their individual DFN sets (bottom row)

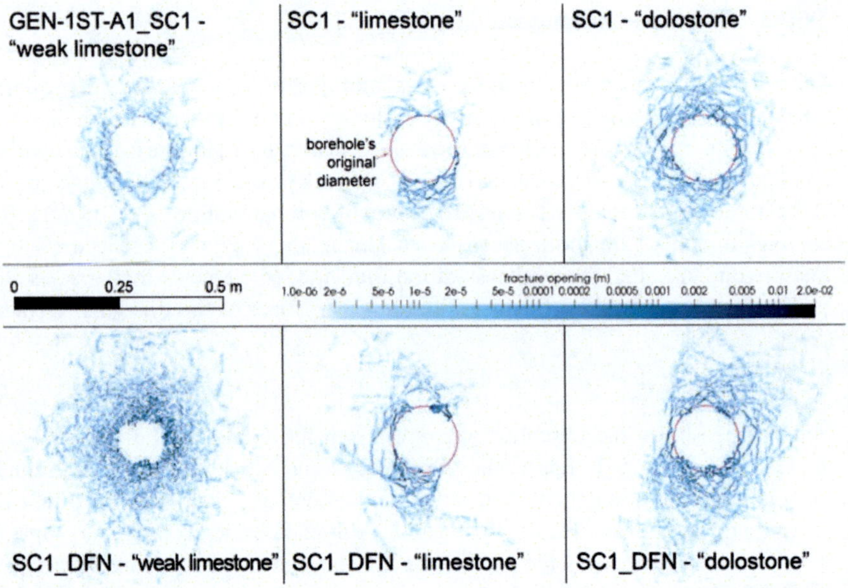

Fig. 5.53 Width of fracture opening [m] at the final time step of the models from GEN-1ST-A1, scenario 1: "weak limestone," "limestone," and "dolostone" (top row) with their individual DFN sets (bottom row)

DFN-2 and -3 seem to concentrate some incontiguous opening channels (Fig. 5.52, SC1_DFN "weak limestone" and "limestone"). The DFN-2 and -3 in the "dolostone" disperses the opening laterally but do not constitute coherent lines (Fig. 5.52, SC1_DFN "dolostone").

In GEN-1ST-A1, fracture openings with a maximum of 2 μm restrict half a meter square around the borehole, as illustrated in Fig. 5.53. Again, the pattern of the opening fractures orients along the EDZ, with a decline in opening from 20 mm to 2 μm away from the borehole. In contrast to GEN-1, a preferred opening along DFN lines is not visible in the DFN models (Fig. 5.53, bottom row).

Although different in fracture pattern, the fracture opening from the borehole to the model's rim develops uniformly. Hence, in GEN-1 and GEN-1ST-A1, the openings cannot be traced beyond the 1 m and 0.5 m square, respectively, independent from the material.

Fracture Slip

Figure 5.54 illustrates the slip distance in the models of scenario 1 for GEN-1. In the prematurely terminated "weak limestone" model, the slip is consistently below 0.1 mm over the whole excerpt. A pattern of multiple conjugated shear planes shows. For the "limestone" scenario, the greatest slip movements accumulate at the borehole's top and bottom, indicating potential cone-shaped areas for borehole breakouts. The "dolostone" model lacks this distinct pattern, and the amount of slip exceeds

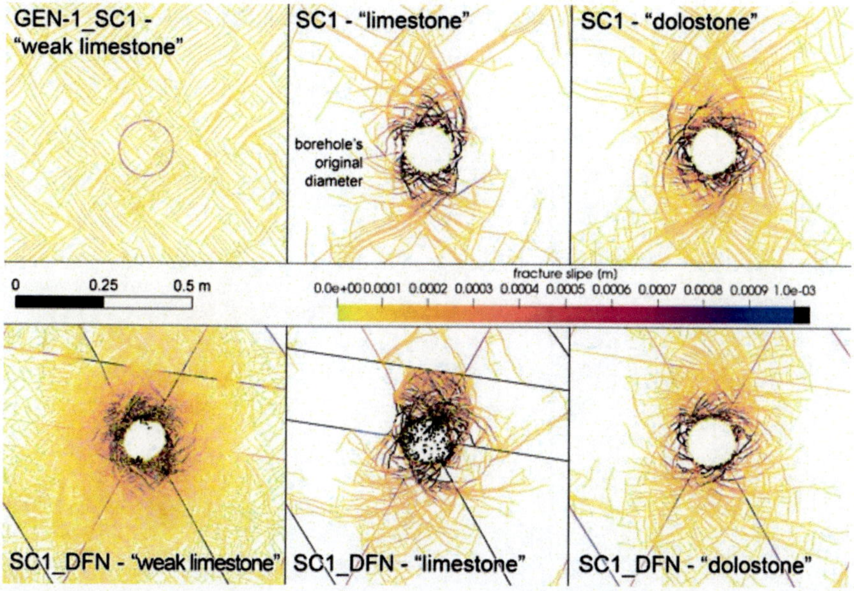

Fig. 5.54 Length of fracture slip [m] at the final time step of the models from GEN-1, scenario 1: "weak limestone," "limestone," and "dolostone" (top row) with their individual DFN sets (bottom row)

1 mm all around the borehole. However, the interacting network of shearing fractures is denser at the top and bottom than at the walls.

Regarding the models with DFN, the amount of slip decreases with distance to the borehole and uniformly around the borehole in the "weak limestone." All DFN lines, apart from the one intersecting the borehole, slip by more than 1 mm (Fig. 5.54, SC1_DFN – "weak limestone"). In the "limestone" with DFN, the largest slip accumulates at the top between the two DFN lines from set 2 and 3. It seems as if these two lines plus the broken line from set 1 confine and concentrate the fracturing within. Equally to the "weak limestone" + DFN model, the DFN lines also move more than 1 mm, also the one intersecting the borehole. The bottom develops about the same as the model without DFN (Fig. 5.54, SC1_DFN – "limestone"). In the "dolostone" + DFN model, the slip develops similarly to the one without DFNs. However, the DFN-2 and -3 seem to concentrate the amount of slip even denser at the top and bottom. Concurrently, the depth at the top and bottom is reduced, while at borehole walls, the slip extends evenly strong and wide than without DFNs.

The difference that the "weak limestone" model for GEN-1ST-A1 ran longer than for GEN-1, does not change the amount of slip, nor the fact that the slip decreases with distance from the borehole and that the DFN sets involved, comparably, slip even more than the emerging fractures (Fig. 5.55, "weak limestone" ex- and including DFNs).

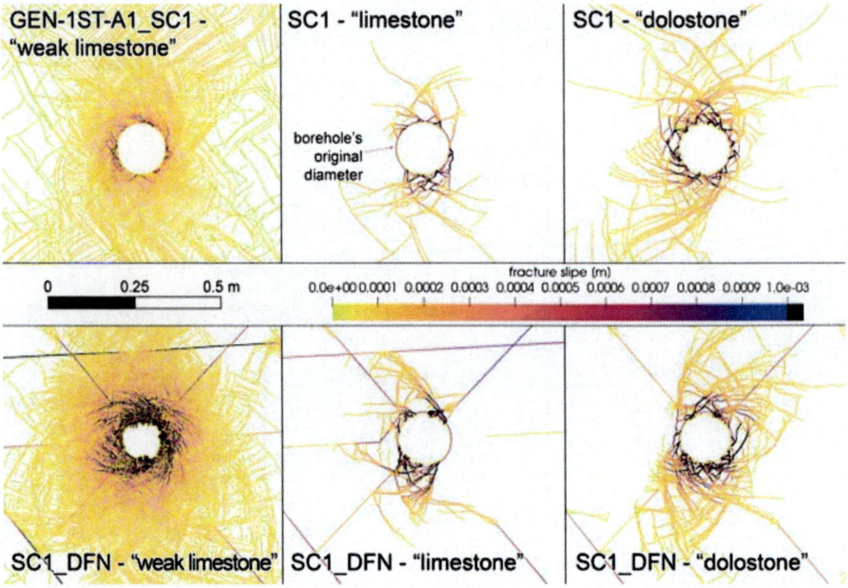

Fig. 5.55 Length of fracture slip [m] at the final time step of the models from GEN-1ST-A1, scenario 1: "weak limestone," "limestone," and "dolostone" (top row) with their individual DFN sets (bottom row)

In the "limestone," the fracture slip is only above and close to 1 mm at the borehole margin and rapidly drops to zero within 0.25 m from the borehole. Equally, the "dolostone" shows this effect, although the occurrence of single larger slips is higher around the borehole (Fig. 5.55, "limestone" & "dolostone"). Fracture slips decrease within a distance of half a meter to 0.2 mm and less.

Concerning the DFN models, large slip movements from newly emerging fractures orient towards the DFN-2 and -3. In addition, if a DFN set is either decoupled from the DFN network or intersected by a newly emerging fracture, the slip decreases. However, the overall slip distance for both "limestone" and "dolostone" models is larger than in models without DFNs. (Fig. 5.55, "limestone" & "dolostone" + DFN). The amount of slip decreases with distance from the borehole. However, in the "dolostone" it decreases faster than in the "limestone".

5.6.2.4 Scenario 2—Multiple Lithologies

Scenario 2 (SC 2) is a stratigraphic sequence of all three materials, "weak limestone", "limestone", and "dolostone" with the model setup after Sect. 4.4.4.2. The sub-scenarios one to six in SC2 present the excavation of boreholes in slices of a 3D volume, representing a typical geological setup of the reservoir based on the results from this work (This chapter) and discussion in Sect. 6.5.1.2, simulating, on the one hand, running a borehole in the reservoir and on the other hand identifying different fracture patterns in various lithologies. Visualization of the results is similar to the previous Chapters.

SC2-1: Borehole in "limestone" in between "weak limestone"

Scenario 2–1 simulates a borehole excavation in "limestone" betwixt between two layers of "weak limestone." Noticeably, during the reduction of the Young's modulus in the first 150 k time steps, fractures already emerge in the remote layers of "weak limestone," not connecting to the borehole. However, the shifting of stresses leads to intense fracturing of the two weak layers even before fractures on the borehole margin emerge.

Only after 130 k time steps, fracturing d branches off as a wing-shaped crack at the borehole's top to the south and at the bottom to the north, with two branches to the south. After reaching the "weak limestone," they connect with the pre-existing cracks and intensify (Fig. 5.56, left, Time: 200).

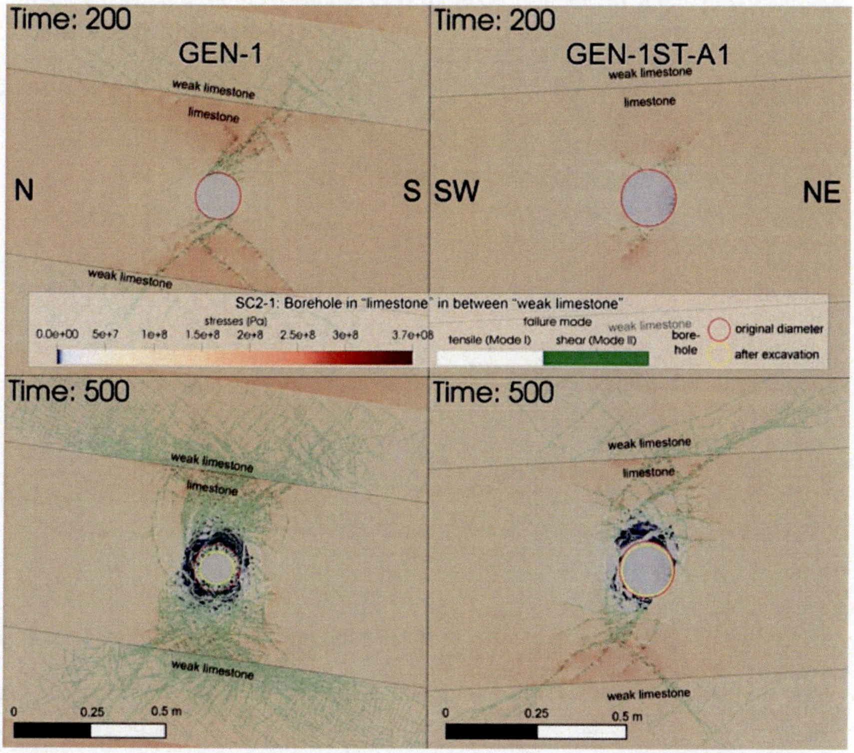

Fig. 5.56 Scenario 2-1, Borehole in "limestone" in between "weak limestone": modelled fracture initiation and propagation around the borehole with absolute stresses and failure mode of fractures. Left: GEN-1 at the time steps 200 k and 500 k, right: GEN-1ST-A1 at the time steps 200 k and 500 k

With progressing time steps, fractures propagate vertically towards the base and hanging weak layers, where they connect with the existing cracks in these layers, which does not restrict the horizontal extension of fractures in the limestone. The transgressing fractures cause further shear fracturing in the weak layers. The EDZ extends over 0.3 m, and the accompanying deformation shrinks the borehole diameter to less than 0.1 m. However, the margin elements are not completely free of stress (dark blue) but somehow interlock at a low stress magnitude (light blue) (Fig. 5.56, left, Time: 200).

The fracture initiation in the first time steps of GEN-1ST-A1 only occurs at the borehole perimeter and spread around 10 cm from the borehole. The weak layers, which are heavily fractured in GEN-1, are not damaged (Fig. 5.56, left, Time: 200). The excavation and the reduction of support pressure leads to a fracture propagation similarly channeled as in GEN-1. However, only after the first fractures within the "limestone" reach the weaker layers, fracturing within the limestone stops, and spreading continues at the roots of the "limestone" fractures in the "weak limestone." Although the borehole diameter is hardly altered, one larger segment at the top with around 10 cm depth and the right bottom with around 5 cm detaches from the rock.

In contrast to Fig. 5.47, where the fractures spread fan-like from the borehole, the development of fractures in Fig. 5.56 is more focused on the pre-existing fractures, as already seen in the DFN-models before (Sect. 5.6.2.2). The extend of fractures exceeds the SC 1 "limestone" but is inferior to SC 1 "weak limestone."

SC2-2: Borehole in "weak limestone" in between "limestone"

The borehole in SC2-2 locates in the "weak limestone" only. In GEN-1, fractures pierce the whole layer of "weak limestone" instantly by reducing the Young's modulus in the borehole. Fracturing condenses to the top and bottom of the borehole and also penetrate through the borehole. The fractures propagate in the lying and hanging boundary of the "weak limestone" into the "limestone," where they extend distinctly (Fig. 5.57, left, Time: 200). Further deformation due to excavation and removal of the borehole squeezes the borehole diameter by 35% to 10 cm, and some deep blue mesh elements push even further, showing no final ring support. Shear fractures over an extension of almost 1.5 m densely pierce the weak layer, and shearing continues in the "limestone" above and below (Fig. 5.57, left, Time: 500). Not illustrated in Fig. 5.57, distinct fractures progress from the "limestone" below to the "dolostone," where they spread fan-like for another 1.5 m away from the borehole.

In GEN-1ST-A1, the shear fractures stay within the "weak limestone" and do not transit to the upper and lower bounding "limestone" (Fig. 5.57, left, Time: 200 & Time: 500). Fractures arise at the top and bottom of the borehole and advance simultaneously vertical and parallel with the bedding. They move NE and SW from the borehole's top and bottom until they connect in a flat bow to the "limestone" boundary. Remarkably, no fracturing within the borehole area happens (Fig. 5.57, left, Time: 200). For the next 300 k time steps, fracturing progresses in the "weak limestone" only and creates an EDZ that circularly surrounds the borehole with a diameter of the weak layers' thickness (0.3 m). The borehole diameter reduces by 2 cm, while a

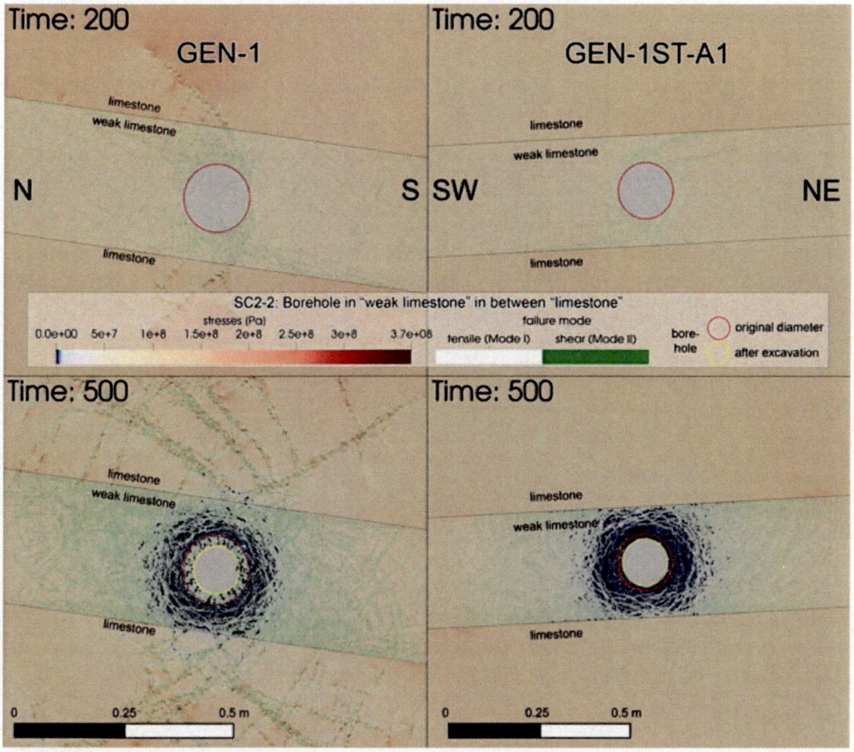

Fig. 5.57 Scenario 2-2, Borehole in "weak limestone" in between "limestone": modelled fracture initiation and propagation around the borehole with absolute stresses and failure mode of fractures. Left: GEN-1 at the time steps 200 k and 500 k, right: GEN-1ST-A1 at the time steps 200 k and 500 k

dense fracture network extends over half a meter from the borehole (Fig. 5.57, left, Time: 500).

SC2-3: Borehole intersecting "limestone" at the Base and "weak limestone" in the Hanging Wall

During the numerical preparation for the excavation in SC2-3, fracturing starts likewise to SC2-1 in the weak layers and afterward initializes at the borehole's bottom margin. While several shear fractures fully disintegrate the rock around the borehole's roof in the "weak limestone," a single fracture emerges at the borehole's bottom in the "limestone" and pierces it straight to the "dolostone," where it splits up (Fig. 5.58, left, Time: 200).

Although minor cracks already transit to the "limestone" above the "weak limestone" before excavation, an intense fracturing develops after removing all restraining borehole support. As a consequence, fractures spread even further both up and down. Downwards, fracturing creates a second main branch from the bottom of the borehole perimeter, developing conjugated shear planes. These conjugates shear planes can

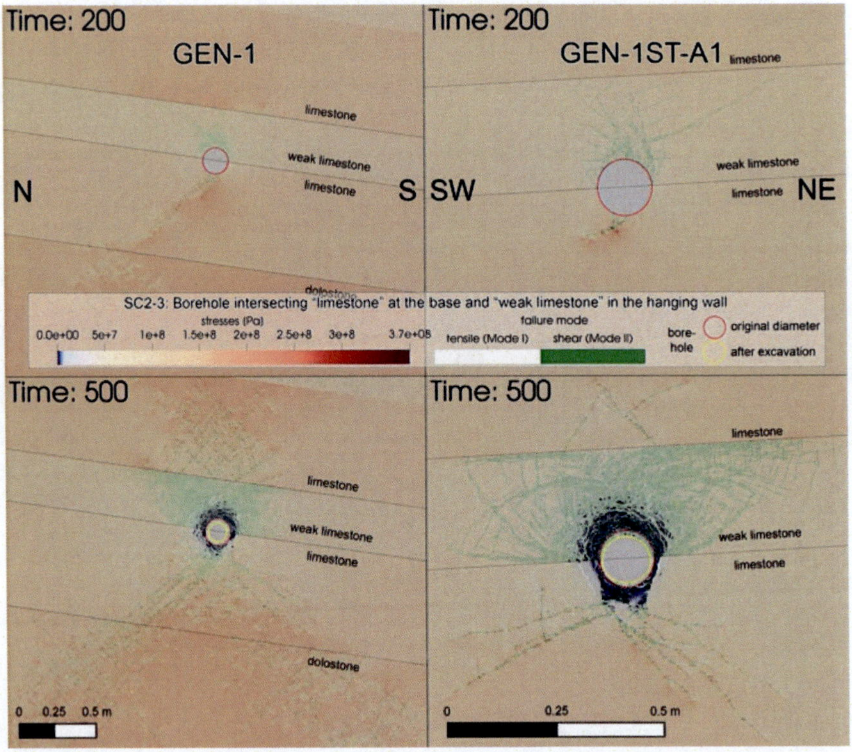

Fig. 5.58 Scenario 2-3, Borehole intersecting "limestone" at the base and "weak limestone" in the hanging wall: modelled fracture initiation and propagation around the borehole with absolute stresses and failure mode of fractures. Left: GEN-1 at the time steps 200 k and 500 k, right: GEN-1ST-A1 at the time steps 200 k and 500 k

be distinguished in distinct branches in the "limestone" but spread unorganized in the "dolostone" up to 2.5 m and more away from the borehole. Above the borehole, shear fractures spread laterally in the "weak limestone" and trespass in the hanging "limestone" layer, where they further connect sporadically to the second weak layer. However, they extend only 1.5 m to the top, which is 1 m shorter than in the opposite direction. The local stress field mirrors the strong fractioning of the weak layer. Hence, the EDZ above the borehole extends by 10 cm and more, and tensile fractures that further dissect the elements accompany shear failures. In the "limestone," the EDZ is less than 10 cm (Fig. 5.58, left, Time: 500).

GEN-1ST-A1 copies the mechanical failure behavior of GEN-1 at the beginning of the simulation, but with a decrease fracturing degree that scratches the transition to "limestone" in the hanging wall with minor stress peaks. At the borehole's base, the emerging crack does not reach the dolostone but ends after 10 cm (Fig. 5.58, right, Time: 200). Further deformation with distinct fracture growth proceeds half a meter with the "limestone." However, many shear fractures penetrate the "weak limestone" and loosen the rock up to a depth of 15 cm around the borehole's top half perimeter. Single fractures also proceed in the "limestone" above the "weak limestone." The EDZ becomes a funnel-shaped form that squeezes the borehole by a few centimeters from the top (Fig. 5.58, right, Time: 500).

SC2-4: Borehole in "limestone" with "dolostone" at the Base and "weak limest." in the Hanging Wall

Equally to the previous models, fractures in the "weak limestone" already emerge with reducing the Young's modulus of the borehole in the "limestone" in GEN-1. Besides these early fractures, further branches initiate in the top and bottom of the borehole. One branch each moves in a wing-shaped deflection to the "weak limestone," where it connects with pre-existing fractures, and the other to the boundary of "dolostone," where it progresses from a distinct fracture to a widespread network of thin fractures (Fig. 5.59, left, Time: 200). With progressing elimination of support pressure, the fracture network grows continuously around the borehole, where it

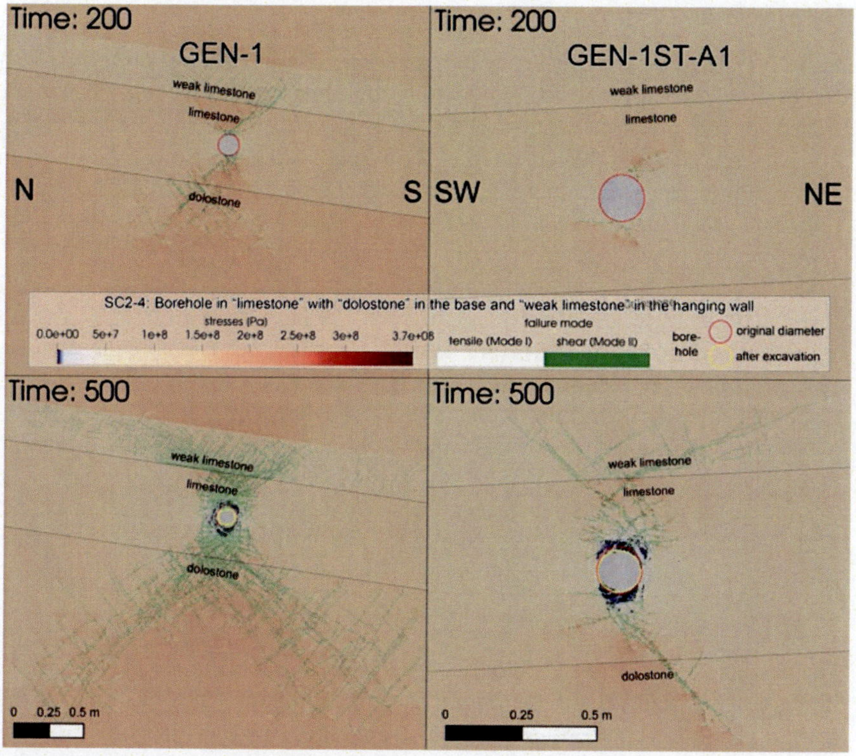

Fig. 5.59 Scenario 2-4, Borehole in "limestone" with "dolostone" at the base and "weak limestone" in the hanging wall: modelled fracture initiation and propagation around the borehole with absolute stresses and failure mode of fractures. Left: GEN-1 at the time steps 200 k and 500 k, right: GEN-1ST-A1 at the time steps 200 k and 500 k

loosens the rock mass up to 0.3 m uniformly around the borehole and squeezes it to a diameter of 12 cm. Vertically, fracture density accumulates, where the fractures extend over half a meter along the boundaries to the other lithologies. From there, fracturing continues severely at the base of the "weak limestone." In the "dolostone," fractures extend by more than 3 m as a conjugates shear fracture. These shear fractures split and partly propagate back to the "limestone," where they stop (Fig. 5.59, left, Time: 500).

During the reduction of the Young's modulus Gen-1ST-A1, shear fractures only emerge for 10 cm in the "limestone" (Fig. 5.59, right, Time: 200). In contrast to GEN-1, fractures at the top and bottom do not develop evenly. While the top fractures propagate to a cone-like network and have multiply transits to the "weak limestone," the fractures at the bottom focus on one distinct branch, where the branch continues almost unaltered, although one branch pierces back to the "limestone." Although the fracture development is different vertically, two notches show at the top and bottom with a depth of approximately 5 cm depth (Fig. 5.59, right, Time: 500).

SC2-5: Borehole intersecting "dolostone" at the Base and "limestone" in the Hanging Wall

In SC2-5, the borehole locates at the stratigraphic boundary of "limestone" and "dolostone." According to Fig. 5.47 (SC1 "limestone") and Fig. 5.48 (SC1 "dolostone), fracture initiation behaves in the same way, where a conjugated shear fracture pair emerges from the reduction of the Young's modulus. However, even in the pre-excavation stage, fractures extend more than in the lithology-individual models (Fig. 5.60, left, Time: 200).

Again, the "weak limestone," even more remote than in Fig. 5.56 (SC2-1), is slightly fractured. However, with progressing removing the borehole's support pressure, fracturing increases laterally and in length, leading to a denser fracture network, which keeps the conjugated fracture network's appearance. The borehole squeezes vertically for a few centimeters, but a light blue element color shows that the remaining stresses support the borehole (Fig. 5.60, left, Time: 500).

The development in the first stage of pre-excavation in GEN-1ST-A1 corresponds to the equivalent rocks of SC1 (Figs. 5.47 and 5.48). In the following time steps, the fractures in the "dolostone" increase faster and spread further than in the "limestone". With an advanced fracture network, the EDZ extends up to 10 cm in the "dolostone," whereas only scattered loose segments of a few centimeters form in the "limestone."

SC2-6: Borehole in "dolostone" with "limestone" in the Hanging Wall

The borehole of SC2-6 is in "dolostone" with a distance of 0.5 m to the "limestone" in the hanging wall. Again, the preliminary fracture initiation forms a conjugated fracture network with almost orthogonally structures. However, in contrast to SC1 – "dolostone" (Fig. 5.48), the fractures extend further, and more off-branches establish. All fractures reach the lithological boundary but hardly transit in the "limestone" (Fig. 5.61, left, Time: 200). At the time step 500 k (Fig. 5.61, left, Time: 500), the fracture network becomes even denser but does not extend its reach. Around the borehole, the shear failures accumulate in a diameter of 0.5 m, creating an EDZ of

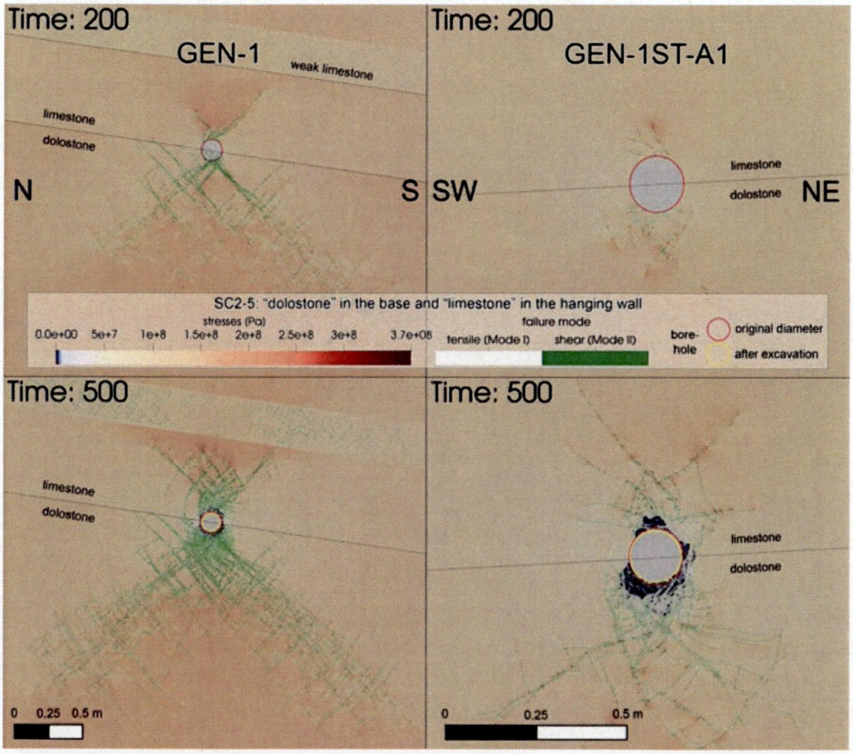

Fig. 5.60 Scenario 2-5, Borehole intersecting "dolostone" at the base and "limestone" in the hanging wall: modelled fracture initiation and propagation around the borehole with absolute stresses and failure mode of fractures. Left: GEN-1 at the time steps 200 k and 500 k, right: GEN-1ST-A1 at the time steps 200 k and 500 k

at least 0.1 cm. The final extend of fractures along the lithological boundary is ca. 1.5 m.

For GEN-1ST-A1, the initiating crack branches are less und shorter than in GEN-1 and more distinct. They develop similarly to the ones in SC1 – "dolostone" (Fig. 5.48) but extend twice as far (Fig. 5.61, right, Time: 200). Remarkably, the initial three large fracture branches limit the lateral extent of the final fracture network, extending horizontally not more than 1 m and along the lithological boundary by less than 0.8 m. While a dense fracture network entangles the borehole of GEN-1, there are almost no lateral fractures to the side of GEN-1ST-A1's borehole. However, the EDZ extends up to more than 0.5 m vertically and 0.3 m horizontally. To the borehole's sides, larger pieces detached in a tensile mode in large coherent elements (Fig. 5.61, right, Time: 500). The area of the EDZ, especially to the margins inside the borehole, is deep blue. However, the borehole diameter only decreases hardly.

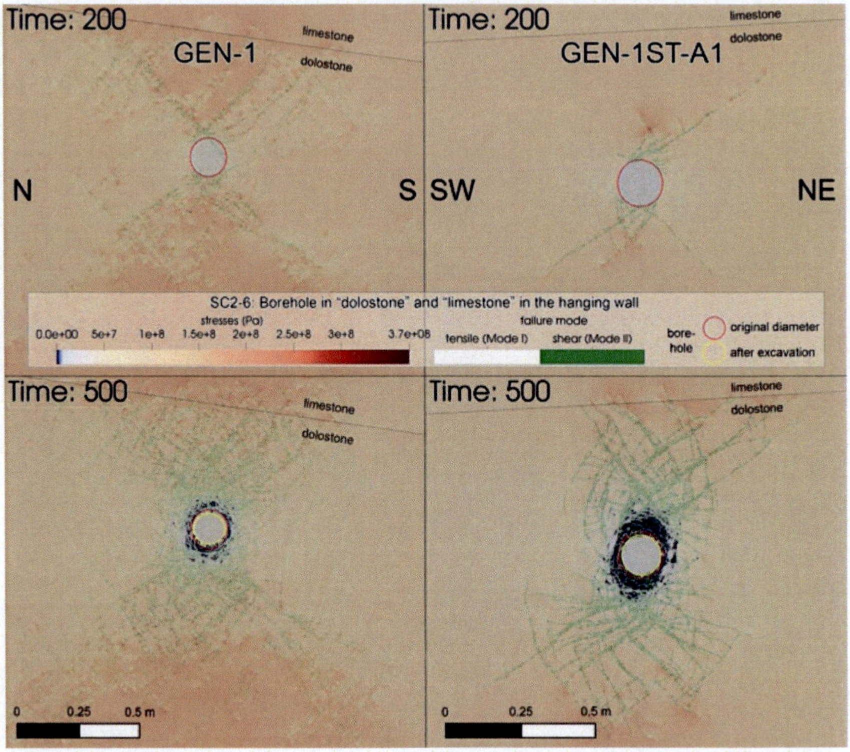

Fig. 5.61 Scenario 2-6, Borehole in "dolostone" with "limestone" in the hanging wall: modelled fracture initiation and propagation around the borehole with absolute stresses and failure mode of fractures. Left: GEN-1 at the time steps 200 k and 500 k, right: GEN-1ST-A1 at the time steps 200 k and 406 k

5.6.2.5 Scenario 2 with Discrete Fracture Networks (DFNs)

Scenarios 2-3 and 2-5, which split the borehole into two different lithologies, are remodeled with the DFNs from Sect. 5.6.1.3.

SC2-3_DFN: Borehole intersecting "limestone" at the Base and "weak limestone" in the Hanging Wall + DFN

Fractures start propagating along the DFN sets DFN-2 and DFN-3. As in SC3-2, a transition of fractures is stopped by DFN-1 and advances along DFN-2 and -3. Hence, DFN-1 also acts like a lithological boundary. Still, fractioning in the "weak limestone" is severe but not as strong as in SC3-2. The distinct fractures in the "limestone" almost vanish and appear only along the DFNs (Fig. 5.62, left, Time: 200). After removing all support pressure, the borehole squeezes to half its diameter while fracturing disintegrates the "weak limestone." From the boundary to the "limestone" above, some fractures transit and develop distinct paths that aim for DFNs or their crossings, where stresses magnitudes peak. In contrast to SC3-2 (Fig. 5.58), wherein

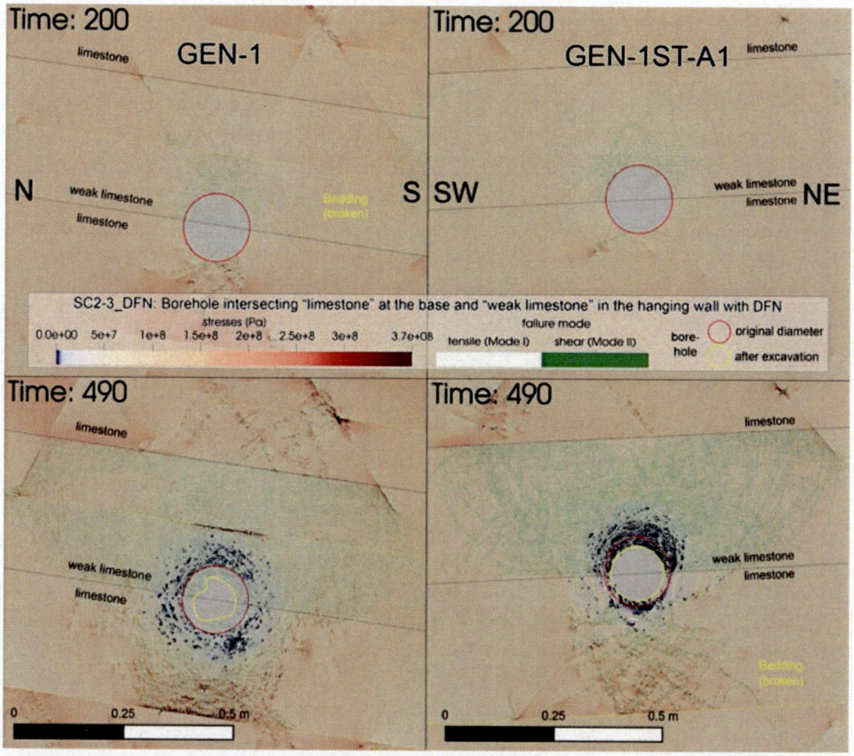

Fig. 5.62 Scenario 2-3_DFN, Borehole intersecting "limestone" at the base and "weak limestone" in the hanging wall + DFN: modelled fracture initiation and propagation around the borehole with absolute stresses and failure mode of fractures. Left: GEN-1 at the time steps 200 k and 490 k, right: GEN-1ST-A1 at the time steps 200 k and 490 k

the bottom "limestone" fractures stay distinguishable, a dense fracture network—comparable to the "weak limestone" disintegrates the "limestone", resulting in a uniform EDZ of 5 cm around the borehole, significantly less than in SC3-2 (Fig. 5.62, left, Time: 490). For GEN-1ST-A1 applies the same fracture development as in GEN-1 (Fig. 5.62, right, Time: 200) with a lower extend and a lower degree of fracturing. Although the "weak limestone" misses the broken DFN, which reduces fracturing in GEN-1, it only disintegrates over 1 m. Fractures from this weak layer, except along DFNs, do not transit into the "limestone." Also, the limestone's fractures do not skip the broken DFN (bedding in Fig. 5.62, right, Time: 490); Single fracture branches are still distinguishable. As a result, the EDZ around the borehole's roof more than 10 cm, while the floor only loosens up for a couple of cms.

SC2-5_DFN: Borehole intersecting "dolostone" at the Base and "limestone" in the Hanging Wall + DFN

For both wells, GEN-1 and GEN-1ST-A1, fracturing before excavation decreases in this scenario, compared to SC2-5 (Fig. 5.60) and concentrates solely along the DFNs or between them (Fig. 5.63, top row). The conjugated shear fractures from GEN-1 in SC2-5 disappear completely (Fig. 5.63, left, Time: 500).

Hence, in the "dolostone," fractures develop in a fan-shape, macroscopically independent from the DFN half a meter around the borehole's bottom. Although the lithological boundary between "dolostone" and "limestone" limits fracture propagation at first, further decrease in support pressure leads to an upwards turning of the initially "dolostone"-fractures into the "limestone." The fractures initiated at the top in the "limestone" stay independent, do not coalescent with the ones from the "dolostone," and accumulate in the rock's severe disintegration along the broken DFN. Thus, two separate EDZs show in the top and at the bottom, where the rock mass loosens by more than 10 cm. In the "limestone" this zone is a 10 cm broad, while it covers

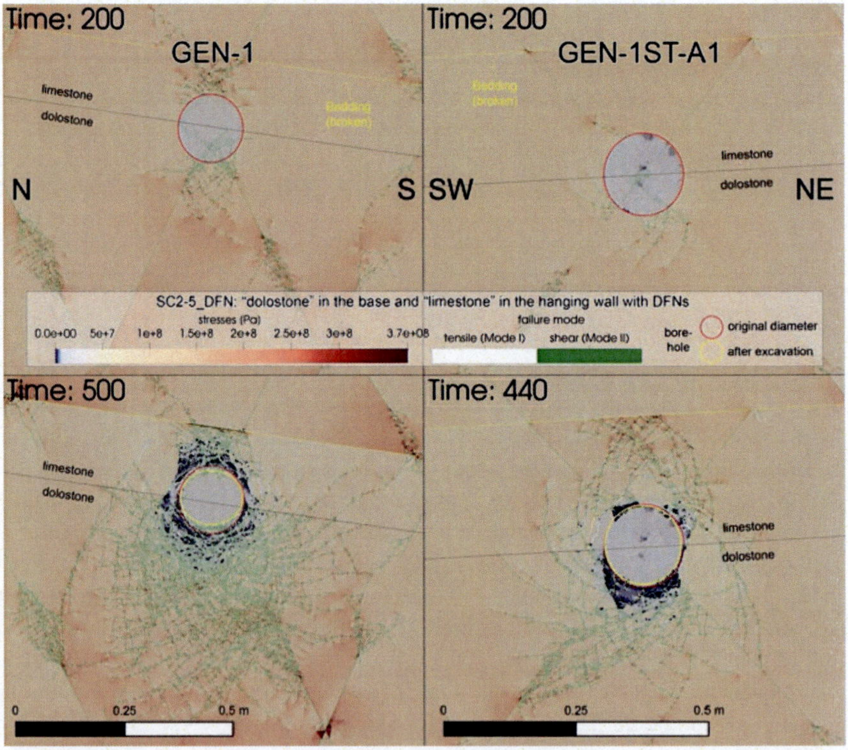

Fig. 5.63 Scenario 2-5_DFN, Borehole intersecting "dolostone" at the base and "limestone" in the hanging wall + DFN: modelled fracture initiation and propagation around the borehole with absolute stresses and failure mode of fractures. Left: GEN-1 at the time steps 200 k and 500 k, right: GEN-1ST-A1 at the time steps 200 k and 440 k

half the perimeter at the bottom for the "dolostone". In between the right and left abutment detaches only a few centimeters (Fig. 5.63, left, Time: 500).

GEN-1ST-A1 shows a similar fracture pattern with the detachments at the bottom but to a lesser extent and a lesser density as in GEN-1. Noticeably, in Fig. 5.63, two separate areas concentrate at the tip of DFN-2 and -3, which connect to the edge of the borehole diameter (Fig. 5.63, left, Time: 440). However, the simulation abort due to numerical instability, but the borehole did not deform any longer. In conclusion, the borehole's final perimeter does not deviate much from the original one.

Notably in all Scenarios, shear failures occur predominantly, while tensile fractures emerge secondarily between the shear fractures.

5.6.2.6 Fracture Opening and Slip in Scenario 2

Identical to Sect. 5.6.2.3, this Chapter presents the fracture apertures and slip values of the newly formed fractures.

Fracture Opening

Figure 5.64 shows the opening width of fractures from all eight models. All excerpts of the models have the same scale. As already stated in Sect. 5.6.2.3, fracture opening is at the maximum (20 mm) at the borehole's margin and declines to the far-field (<2 μm). In the scenarios SC2-2, SC2-3, SC2-3_DFN, and SC2-5_DFN (Fig. 5.64), the layered sequence is visible in aligned, discontinuous opening paths. In contrast to

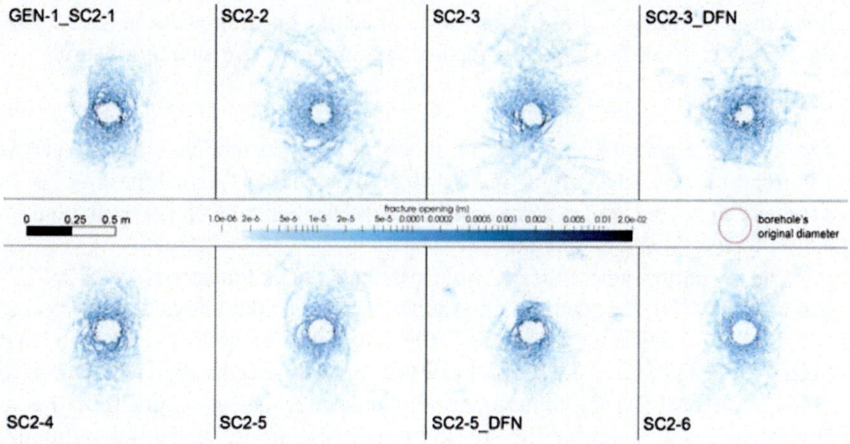

Fig. 5.64 Width of fracture opening [m] at the final time step of the models from GEN-1, scenario 2

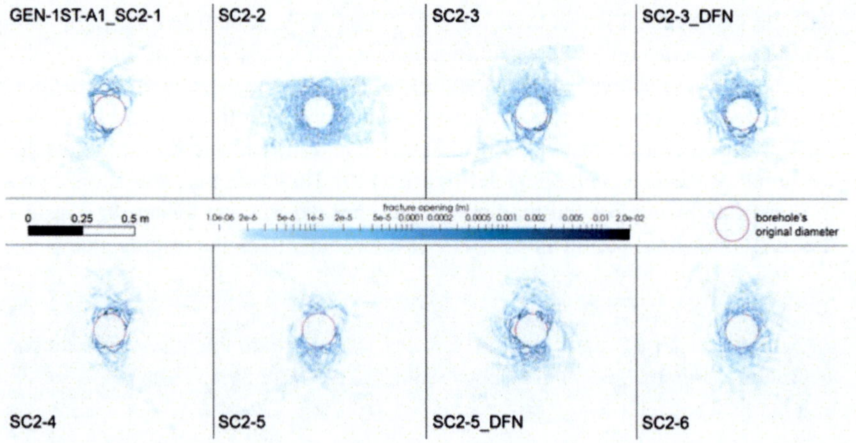

Fig. 5.65 Width of fracture opening [m] at the final time step of the models from GEN-1ST-A1, scenario 2

the openings in the "Scenario 1" models, the extent of the opening with the distance to the borehole.

Hence, SC2-1 and SC2-4, both in "limestone," open horizontally and spread vertically, while a close sequence of different materials leads to a diffuse and heterogeneous opening of fractures (SC2-2, SC2-3, SC2-3_DFN). Materials with small differences in parameters develop similar openings (SC2-5 and SC2-6). However, different DFN sets may alter and abolish the similarities (DFN2-5).

Figure 5.65 shows the openings for GEN-1ST-A1 in Scenario 2. Comparably, the openings develop in the same pattern as in GEN-1, to a lesser extent. Still, the anisotropic features in SC2-2, SC2-3, SC2-3_DFN, and SC2-5_DFN are visible. The openings of SC2-1 and SC2-4 are similar, while SC2-2 opens only within the "weak limestone." Although DFN-1 does hardly affect the opening width in SC2-5_DFN, the DFN sets 2 and 3 change the fracture opening in a more lateral direction.

Fracture Slip

The value of fracture slip in GEN-1 shows in all eight models offsets, lager than 1 mm at the borehole's margin. SC2-1, which is betwixt by "weak limestone", shows a fracture slip of > 1 mm in 25 cm radius around the borehole that is similar but even denser than in SC2-4 (Fig. 5.66).

In the scenarios, where the borehole intersects "weak limestone" (SC2-2, SC2-3, and SC2-3_DFN), the fractures slip radially surrounds the borehole with 25 cm. A transit of the fracture to "limestone" is also followed by an increasing slip length. SC2-3_DFN and SC2-5_DFN also show that long slip lengths align with DFN-2 and -3. SC2-5 and SC2-6, in "limestone" and "dolostone" show again similar fracture slip extends. with a denser slip spacing in the "dolostone" at the top and bottom. However, as already seen in Fig. 5.54, the "dolostone" has a denser slip spacing at

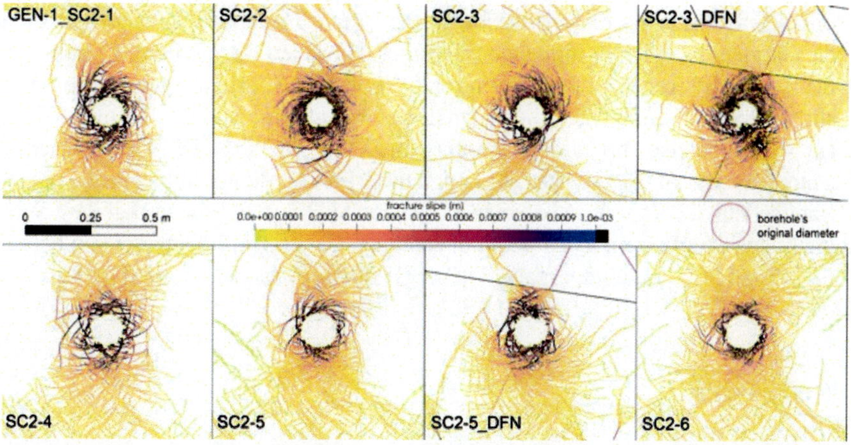

Fig. 5.66 Length of fracture slip [m] at the final time step of the models from GEN-1, scenario 2

the top and bottom, whereas laterally at the walls, long reaching fracture slip lengths form a wedge-shaped form.

Compared to GEN-1, the fracture pattern in GEN-1ST-A1 is less widespread, and so are the slip movements (Fig. 5.67). However, the patterns of individual scenarios are comparable, just as in GEN-1. Hence, SC2-1 and SC2-4 show equal fracture slip over 1 mm at the borehole's margin than becomes less as soon as the fractures approach another material. The fracture slip length arranges gyroscopically around the borehole in SC2-2. Interestingly, two slip-free notches in the vertical distance to the neighboring lithologies develop, while to the sides of it, the slip is over 1 mm at the boundary to the "limestone." In SC2-3 and SC2-3_DFN, the amount

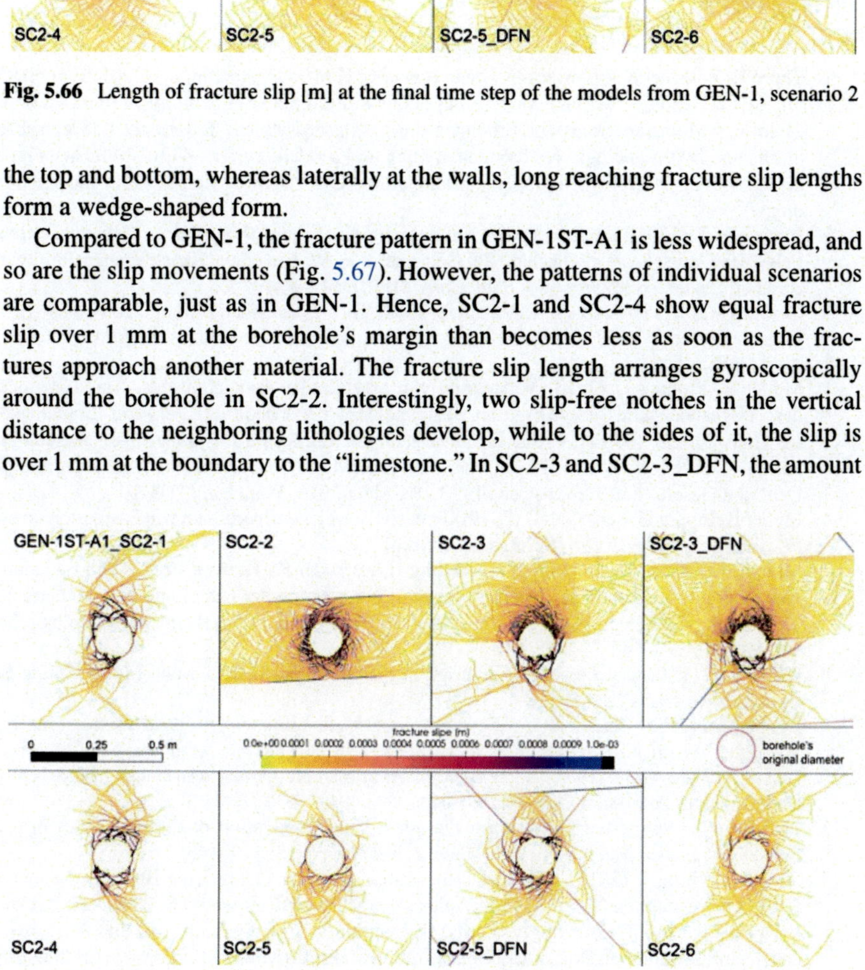

Fig. 5.67 Length of fracture slip [m] at the final time step of the models from GEN-1ST-A1, scenario 2

of fracture slip is not distinguishable in the "weak limestone," where the DFN sets do not macroscopically influence fracturing. However, the amount of fracture slip distributes differently in the "limestone." While SC2-3 forms larger cohere fracture slip intervals, SC2-3_DFN shows small-scaled but large slips at the tip of DFN-2 and -3, which also slip at large amounts. In SC2-5_DFN, the DFN sets concentrate a large amount of fracture slip. Although the DFN sets do not influence the range of fracture slip, a comparison between SC2-5 and SC2-5_DFN shows that they control the fracture movement and amount of shearing. Generally, the fracture slip between SC2-5 and SC2-6 is comparable to its single lithology models from SC1 (Fig. 5.55).

References

1. Thuro K, Zosseder K, Bohnsack D, Heine F, Konrad F, Mraz E, Stockinger G (2019) Dolomitkluft - Erschließung, Test und Analyse des ersten kluftdominierten Dolomitaquifers im tiefen Malm des Molassebeckens zur Erhöhung der Erfolgsaussichten. Teilprojekt B: Geomechanische und hydro-geologische Parametrisierung und Modellierung. – 413p., München (Technische Universität München - Ingenieurfakultät Bau Geo Umwelt - Lehrstuhl für Ingenieurgeologie)
2. Mutschler T (2004) Neufassung der Empfehlung Nr. 1 des Arbeitskreises "Versuchstechnik Fels" der Deutschen Gesellschaft für Geotechnik e. V.: Einaxiale Druckversuche an zylindrischen Gesteinsprüfkörpern. – *Bautechnik* 81(10): 825–834
3. Gercek H (2007) Poisson's ratio values for rocks. Int J Rock Mech Mining Sci 44(1):1–13
4. Gegenhuber N, Pupos J (2015) Rock physics template from laboratory data for carbonates. J Appl Geophys 114:12–18
5. Müller B, Pipping U (2013) Vorstellung von praktikablen geotechnischen Klassifikationen der Festgesteine und Festgebirge an Beispielen aus der Praxis des Fels-und Tunnelbaues. In: Thuro K (ed) 19. Tagung für Ingenieurgeologie mit Forum für junge Ingenieurgeologen und Fachausstellung, 67–73, Technische Universität München, Fachsektion Ingenieurgeologie, Deutsche Gesellschaft für Geotechnik, 13.-16. März 2013, München (TUM)
6. Müller B, Pippig U, Sebastian U (2019) Geotechnische Klassifikationen von Festgesteinen und Festgebirgen. 197p. Berlin (Springer Spektrum)
7. Mraz E, Bohnsack D, Stockinger G, Käsling H, Zosseder K, Thuro K (2018) Die Bedeutung von Analogaufschlüssen des Oberjura für die Interpretation der Lithologie der geothermalen Tiefbohrung Geretsried. – Jahresberichte und Mitteilungen des Oberrheinischen Geologischen Vereins 100: 517–548
8. ISRM (1981) Basic geotechnical description of rock masses. Int J Rock Mech Mining Sci Geomech Abstracts 18: 85–110
9. Dunham RJ (1962) Classification of carbonate rocks according to depositional texture. In: Ham WE (ed) Classification of carbonate rocks, vol 1, pp 108–121 (Am Assoc Pet Geol Mem)
10. Embry AF, Klovan JE (1971) A late Devonian reef tract on northeastern Banks Island, N.W.T. Bull Canadian Petroleum Geol 19: 730–781
11. ISRM (1978) Suggested methods for the quantitative description of discontinuities in rock masses. Int J Rock Mech Mining Sci Geomech Abstracts 15: 319–368
12. Steiger T, Uhlig S (2018) Bio- und Lithostratigraphie der Geothermie-Bohrung Geretsried GEN- 1 mit Sidetrack GEN-1ST-A1. (Anlage 1). In: Thuro K, Zosseder K, Bohnsack D, Heine F, Konrad F, Mraz E, Stockinger G (eds) Dolomitkluft - Erschließung, Test und Analyse des ersten kluftdominierten Dolomitaquifers im tiefen Malm des Molassebeckens zur Erhöhung der Erfolgsaussichten. Teilprojekt B: Geomechanische und hydro-geologische Parametrisierung und Modellierung, München (Technische Universität München - Ingenieurfakultät Bau Geo Umwelt - Lehrstuhl für Ingenieurgeologie)

13. Stockinger G, Käsling H, Menschik F, Thuro K (2019) 3D rotation applied to in situ stress fields for 2D numerical modelling, borehole stability and drill core recovery in deep geothermal wells. In: Fontoura SD, Rocca RJ, Mendoza JP (eds) Rock mechanics for natural resources and infrastructure development—Full papers. CRC Press, London, pp 3144–3151
14. Koch R (2018) CT-Analyse von drei ausgewählten Kernproben der Bohrung GEN-1ST-A1. (Anlage 2.3). In: Thuro K, Zosseder K, Bohnsack D, Heine F, Konrad F, Mraz E, Stockinger G (eds) Dolomitkluft - Erschließung, Test und Analyse des ersten kluftdominierten Dolomitaquifers im tiefen Malm des Molassebeckens zur Erhöhung der Erfolgsaussichten. Teilprojekt B: Geomechanische und hydro-geologische Parametrisierung und Modellierung, München (Technische Universität München - Ingenieurfakultät Bau Geo Umwelt - Lehrstuhl für Ingenieurgeologie)
15. Mraz E, Moeck I, Bissmann S, Hild S (2018) Multiphase fossil normal faults as geothermal exploration targets in the Western Bavarian Molasse Basin: case study Mauerstetten. Zeitschrift der Deutschen Gesellschaft für Geowissenschaften 169: 389–411
16. Terzaghi, R.D. (1965): Sources of Error in Joint Surveys. – *Géotechnique, 15*: 287–304.
17. Seithel R, Steiner U, Müller B, Hecht C, Kohl T (2015) Local stress anomaly in the Bavarian Molasse Basin. Geothermal Energy 3(1): 22
18. Wolfgramm M, Thiem S, Zimmermann J, Budach I, Buse C, Kabus F (2018) Forschungsbericht GTN: Erschließung, Test und Analyse des ersten kluftdominierten Dolomitaquifers im tiefen Malm des Molassebeckens, 93p. Berlin (Geothermie Neubrandenburg GmbH)
19. Drews MC, Seithel R, Savvatis A, Kohl T, Stollhofen H (2019) A normal-faulting stress regime in the Bavarian Foreland Molasse Basin? New evidence from detailed analysis of leak-off and formation integrity tests in the greater Munich area, SE-Germany. Tectonophysics 755: 1–9
20. Backers T, Meier T, Gipper P, Munsch P, Bücken D, Nokar K (2017) Abschlussbericht zum Teilprojekt B: Struktur- und Spannungsfeld im Verbundprojekt MAFA: Parametrisierung von Fazies, Diagenese, Struktur- und Spannungsfeld sowie Optimierung der Testabläufe im Malm zur Verringerung des Erfolgsrisikos. 44p. Berlin (geomecon GmbH)
21. Kirsch (1898) Die Theorie der Elastizität und die Bedurfnisse der Festigkeitslehre. Zeitschrift des Vereines Deutscher Ingenieure 42: 797–807
22. Zang A, Stephansson O (2010) Stress field of the earth's crust. 316p. Springer, Dordrecht
23. Zoback MD, Moos D, Mastin L, Anderson RN (1985) Well bore breakouts and in situ stress. J Geophys Res: Solid Earth 90(B7): 5523–5530
24. Hoek E, Martin C (2014) Fracture initiation and propagation in intact rock—a review. J Rock Mech Geotech Eng 6(4): 287–300
25. Perras MA, Diederichs MS (2016) Predicting excavation damage zone depths in brittle rocks. J Rock Mech Geotech Eng 8(1): 60–74
26. Lisjak A, Figi D, Grasselli G (2014) Fracture development around deep underground excavations: insights from FDEM modelling. J Rock Mech Geotech Eng 6(6): 493–505
27. DIN EN ISO 14689 (2018–05) Geotechnische Erkundung und Untersuchung – Benennung, Beschreibung und Klassifizierung von Fels (ISO 14689:2017); (Deutsche Fassung EN ISO 14689:2018). Beuth Verlag GmbH, Berlin, 2018–05

Chapter 6
Discussion

6.1 Evaluation of Rock Mechanical Parameters

6.1.1 Dependability on the Parameters of the Tested Rocks

Analog samples show huge discrepancies within their range of rock mechanical parameters. Therefore, a proficient comparison of the analog samples with the in situ cores and their utilization as input parameters for the modeling needs a profound interpretation and discussion of all sampled parameters, which is dealt with in the following Chapters.

6.1.1.1 Quality and Reliability of Rock Mechanical Properties from Analog Rocks

The different sampling methods of the analog rocks impact the rock's quality and condition at first hand even before being prepared for laboratory testing. As introduced in Sect. 4.1.1, the limestones BK, BO, and OKL (refer to Fig. 4.2, p. 50 or Table 6.1, p. 167 for the complete rock names) serve as flux limestones for concrete production. Hence, they are mined destructively, where the rock and rock mass are shattered in mining operations (BK and BO) or surface excavations (OKL) by blasting. Although the blasting damage cannot be quantified, blasting induced microcracks deteriorate the overall strength. Noticeably, BK, and OKL show the lowest dynamic properties of all rock mechanical parameters (Fig. 5.1) with the largest grain sizes and highest porosities (Figs. 5.13 and 5.14). While BK also shows relatively week static values, OKL has comparably high static elastic values and high tensile strength, with a low uniaxial compressive strength (Sect. 5.1.2). An increased porosity that degrades the dynamic properties simultaneously might correlate with an advanced weathering rate. For OKL, the large grain size and the combination from the matrix- and grain-supported skeleton might lead to stiffer and more elastic deformation, which results in a larger static Young's modulus and a larger indirect tensile strength. However, it

G. M. Stockinger, *Fracturing in Deep Boreholes*, Springer Theses,
https://doi.org/10.1007/978-3-030-94569-5_6

Table 6.1 Analog rocks assigned to matching properties of the in situ overcored specimens (DC1, DC2, DC3-1, DC3-2, DC3-3, DC4-1, DC4-2, and DC5), with the analog rocks after Fig. 5.1

Properties	Overcored specimens from in situ rock cores							
	DC 1	DC 2	DC 3			DC 4		DC 5
	★	⬢	(1)	(2)	(3)	(1)	(2)	⬢
Dynamic Young's modulus	see DC 3 & 4	BK OKL	DD PFD					
Dynamic Poisson's ratio	QK	div.	QK					div.
P-wave velocity	see DC 3 & 4	BK OKL	BO DD SPK QK					DD PFD
S-wave velocity	see DC 3 & 4	BK	div.					
AI Z_P	BO DD	BK OKL	BO DD		QK	QK	DD	BO (SPK DK)
AI Z_S	BO	< BK OKL	DD PFD			DD	DD	DD
v_P/v_S	BO DD	BK OKL	BO DD		QK	QK	DD	BO (SPK DK)
Derived σ_u	260 MPa SPK	155 MPa	250 – 280 MPa SPK					SPK
Derived τ_S	PFD	BK	40 – 45 MPa					
Porosity	QK	BK	QK					
Grain size	QK SPK							

BK (Bankkalk) BO (Basisoolith) DD (Diefturter Dolomit) DK (Dietfurter Kalkstein)
QK (Quintner Kalk) SPK (Solnhofener Plattenkalk) PFD (Pfraundorfer Dolomit) OKL (Obere Krumme Lage)

affects the uniaxial compressive strength negatively due to the vast amount of occurring imperfections, which after Griffith [16] form an intrinsic zone that concentrates stresses. An inhomogeneous stress distribution leads to stress peaks and, consequently, a premature failure of the rock—i.e., a lower uniaxial compressive strength. BO opposes these two with high dynamic elastic values, with a median porosity of lower than 6% and a maximum grain size of 0.3 mm. Remarkable is the conchoidal fracture pattern of the rock, comparable to SPK and QK. However, the first difficulties with this rock emerged during drilling, where the specimens broke along preexisting fractures during drilling. These preexisting fractures do not show in the dynamic and static elastic values, where BO has the highest values of all limestones. However, they show in the final strength parameters, where σ_u is as low as 100 MPa and as high as 200 MPa, and σ_t varies between 5 and 11 MPa, which indicates a high heterogeneity –not as previously caused by the skeleton of heterogenous grain sizes—but due to the preexisting fractures, presumably caused by blasting.

The drilling during the preparation of QK posed the same problems, also caused by preexisting microfractures. However, these fractures arise naturally from the allochthone shifted deposition and the highly faulted condition (Fig. 4.3). Despite pervaded by microfractures, all specimens show v_P close to pure calcite. Consequently, QK shows high dynamic elastic parameters, with the highest values of v_P/v_S-ratio and the acoustic impedance of all analog rocks. Although QK's E_{stat} is the highest value of all limestone, its strength parameters are the lowest among all analog samples.

KAK has a comparable E_{dyn} and ν_{dyn} and a slightly decreased E_{stat} to other limestones, e.g., DK, QK, and SPK. However, the lithological heterogeneity, reflected by the grain size range from 0.2 mm up to centimeters and the elongated, isolated occurring pores that make up to 9%, lead to a high ν_{stat}, which is presumably due to spalling close to the elongated pores and a consecutive exaggerated dilatation of the radial LVDT, and consecutively to enormously low strength values.

This heterogeneity also causes these comparably low strengths for WAD and WD. However, the kind of heterogeneity is different. For WAD, the main cause of heterogeneity is a highly increased vuggy volume. Compared to the other dolostones, these vugs are distributed irregularly [27], and conspicuously the crystalline size of the dolomite crystals covers a larger range (Fig. 5.15). The dynamic properties result from a mixture of porosity and an individual set of elastic parameters. Apparently, multiple minerals in a rock, here WD with 70% calcite and 30% dolomite [27] with two majorly different elastic properties (compare Sect. 4.2.2, Table 4.1), attenuate the strength of the rock structure itself.

The remaining limestones DK and SPK are in the same range for various parameters: e.g., the dyn. Young's modulus, the static Young's modulus, the static Poisson's ratio, and porosity. Although they match in more points than all others, specific parameters are lower in DK than in SPK: the average Young's modulus, the uniaxial compressive strength, and the indirect tensile strength. A look at the grain size and skeleton support consents an interpretation why these values differ. DK is grain-supported with a grain size between 0.7 and 1.63 mm, while SPK is matrix-supported. Hence, during the first static elastic deformation of the rock, differences show in a shallower stress–strain slope. While SPK has a difference of 3 GPa between the initial slope and the loading cycle, DK shows a difference of almost 10 GPa, a three times larger share of plastic deformation (or fracturing or yielding) that must occur along the grain contacts. Concurrently, since fracturing already occurs in such an early stage, fracture propagation during the third stage of the uniaxial compressive strength can propagate on many more predetermined breaking points, which leads to a faster coalescent of fractures and a reduced strength, which is 80 MPa lower for DK than for SPK. This argumentation also works for the indirect tensile strength, since grain contacts (punctual contacts) break and coalescence faster than an adherend matrix (areal contact). Therefore, σ_t of SPK is more than 4.5 MPa higher than DK.

The remaining dolostones DD and PFD share the same range of the dyn. Young's modulus, static, and average Young's modulus, and static Poisson's ratio. Although the porosity median is almost identical, PFD is likely between 2 and 4% porosity, while DD is mainly between 4 and 6%, which results in a minor larger density of PFD at 2.71 g/cm^3 compared to 2.67 g/cm^3 for DD. Compared to DD, with a crystal size between 0.16 and 0.23 mm, PFD has even smaller crystals with a size of 0.1 to 0.14 mm. The lower porosity and finer grain size might indicate the difference of uniaxial compressive strength by more than 50 MPa, being consistent with the argumentation on WAD. Apparently, from the results, both parameters affect the indirect tensile strength, which is almost equal and where DD even has larger values above 15 MPa. However, the IQR of σ_t for PFD is small and consistent for PFD, whereas the IQR of DD is twice the range and uneven. The IQR of WAD for σ_u and σ_t supports this argumentation, where a higher porosity and larger crystal size leads to a broader distribution of values and a reduced strength overall.

6.1.1.2 Influences on Dynamic and Static Parameters of Carbonates

The dynamic Young's modulus is independent of structural features, such as microfractures. The quasi longitudinal wave v_D shows no correlation with the lithology and is between 4900 to 5100 m/s for the lime- and dolostones BO, DD, DK, QK, KAK, SPK, and PFD. For the remaining rocks, it is 4000 to 4400 m/s. Considering that the mineral density of lime- and dolostones do not differ, the porosity must also reduce the resonance frequency of the body, and due to the multiplication with the bulk density—which also depends on porosity—its factor is doubled. An exception is KAK, which, despite its high porosity, also has a high Young's modulus. In conclusion, E_{dyn} is a property that reflects the perfect elasticity of the rock without considering microfractures and eventually overrating porosity (compare OKL's dynamic and static Young's modulus). Therefore, it should **not** be used as an input parameter for numerical models.

High values of dynamic Poisson's ratio indicate a low porosity and a high degree of cementation. Overall low gas contents associate with low porosity (Fig. 4.9). For instance, with a value of 0.38, QK has less than 1% porosity (Fig. 5.15) and is supposed to have the highest derived strength (Fig. 5.11).

The static Young's modulus for BO, DK, KAK, SPK, WAD, and OKL is around 50 ± 2.5 GPa, except for BK at 33 GPa. However, the diverging average modulus delivers important information. For the rocks, BO, BK, DK, KAK, QK, and OKL, the differences of around 8–10 GPa indicate plastic deformation in the elastic area where existing cracks emerge. Particularly, QK shows differences above 10 GPa, enforcing the impression of large plastic deformation during the linear deformation, presumably on widely existing microcracks. SPK has a difference of 3 GPa, indicating only minor plastic deformation. A mixture of minerals, as in WD, seems to reduce the static Young's modulus. Pure dolostones with low porosity and a fine crystal size have static values of around 60 GPa. Concisely, the more flaws (e.g., porosity, microcracks, heterogeneities, grain-supported skeleton, polymineralic), the larger the difference slope gradient of the initial and second cycle of the linear stress–strain curve.

The static Poisson's ratio is mostly large for weak rocks (BK, KAK, and OKL) with values around 0.30. These rocks have high porosity, large grain sizes, and a grain-supported skeleton. Under loading, these grain contacts fail in the elastic region (as previously discussed) and dilate laterally—resulting in a large radial strain and a high Poisson's ratio. This is a little different for fine-grained rocks with existing microcracks (BO and QK), where the radial strain presumably takes place along these cracks while the remaining strain can be absorbed with the elastic region of the small grains. This stress absorption also applies to the limestones DK and SPK and for the dolostones DD, PFD, and WAD.

The following considerations apply for the uniaxial compression strength results: the more heterogeneities, the lesser the strength. Hence, an interaction of porosity, large grain size, or multiple minerals decreases the strength (BK, KAK, WAD, WD, and OKL). A low porosity, but microfractures or larger grain sizes reduces the strength (BO, DD, DK, QK, and PFD). A homogenous rock, where the loading applies on a randomly distributed fine grain size, has high strength values. The effect of homogeneity also applies to tensile strength. However, an excessive preexisting fracture network may produce non-valid results that lead to a misinterpretation of the real rock properties (QK).

A special case embodies SPK; while the acoustic impedance overestimates the other ten carbonate rocks' strength (Fig. 5.11), SPK is the only rock that exceeds the predicted strength. Therefore, the SPK's equigranularity or a matrix-supported skeleton with a homogenously distributed porosity and no microfractures indicates that higher strengths than anticipated by the acoustic impedance are possible. To what extent this affects the shear strength would require data from shearing tests.

Therefore, the distribution of porosity, grain size, geological history and sampling must be described to enable a transfer of analog samples' strength properties to in situ rocks, as applied in Sect. 6.1.2.

6.1.1.3 Condition of in situ Overcored Drill Cores

The first 5 cm drill core (DC 2) has an effective gas porosity of 7.63%. According to this porosity, the P-wave velocity must be at least 3000 m/s for intact rock (after the equation from Sect. 4.2.2.1). Its P-wave velocity is 3900 m/s at its lowest and 4820 m/s at its highest. This either indicates that some rock pores are still filled with fluids or that the P-wave velocity is higher due to the increased mineral density of 2.73 g/cm^3, which could indicate ferrous minerals, such as pyrite as described by GeoService GmbH ([14]: 7) in the lithological cutting description. Three different P-wave velocities indicate anisotropic behavior, with a weakening of the joint plane intersection the specimen. However, this joint does not affect the dyn. Young's modulus, as previously realized on the analog samples. According to the collected information, this rock would suite well the expression "Fäule" (Sect. 1.4.2.2). This term applies for an intercalary, clayey laminated layer in the "Solnhofener Plattenkalk" (SPK) with a porosity of around 8% and a porosity over 0.019 mD. Investigations from Thuro et al. [38] confirm that the permeability decreases to 28 MPa confining pressure from 0.091 to 0.047 mD in this sample matching the definition of "Fäule".

The second 5 cm drill core (DC 5) has a particle density of calcite (Sect. 4.2.2, Table 4.1) and a porosity below 1%. With this porosity, the calculated v_P is 5800 m/s, and correlates well with the measured 5955 m/s.

The 2 cm specimens (DC 1, DC 3–1, DC 3–2, DC 3–3, DC 4–1, and DC 4–2) have porosities between 1.03 and 2.13% with a density of 2.68 to 2.73 g/cm^3, indicating almost pure calcite. These specimens would correspond to "Flinz", matching a calcite content of 95 to 98%, a porosity of 0.2 to 6% (Sect. 1.4.2.2), and the lithological description after [34]. Although several minor calcite-healed veins intersect the specimens, all v_P exceed the v_P of calcite. These discontinuities withstood overcoring, although, in contrast to DC 2 (Fig. 5.16), they sever the specimens' hole diameter. In conclusion, the discontinuities are not open since this results in a dampened P-wave velocity. Furthermore, either secondary minerals or fluid inclusions lead to velocities exceeding pure calcites' velocity. Radial measurements were not possible due to the short lateral distance, but microcracks parallel to the specimens could not be detected macroscopically.

6.1.2 Matching Attributes of Analog Samples with in situ Rock Cores

This Chapter discusses and concludes on the comparability of the analog rock parameters within situ rock parameters. From the eleven analog rocks, KAK, WAD, and WD withdraw prematurely since they are either too porous, to heterogeneous, do not correspond with the lithology (WD), or they do not overlap with any of the results of the in situ rock cores. Hence, eight analog rocks remain in discussion as a proper fit for the in situ rock cores. The matrix in Table 6.1 presents the properties of the analog rocks that match to the values of the in situ rock cores.

The analog rocks that match with DC 1, 3, and 4 are decently similar. The elastic properties, e.g., the dynamic Young's modulus, the P- and S-wave velocity, and its derivatives are typified predominantly by BO, DD, QK, SPK, and PFD, but differ in what was found crucial for the static mechanical behavior in Sect. 6.1.1.1: porosity, grain size, and preexisting fractures. QK and SPK cover these attributes from DC 1, 3, and 4 perfectly since they are micritic limestones, where QK has hardly any gas porosity and SPK a very high strength. Also, the v_P/v_S-ratio plotted versus the acoustic impedance Z_P of the coinciding positions from QK and DC 1, 3, and 4 show strong cementation and low porosity. If QK was free of microcracks, its strength would be at least similar to SPK or higher. The derived uniaxial compressive strength from Z_P and the reduced porosity, which exclusively accounts for **SPK's** heterogeneity, indicates a rock strength (σ_u) higher than 250 MPa.

DC 2 agrees with the properties of BK and OKL, except for the grain size. Although DC 2 shows anisotropic properties like OKL. However, in contrast to OKL, the axial elastic waves in DC 2 move through one material only and do not intersect the anisotropic plane. Therefore, the axial P-wave velocity and the quasi-longitudinal wave from the resonance frequency represent the properties derived from the fastest waves through the specimen. This assumption matches with the isotropic properties of BK and its porosity. According to this argumentation, the properties of DC 2 adapt well with **BK's** properties. Therefore, this work assumes this material is isotropic. The anisotropic mechanical behavior, which is demonstrated by the US measurements (Sect. 5.2.4), is implemented differently for the subsequent modeling (see Sect. 4.4.1).

DC 5 also shares dynamic attributes with DD and PFD, although the derivatives of these values match better with the limestones BO, DK and SPK. Enforced by the v_P/v_S-ratio plotted versus the acoustic impedance Z_P, the results confirm a highly cemented material with a low porosity and an increased gas saturation, compared to DC 1, 3 and 4. A derived uniaxial compressive strength is at around 230 MPa and matches well with SPK. However, a macroscopic view of the overcored specimen DC 5 (Fig. 5.17) shows some larger components in a micritic matrix and several calcite-healed veins that partly opened, indicating a micro-fracturing due to a heterogeneous material. Therefore, the actual rock strength is probably less than from the specimens DC 1, 3, and 4 and could correspond to **DK**.

None of the overcored samples and none of rock cores equals the lithology and porosity from the dolomites DD and PFD. However, the lower section of Core Run 7 and successive cuttings show a transition to dolostone. Steiger and Uhlig [34] describes these cuttings as parts of a microcrystalline dolostone, irregularly crystalized, with larger single dolomite crystals that break apart along their cleavage planes. Partly, compact dolomite crystals allow only small inclusions of secondary minerals and a secondary crystallization with coarse calcite crystals closes existing primary porosity [34]. From the tested dolostones, this description fits best with the PFD. Mraz et al. [27] describe PFD as a microcrystalline, compact dolostone (99%) that contains coarse-crystalline calcite and other secondary minerals (clay, ferrous/ferric hydroxide and quartz). Additionally, **PFD** is the dolostone with the lowest porosity, which comes closest to the impression of the host rock as highly diagenetic compacted and sealed.

Concluding Remarks on Substitutability of Analog Samples

Due to the previous interpretation and discussion, the following analog rocks represent the in situ reservoir rock best: **BK (Bankkalk)** as the weakest rock possible when porous rocks prevail, **DK (Dietfurter Kalkstein)** as an intermediate carbonate when heterogeneities determine the mechanical behavior, **SPK (Solnhofener Plattenkalk)** as the strongest representative when the micritic rock is in flawless condition and finally **PFD (Pfraundorfer Dolomit)**, when the reservoir rock is massive dolostone.

6.2 Conclusions on the Rock Mass from in situ Drill Cores

Although the drill cores did not fulfill their intended use to directly determine the host rock's strength parameter, they still delivered plenty of information on the rock mass. While the previous Sect. 6.1 adapted them in the best possible way to analog samples, the following Chapter discusses the difficult way to derive a realistic fracture network from these cores. This Chapter discusses the US measurement results and develops new insights on the rock mass.

6.2.1 Time-Dependent Effects

Thuro et al. [38] and the newly evaluated data from Sect. 5.2.4.2 show no time-dependent effects and hence no progressing fracture initiation in the drill cores with time. In contrast to Wieser [41], who measured decreasing P-wave velocities of 1000 m/s over 500 h, this effect did not show in any tested drill cores from GEN-1ST-A1. Although pulling the core drill string took more than 12 h, this effect should still be measurable. Hence, several conclusions may be drawn from that:

- Fracturing, relaxation, or plastic deformation in carbonate rocks (monomineralic) evidentially wears off quickly. It thus describes a completely different mechanical behavior of magmatic rocks (polymineralic) as described by Wieser [41] (Sect. 3.3.3) or in Gischig et al. [15] with the continuing spalling in the Bedretto adit (Switzerland).
- Disintegration/Relaxation of the drill cores happened right after drilling or still in the core barrel.
- Fracture growth in carbonates is independent of a temporal component. Hence, borehole stability and fracture propagation only depend on mechanical factors or thermal stresses, not on a temporal component.
- The rock is very brittle, and the limit of unstable crack growth (σ_{cd}) is close to the rock's short-term strength (σ_u). In Thuro et al.[38], all analog carbonate rocks' stress–strain curves are purely linear up to failure, which confirms this theory. However, real proof could only be achieved by acoustic emission testing or evaluating the volumetric strain (Fig. 2.3).

US measurements identify active discontinuities perfectly, but US velocities do not change across the discontinuities in the long term. Although the drilling mud's influence and the temperature may initially cause some outlying values (Fig. 74, t = 0 h), the discontinuity properties do not change over time.

Finally, the long-term measurements LT-US 4 and LT-US 5 initiated a closer investigation of the P-wave velocities over all core runs (Sect. 5.2.4.1 and 5.2.4.3). Although there is no temporal development of the US velocities, the axial velocities become smaller than the radial velocities, and radial velocities split from equal or overlapping velocities in LT-US4 (Fig. 5.24) into two distinct radial velocities in LT-US5 (Fig. 5.25). This sudden change in the P-wave velocity pattern can be due to scale- or stress-related effects, discussed in the next Chapters.

6.2.2 Scale-Related Effects

Scale-related effects, i.e., the dependence of rock-mechanical properties on geometric dimensions [12], pose large problems of determining reliable rock parameters. On the one hand, a larger sample contains more flaws and is an intermediate between rock and rock mass [5], and on the other hand, a small sample may not be representative for a rock regarding grain size or pore size [37]. The US measurements through Sects. 1 to 3 (Sect. 5.2.4.1, Fig. 5.22) explicitly prove that the originally recovered drill cores have lower P-wave velocities than the overcored specimens. As these drill cores are monomineralic [34] and barely porous (Sect. 5.2.2), fractures are the only plausible cause for the decreased P-wave velocity. Visible fractures partly show as crescent-shaped stress-induced cracks (Sect. 5.3.1.2). Hidden fractures activate in a 2 cm spacing during overcoring or due to mechanical impact. Actually, force induced by careless handling is sufficient to break the cores apart. These fractures are in no relation with the mapped in situ fracture network (Fig. 5.37). Therefore, stress-related relaxation (e.g., core disking) or mechanically induced damage from drilling may have caused these cracks artificially.

Hence, none of the drill cores, recovered in its original condition, would be suitable for static rock-mechanical testing. However, the overcored specimens are representing the rock well. Firstly, their P-wave velocity resembles the P- and S-wave velocity of calcite and the upper limit of limestones (Sect. 4.2.2, Table 4.1). Secondly, the fluctuation of the S-wave velocity diminishes, indicating a homogeneous and undisturbed material. For mechanical testing, the small specimen size should be sufficient due to the small grain size of the material.

6.2.3 Stress-Related Effects

The previous Sect. 6.2.2 mentioned superficially whether the pre-damaging of the drill cores is stress- or mechanically induced. However not contradicting each other, both presumable causes need a certain separate consideration. First of all, for Sects. 1 and 2 (Sect. 5.2.4.1, Fig. 5.22), the axial P-wave velocities are higher than the radial P-wave velocities. Hence, fracturing is more extensive laterally. This is logical when considering that JS1, as a preexisting discontinuity, opens first and stress already depletes in plastic deformation or strain. However, the core barrel limits the lateral expansion of the core, and the remaining stress may deplete in axial fracturing, although this is limited by the bottom of the drill rig and the skin friction of the barrel. Noticeably, all core runs, except CR5 and CR6, contain this steeply inclining joint, which is almost congruent with the shear plane of τ_{xy}/τ_{yx} from the stress rotation in Sect. 5.5. As a hypothesis, the high shear stresses might have caused the core jamming, which led to the abortion of all core runs. Hence, due to the coring, the stress state within the rock changes from the in situ environment to a new stress regime defined by the rotated stresses in the core barrel. The shear stress development in Sect. 5.5.2 (Figs. 5.40, 5.41 and 5.42) shows a climax for the τ_{xy}/τ_{yx}-stresses in the well GEN-1ST-A1 in the coring section. Depending on the stress regimes, their values range from 24 to 55 MPa, where σ_{zz} acts as normal stress to these shear planes (see absolute values in Tables 5.9, 5.10 and 5.11). In a simplified 2D consideration of the Mohr–Coulomb criterion ($\tau = \sigma_n * \tan(\varphi)$), where no cohesion and dilatation apply, there would be a minimum required angle of inner friction for the plane **not** to fail in shear and to prevent jamming. Hence, the friction angle of JS1 has to be at least

23.9° (strike-slip, Seithel et al [32]), 39.4° (strike-slip, Backers et al. [4]), or 20.9° (normal faulting, Wolfgramm et al. [43]). Santarelli and Dusseault [30] compares the cause for core jamming with the stress state within a triaxial compressive test at, which would be promoted by a weak shear plane. In contradiction to this hypothesis are the results from Byerlee [11], who claims that rock surfaces do not fail in shear up to a friction angle of 41° ($\mu = 0.85$). Hence, the results from the simplified calculations would be too low in order to cause shear failure. However, five reasons may still be in favor of the hypothesis:

1. Clay minerals on joints reduce the angle of friction [11]. Barthel [6] describes an accumulation of clay minerals on the layer boundaries of the Solnhofener Plattenkalk.
2. Barton ([8]: 260) describes for "Rock wall contacts" of slickensided, planar surfaces an angle of friction between 34° and 7.1°. Hence, for 34°, only the strike-slip regime after Backers et al. [4] would cause jamming.
3. A differential stress regime, where the difference between S_{Hmax} and S_{hmin} is even larger than in the gradient after Backers et al. [4], results in a higher friction angle and exceeds the friction coefficient of Byerlee [11].
4. A local change of stresses, due to core drilling, causes a redistribution of stresses, as the numerical models show (Sect. 5.6.2), and the shearing motion along JS1 leads to a jamming of the core barrel.
5. The successful Core Run 6, with a TDR of 90% and an RQD of 53%, lacks JS1. The reason for core jamming in CR 5 could not be identified, and JS1 was not detected in the strongly disintegrated drill core. However, in CR 7, at the footwall of CR 6 (where the core run stopped due to core jamming), JS1 reappears.

The effect of the stresses in CR 6 is visible, but a lack of JS1 did not lead to core jamming. Differently than in Sects. 1 and 2, Sect. 3 shows larger radial P-wave velocities (Sect. 5.2.4.1, Fig. 5.22). The absence of JS1 eradicates the possibility of lateral relaxation of the drill core and eliminates the possibility that stresses concentrate and dissipate at existing discontinuities, and consequently, secondary fractures cannot form. Therefore, relaxation and plastic deformation pursue uniformly in the path of least resistance, parallel to the drill string. Fractures emerge perpendicular to the core barrel with distinct spacing, defined by the major effective principal stress, dividing the drill core into discs. Due to the fractures' predetermined radial propagation, the axial P-wave velocities decrease strongly, whereas the radial velocities experience an only minor decline. This minor decline also proves minor fracturing in the axial direction since the median of the radial velocities in Fig. 5.22 (Sect. 3) is over 1500 m/s below pure calcite's velocity. While LT-US 4 shows an almost congruent development of the radial velocities (Fig. 5.24), 1.2 m further, the radial velocities suddenly differ by 1000 m/s in LT-US 5 (Fig. 5.25). Although the velocities are reproducible and steady, a difference in radial velocity can be measured through the entire CR 6 (Fig. 5.26) but cannot be tied to any macroscopic properties. Until, in 5382 m MD, so-called saddle-shaped discs appear (Sect. 5.2.4.3). These saddle-shaped discs develop when differential stresses cause core disking (Sect. 2.4.2.3).

From this depth, the axis through the troughs (direction of σ_1) and the saddles (direction of $\sigma_{2/3}$) define the radial measurements' direction (Fig. 2.12). Throughout all these measurements, the velocities along the troughs are faster (red line in Fig. 5.26) than along the saddles (black line in Fig. 5.26), which is logical when reconsidering that fracturing always happens parallel to σ_1 and perpendicular to the least principal stresses. These results prove a differential stress field.

Nevertheless, complementing the orientation of the troughs' and saddles' linear (Sect. 5.4.3) and taking the stress development along the drill trajectory (Sect. 5.5.2) into consideration, it becomes possible to deduce the stress field and its least possible gradient. Therefore, when the saddles develop at the top and bottom of the borehole, then $\sigma_{2/3}$ is acting in this direction, while σ_1 is acting horizontally. The horizontal stresses along the borehole equalize (Fig. 5.39, σ_{xx} and σ_{yy}) because it strikes ~ 45° to the in situ stress field. Hence, S_{Hmax} and S_{hmin} become equal, and the resulting stress is the horizontal stress acting perpendicular to the borehole, here σ_1. The only remaining stress is σ_{zz}, which acts vertically on the borehole and replaces $\sigma_{2/3}$. Thus, the normal faulting regime cannot apply since σ_{zz} is larger than $\sigma_{xx} \approx \sigma_{yy}$ (Sect. 5.5.2.2, Fig. 5.42b). Likewise, the strike-slip regime after Seithel et al. [32] cannot apply since all principal stresses equalize in the core run section $\sigma_{xx} \approx \sigma_{yy} \approx \sigma_{zz}$ (Sect. 5.5.2.1, Fig. 5.40b), which would contradict the effective differential stresses. Finally, the strike-slip regime after Backers et al. [4] (Sect. 5.5.2.1, Fig. 5.41b) agrees qualitatively with the observations and P-wave measurements, where the horizontal stress $\sigma_{xx} \approx \sigma_{yy}$ is larger than the vertical stress σ_{zz}. All considerations neglect the effect of the occurring shear stresses.

In Fig. 5.26, area 2, two discontinuities intersect the core at 5380.3, and 5381.5 m MD and the axial velocities increase. Considering that relaxation along JS1 causes minor axial fracturing and a higher v_P, the same could apply to other joint sets. Although not assignable to a specific joint set, some of the joint sets from JS2 to JS5 may be active or activated by drilling and represent weak zones in the rock mass. Contrarily, there are also sections where axial velocities and radial velocities decrease below average (Fig. 5.26, 5382 m MD). Hence, stress-related discing might even be severe in some regions. The discs of the overcored specimen match pretty well with the thicknesses measured in CT 2 (refer to Fig. 5.26 and to Sect. 5.3.1.1 for the results). This agrees with the consistent P-wave velocities in this section. However, in this section, the P-wave velocities are significantly above the average, indicating that the spacing in sections with lower P-wave velocities must be narrower.

Concluding Remarks on the Rock Mass:

The high shear stresses τ_{xy}/τ_{yx} are the most probable cause for core jams by activating JS1 in a shear motion. Hence, by choice of the drill trajectory for intended coring, high shear stresses, and consequently, core jams could be avoided, and the core quality could be improved.

JS1 dissipates the in situ stresses and reduces damage to the rock. Hence, a retrieval of specimens is possible (Figs. 5.16 and 5.18), which allows the measurement of the true dynamic rock properties of the rock cores (Sect. 6.2.2). Where JS1 is missing, other joint sets cannot or only hardly reduce fracturing of the rock core, and stress-related fractures progressively disintegrate the drill cores into discs. Hence, the rock cores indicate that either **structural elements** (JS1), or **in situ stresses**, as in CR 6, locally control fracturing. The joint sets JS2 to JS5 do not evidently change or influence fracturing. Core disking suggests that the rock is under a strike-slip stress regime with a strong differential stress gradient.

6.3 Fracture Network

6.3.1 Bedding

The geometrical method from Sect. 5.4.1 (Fig. 5.34) proves that the steeply inclining joints in the drill cores have an orientation of 162/09. The fracture picking in the HMI log of GEN-1ST-A1 (Sect. 5.4.2) also shows these shallow southward dipping planes. At the beginning of the open hole, these planes are a little steeper (Fig. 5.35) and become shallower to the end of the HMI log (Fig. 5.36), where its spatial orientation of 165/09 is almost congruent to the one in the rock cores. Compared to Lemcke [21], who claims that the Weißjura-Group dips with an angle of 8° towards the south (Sect. 1.4.1) and, more recently, to the investigations of Schubert [31] in Sect. 3.3.1, this fits perfectly with this joint set. Comparable to Schubert [31] in GEN-1, the decrease in dip from 12° to 9° might indicate an apparent shallower transition of GEN-1ST-A1 into a basin depositional environment from the Purbeck to the Jurassic platform carbonates after Steiger and Uhlig [34]. Further argumentation also allows deducing the spacing of this joint set. Hence, after Steiger and Uhlig [34], the comparable equivalent rock is from the Solnhofen-Fm., deposited in a basin environment with bedding thicknesses from < 1 mm and 300 mm (Sect. 1.4.2.2, Barthel [6]). The lithologically and dynamically comparable Quintner Limestone (QK) shows layers thicknesses up to 600 mm (Fig. 4.3). The HMI log from 4720 to 4888 m MD shows a mean spacing distance of 450 mm (Sect. 5.4.2). An extreme to very close spacing (after ISRM [44]) show the drill cores from the Core Runs 3 and 4, whereas the small core diameter might not resolve possibly larger spacing distances.

6.3.2 Fracture Network from the HMI Logs

Regarding the remaining joint sets, the consideration of possible orientation and planes becomes more diverse and complex. The stereographically plotted fracture traces from the HMI logs posed the issue that the shallow dipping JS1 predominates over the whole logged distance with 463 traces compared to 63 vector entries for the remaining joint sets. Hence, the unweighted density plot does not capture other joint sets. A weighted density plot reveals clusters of pole vectors for JS2, JS4, and JS5 (Fig. 5.35) up to 4720 m MD. 49 individual pole vectors form these three joint sets, where one dips with 67° to the north (Fig. 5.35, JS5), and two strike predominantly NS (JS2 and JS4). While an explicit cluster of pole vectors defines JS5 and limits its strike to an uncertainty of 10°, JS2 combined with JS4, could also form a perfectly NS striking single fracture set with an orientation of 089/86, which may be supported by two sources. Firstly, Backers et al. [4] describe a similar fracture set with an orientation of 284/89 in outcrops of the Franconian Alb, and secondly, the direction under a strike-slip regime would correspond to drilling-induced fractures. In fact, these fractures arise at the beginning of the open hole section, where Fig. 5.44b shows negative minimum tangential stresses and where the fluid pressure still exceeds the positive tangential stresses up to 4720 m MD (Sect. 5.5.4.2). However, traces of this joint set show even deeper than 4720 m MD. Therefore, if DITFs reach that deep, either the S_{Hmax} in given stress field is larger or the pore pressure is less.

Nevertheless, from 4720 to 4888 m MD the fracture pattern changes, and even the statistical weighted plot shows no clusters. Clusters of pole vectors, excluding JS1, define JS4 (110/80) from five pole points, and nine pole vectors form JS5 (337/71) (Fig. 5.36). While JS5 stays unaltered within the HMI logs, JS4 aligns with JS4 from the drill cores (Fig. 5.37). The traces of JS4 and JS5 disappear below 4740 m MD, indicating that they become presumably mechanical inactive. If JS4 represents a set of DITFs, these disappear below 4500 m TVD (4740 m MD).

6.3.3 Fracture Network from the Drill Cores

Although other fracture sets, except for JS1, either disappear (JS2, JS4, and JS5) or do not even develop in the HMI logs (Fig. 5.37), their traces reappear in the drill cores. The pole points (black diamond, red cross, and green triangle in Fig. 5.37) from all three upper sections of the core runs (5036.12–5036.48, 5201.8–5202.0, and 5204.19–5204.64 m MD) are unmistakable due to the geometrical constraints. However, the deeper drill cores (5378.30–5386.69 m MD) lack this marker. Therefore, this newly applied method of congruently matching clustered poles (Fig. 5.37, orange plus sign) needs further discussion to limit the refutability and the possibility that the fracture network changes with depth.

Firstly, the reappearance of JS1 in Core Run 7 indicates an unchanged continuation of the fracture network. Hence, CR 6 revealed an unstratified (massive) facies.

Secondly, the resulting fracture sets JS5 and JS3 agree with the extensional anti- and synthetic EW-striking faults in the Upper Jurassic carbonates [3, 21], which is confirmed from a 3D seismic survey from Shipilin et al. [33] (compare to Sect. 1.4.1 and 3.3.2, Table 3.1). Hence, JS5 represents the antithetic fault with a mean dip of 64° towards the north, which agrees perfectly with the 50° to 70° as described by Bachmann et al. [2]. The corresponding synthetic fault JS3 has a slightly steeper dipping angle of 71°. However, differently than indicated by literature, these faults do not appear as a single line in the reservoir, but more likely, as progressive stepped faults, indicated by 28 fracture traces on a drill core length of no more than 10 m. Hence, JS5 has a joint frequency of 1/m and JS3 of around 2/m.

Thirdly, two rotated clusters from the drill cores represent JS2 and JS4. While these joint systems strike nearly NS in the upper HMI log, they rotate 20 to 30° clockwise. These joint systems fit the description from Unger [39] and Kraemer [20] (Sect. 1.4.1) and Megies and Wassermann [24] (Sect. 2.4.2.2), who described NNE-SSW striking, subvertical strike-slip faults. The indenter theory by Ratschbacher et al. [29] confirms these NNE-SSW striking faults kinematically. The illustrated shear angle of 40° diverging from the thrust force's direction, however, is too shallow for our results (Sect. 1.4.1, Fig. 1.6). Considering JS2 and JS4 are one part of a conjugated shear failure surface, an angle β results from the difference between σ_1, which is NS, and the striking direction of the joint set (2β is the angle between conjugated failure surfaces [7], Sect. 2.1.2.1, Fig. 2.1). For JS2, this is 34°, and for JS4, 21°. Hence, the Eq. $2\,\beta = 90 - \varphi$ (Sect. 2.2.2) after Barton [7] results in a theoretical angle of friction φ for JS2 of 22° and JS4 of 48°. JS4 meets the requirements for a strike-slip fault, since its inclination is almost vertical and the angle of friction of the rock coincides with the friction coefficient μ of limestone of 1 ($\varphi = 45°$, [1] and 1.1 ($\varphi = 48°$, Mogi [26]). Both fracture sets JS2 and JS4 appear in a frequency of 1.5/m. Classifying JS2 as a pure strike-slip fault is difficult for two reasons: an angle of friction of 22° is more likely representative for shale [23], and its dip is too shallow. However, a comparison with data from literature may explain the emergence, affiliation, and importance of JS2, particularly for geothermal use.

6.3.4 Comparison of the in situ Fracture Network with Literature

3138 measurements of discontinuity planes by Backers et al. ([4]: 21) revealed an orthogonal system of three joint sets in outcrops of the Upper Jurassic carbonates in the Franconian Alb, illustrated by Fig. 6.1a, which is an altered excerpt the Geothermie Atlas, published by the Bavarian Ministry of Economic Affairs, Regional Development and Energy [35]. One joint set with 197/02 corresponds to the bedding, which is sub-horizontal, slightly south-wards dipping, and two vertical joint sets with 192/89 (EW-striking) and 284/89 (NS-striking) (Fig. 6.1b).

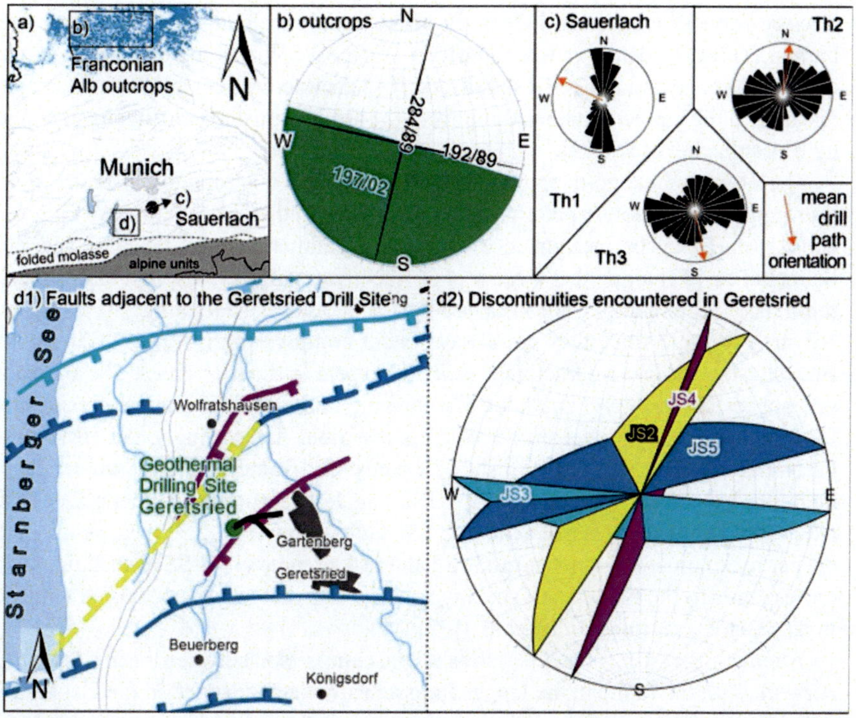

Fig. 6.1 Compilation of the published fracture networks for the Upper Jurassic carbonates in southern Bavaria. **a** Overview map after the Geothermie Atlas from **b** the Franconian Alb in the North, **c** the Geothermal well Sauerlach and **d** the Geothermal well Geretsried with the fracture networks from **b** after Backers et al. ([4]:21), **c** Seithel et al. ([32]:12ff), **d1** Excerpt of the Geothermie Atlas [35] with the location of the geothermal well Geretsried and the faults colored according to **d2** the fracture network, mapped in the drill cores from Sect. 5.4.2, Fig. 5.37 without JS1

Approximately 120 km south, Seithel et al. [32] reveal two mainly ENE-SWS oriented fracture sets in Th2 and Th3 and one N-S, NNE-SSW striking discontinuity set in Th1 of the geothermal wells of Sauerlach (Fig. 6.1c). Although these sets fit the fracture network from Backers et al. [4] (Fig. 6.1b) to a limited extent, the majority of Seithels' [32] values deviate significantly from distinct values of the outcrops. Accordingly, from the comparison of both results, the fracture network changes evidently from the surface to the immerged carbonate layers. Logically, two different mechanical causes distinguish the formation of the joint sets. Firstly, the uplift of the Franconian Alb and its forebulge in the course of the subsidence of the southern sediments might cause the orthogonal joint network in the outcrops (Fig. 1.4). Besides a perfect alignment of the fracture network 192/89 (Fig. 6.1b) to the underground faults in the eastern part of Fig. 6.1a, Boersma et al. [9] support Mode I fractures induced from tectonic uplift that form parallel and perpendicular (192/89 and 284/89) to mechanically layered rocks (bedding with 197/02). Secondly, the fracture network progresses differently with gradual subsidence and increasing overburden. The subsidence causes the Mode I ENE-WSW striking extensional faults, described above and in Sect. 1.4.1. Mode II shear fractures emerge with decreasing distance to the alpine units, resulting from the alpine thrust, mechanically explained by Ratschbacher et al. [29].

The obvious change in fracture patterns from the outcrops to the underground also supports the fracture patterns of the rock cores in this thesis (compare Figs. 6.1c and d1) and may be brought in a context for geothermal use. For Th2 and Th3, Seithel et al. [32] argue that these sets of fractures, which correspond to this thesis' joint sets JS3 and JS5, "are unfavorable for geothermal targeting" in an SS and NF stress regime since they are always under compression ([32]: 19). Purely NS-striking, steeply dipping fracture sets do not appear in the rock cores. These fracture sets appear between two antithetic EW-striking faults and are assumed responsible for a high productivity regardless of the stress field. Interestingly, the HMI log of GEN-1ST-A1 in the first 230 m shows exactly this fracture pattern, where steeply inclining NS-striking fracture planes (JS2 and JS4) and only antithetic EW faults (JS5) appear but **no** traces of synthetic EW faults show (Fig. 5.35). However, in the deeper rock cores, JS3 appears, and JS2 and JS4 become NNE-SSW striking, which corresponds to the NNE-SSW striking fracture sets that show a dilatation tendency in an SS stress regime Seithel et al. ([32]: 19).

Another interesting context establishes the comparison between the in situ fracture network with the faults in the Upper Jurassic carbonates mapped in the Geothermie Atlas (Fig. 6.1d, [35]). The strike and the dip direction allow to color mapped fault lines (Fig. 6.1d1) according to fracture networks (Fig. 6.1d2). Hence, a confirmed synthetic fault system matches JS3 in the north (turquoise line in Fig. 6.1d1). As the only synthetic EW striking extensional fault system, five distinct antithetic faults allocate in the southern area (Fig. 6.1d1, blue lines). The yellow line's strike and dip direction might correspond with JS2, which connects two antithetic EW-striking fault sets at an approximate angle of 45° (Fig. 6.1d1, yellow line). A purple line branches off to the north at a steeper angle from this yellow line, coinciding with JS4. A shallower, but with a generally agreeing direction with JS4, fault line also maps through the wells of Geretsried (Fig. 6.1d1, purple line intersecting the green dot).

Concluding Remarks on the Fracture Network:

JS1 (as it applies in the Sects. 5.4 and 5.6) represents fully persistent bedding planes of the rock, with a spacing distance between 20 and 600 mm. According to the characteristics of the joint sets, JS3 and JS5 represent fossil extensional faults that are under compression and not conductive. JS4 represents and proves the presence of strike-slip faults, but the angle may be too steep for dilatation and conductivity. JS2 likely connects the tips of antithetic extensional faults [28] and may dilate in a SS stress regime, hence JS2 is potentially active and conductive. However, the cores never disintegrated along any of these four mapped joint sets. Therefore, these faults appear as mechanically inactive. A determination of these fracture sets' true spacing distance is impossible since, on the one hand, the fractures are macroscopically indistinguishable. On the other hand, if the drilling triggered some discontinuities, they may be untraceable since they become mechanically corroded during drilling. Nevertheless, the joint frequency on the rock cores allows an estimation of their occurrence, with JS3 most commonly every half a meter, JS2, and JS4 intermediately every two-third of a meter, and JS5 every meter.

6.4 Determining and Quantifying the in situ Stress Field

Sect. 2.4.2.2 summarizes the claim of different authors that a strike-slip stress regime (with transition tendencies to NF) is predominant for the Upper Jurassic carbonates in the Bavarian Molasse Basin. Wolfgramm et al. [43] and Moeck et al. [25] support a strike-slip regime directly linked to the wells of Geretsried in Sect. 3.3.5. A reverse faulting regime is excluded. From all authors, gradients and tendencies differ strongly. Therefore, this Chapter applies the previous results for a more accurate evaluation, whether the regime is a pure strike-slip regime or at a transition and narrows down the stress gradient.

6.4.1 Approvals and Contradictions for possible Stress Regimes

The results of the indicated stress regimes are partly ambiguously and require a joint consideration and discussion. The basics for the discussion on the stress regimes contains Sects. 2.2.2 and 2.4.2.

6.4.1.1 Normal Faulting Stress Regime

Borehole breakout orientations from the reevaluation of the HMI logs from GEN-1 strike NS, in an almost EW striking horizontal borehole, indicating an NF regime, with σ_V as the largest principal stress (Sect. 5.3.2.1, Fig. 5.29), which agrees with the extensional EW-striking joint sets JS3 and JS5 (Fig. 5.37) and micro-tectonic indicators, i.e., stylolites and slickolithes, on the surfaces of JS1 (Fig. 5.28). However, as Sect. 5.3.2.1 already discusses, these are most likely paleo-stress indicators. Contrarily, the curvature of obviously fresh stress-induced cracks (SIC) suggests larger horizontal than vertical stresses, which is confirmed by the investigations of the core disking, combined with US-measurements. Also, the orientation of NS-striking joint sets in Fig. 5.35 could represent DITFs in a NF stress regime (Sect. 6.3). However, the minimum tangential stress never falls below 35 MPa (Sect. 5.5.3.1), making the occurrence of DITFs unlikely even with a larger share for S_v. Likewise, for an NF regime (at the transition to a strike-slip regime), the maximum tangential stresses are low in GEN-1 and negate the development of many BBOs (Fig. 5.43), considering the potential high rock strength (Sect. 6.1.2).

In summary these factors support (+), oppose (−), or are neutral for (•) a NF stress regime:

+ NS-direction of borehole breakouts in GEN-1 (Fig. 5.29)
+ Extensional faults, JS3 and JS5 (Fig. 5.37)
+ (Paleo) micro-tectonic indicators on the bedding surface, JS1 (Fig. 5.28)
− Curvature of fresh stress-induced cracks (SIC) (Fig. 5.28)
− Direction of borehole breakouts in GEN-1ST-A1 (Fig. 5.30)
− Orientation of saddle-shaped core disking patterns (Sect. 6.2.3)
− Differential radial US-velocities of the drill cores (Sect. 6.2.3)
− Low maximum tangential stresses at the borehole wall may not cause BBOs (Fig. 5.43c)
• Orientation of DITFs fits well, but the minimum tangential stress is too high (Fig. 5.44c)

6.4.1.2 Strike-Slip Stress Regime

In general, the pros that speak for a NF regime become cons and vice versa for a strike-slip regime. However, two additional factors add, in favor for a pure strike-slip regime. The orientation of DITFs still fits, but additionally negative hoop stresses allow the formation of tensile fractures. In addition, JS4 in Fig. 5.37 may represent a set of strike-slip faults, which agrees with works of different authors (Sect. 6.3). Also, high maximum tangential stresses are in favor for borehole breakouts (Fig. 5.43b).

In summary these factors support (+), oppose (−), or are neutral (•) for a pure strike-slip stress regime:

+ Curvature of fresh stress-induced fractures (Fig. 5.28)
+ Direction of borehole breakouts in GEN-1ST-A1 (Fig. 5.30)
+ Orientation of saddle-shaped core disking patterns (Sect. 6.2.3)
+ Differential radial US-velocities of the drill cores (Sect. 6.2.3)

+ Large maximum tangential stresses at the borehole wall (Fig. 5.43b)
+ Orientation of DITFs fits well and minimum tangential stresses are negative at the beginning of the open hole section (Fig. 5.44b).
+ JS4 might be the result from a strike-slip stress regime (Fig. 5.37)
– NS-direction of borehole breakouts in GEN-1 (Fig. 5.29)
– Extensional faults, JS3 and JS5 (Fig. 5.37)
– (Paleo) micro-tectonic indicators on the bedding surface, JS1 (Fig. 5.28)

Since the pros prevail and the cons are likely to be fossil remnants, this thesis sees proof for a strike-slip stress regime in the Upper Jurassic carbonates. At the same time, the gradient may still be discussed.

6.4.2 Quantifying the in situ Stress Field

Quantifying the stress field with indirect methods is very inaccurate and should be avoided. Since data of formation integrity tests (FITs) or comparable direct measurement methods for the reservoir rock is not available, narrowing the stress gradient down is an initial approach:

- A strike-slip stress regime with a gradient after Seithel et al. [32] is unlikely as the stresses equalize at the core runs, whereas the cores show differential strain. Besides, DITFs cannot form with this stress regime.
- A strike-slip regime with a gradient after Backers et al. [4] is sufficient to create DITFs. These NS-striking, subvertical dipping joints disappear in 4500 m TVD, while the minimum tangential stress already becomes positive (including p_r) in 4370 m TVD. Whereas Budach's stress gradient (Budach et al. [10], Sect. 2.4.2.2, Table 2.1) fulfills these requirements and causes negative tangential stresses up to the observed depth of 4500 m TVD.

An absolute indirect quantification suggests recent research from Lim and Martin [22] on the Lac du Bonnet Granite, which implies that disking initiates when the maximum principal stress (σ_1) normalized by the Brazilian tensile strength (σ_t) of the rock is 6.5. With an increasing ratio of σ_1/σ_t, the ratio of disk thickness (t) divided by the core diameter (D) decreases, ending up in completely fragmented material when the ratio of σ_1/σ_t exceeds 11. Although all studies on core disking were conducted in magmatic rocks, no publications describe this emergence in carbonates. Therefore, it is questionable if a direct comparison is adequate since carbonates (monomineralic rocks) behave mechanical differently than polymineralic rocks, as Sect. 6.2.1 proves. Yet, a lack of alternatives deploys Lim and Martin [22] method subsequently.

Data on the original core diameter (89 mm), the disc thickness, and the rock's tensile strength allows calculating the maximum principal stress original acted on the core. The thickness of the cores, obtained by overcoring and CT-scanning, is 15.2

Table 6.2 Principal stresses derived from measured core disking thicknesses in relation to the likely and maximum occurring tensile strength of rock

Core diameter D (mm)	Disc thickness t (mm)	Ratio t/D	[1)]Stress condition after [22] $\sigma_1/\sigma_t =$		Maximum principal *in situ* stress for $\sigma_t =$ (MPa)	
					10.9	18
89	7.9	0.09	Max:	11	106	198
			Min:	9	87	162
	15.2	0.17	Max:			
			Min:	8	78	144
	22.5	0.25	Max:			
			Min:	7	68	126

$\pm$ 7.3 mm. Considering that core disking emerged in CR 6 only, a comparable rock would be DK (Sect. 6.1.2). Hence, an applicable mean tensile strength would be σ_t = 9.7 MPa (Fig. 5.8). Flawless specimens, unaffected by stress-induced fracturing, might be comparable with SPK and have a maximum tensile strength of σ_t = 18 MPa (Fig. 5.8). The results of the possible occurring effective maximum principal stresses are listed in Table 6.2. For a tensile strength of 9.7 MPa, the least possible stress is 68 MPa, and the maximum possible stress 106 MPa, while for σ_t = 18 MPa, the stresses range between 126 and 198 MPa. From all effective maximum stresses, σ_{xx}/σ_{zz} that act perpendicular to the drill cores of Core Run 6 from Tables 5.10, 5.11 and 5.12, only σ_{xx} with a value of 98 MPa after Backers et al. [4] is high enough to be considered in Table 6.2.

A look at the specimen DC 5, from Fig. 5.17, reveals a less homogeneous material than expected from SPK. Hence, a tensile strength of 10 MPa is a reasonable value for this rock. The orientation towards the lower ratio (t/D, Table 6.2) confirms the results of the P-wave measurements, which suggest a generally finer disking for the whole Core Run 6 (Sect. 6.2.3). From this point of view, the stress gradient after Backers et al. [4] is a good approximation of the prevailing stress field. While the previous results refute a lower differential stress gradient, a higher differential stress gradient (e.g., after Budach et al. [10]) may be possible but will not be further discussed.

Concluding Remarks on the in situ Stresses:

It is particularly noticeable that—independent from the stress regime—the frequency of occurring borehole breakouts does not relate with the development of the maximum tangential stresses along the borehole in GEN-1 (Fig. 5.32). Unexpectedly, this shows in presumably **structural-controlled** areas, such as the platform-basin carbonates (PBCs, Fig. 5.32) and in prone thick-bedded or massive facies of the platform carbonates (PCs), **lithology** also accounts as a controlling factor for fracturing and borehole stability. Therefore, three main factors are isolated from previous results and discussions that verifiably control rock mass's deformation around a borehole in the Upper Jurassic carbonates: **stress, discontinuities, and lithology.**

6.5 Conclusions from Numerical Models

The numerical simulations' goal is to imitate and understand the rock mass's mechanical behavior when excavating a borehole and widening the knowledge beyond the elastic consideration. Hence, some questions stayed unanswered until analyzed by numerical models:

- Is the occurrence of irregular borehole breakouts controlled by other factors than stress?
- What is the appearance of the Unilateral Patterns linked with?
- How do alternating lithologies influence borehole stability and fracturing?
- What role play discontinuities?
- Is there any naturally occurring opening of fractures?
- Is there a permeability change in the rock mass due to the drilling process?
- How does the borehole hydraulicly connect to the fractured rock mass?

However, before finding answers to the questions above, the input parameters, as presented in Sect. 5.6.2 need individual discussion and interpretation to conclude real rock mass behavior. Since the simulation does not specifically deal with a single identified spot within the well but rather simulates a wide range of possible scenarios with generalized input parameters, these parameters need a review before the models attain a closer examination and discussion.

6.5.1 Reliability of Input Parameters for Numerical Models

6.5.1.1 Material Parameters

In general, the used parameters of "weak limestone" adapts to BK, "limestone" to SPK, and "dolostone" to PFD, as discussed in Sect. 6.1.2. Their values are not the exact median of the corresponding rocks but reflect an even number within their IQR. Therefore, this approach stays within the true rock values and depicts the most critical occurring differences.

Differently than in the models from Thuro et al. [38], only static elastic properties apply. Using static instead of dynamic properties is due to two reasons: Firstly, the borehole excavation and its deformation are static processes. Secondly, as elaborated in this work, static parameters capture flaws in the rock (Sect. 6.1). These flaws cannot be implemented at the model's scale since mesh elements of 5 mm represent the micritic rock. Hence, the input parameters must be adopted and attenuated.

Although several authors [1, 11] suggest a higher friction coefficient for carbonate rocks, this work keeps the default value of $\mu = 0.6$ and might overestimate shearing. Z_S in Fig. 5.21 indicates a maximum cohesion at 40 MPa and a minimum of 24 MPa for the in situ rocks. However, since Z_P (Fig. 5.20) underestimates SPK's strength, its cohesion is artificially increased to 50 MPa. Whereas it overestimates the strength for BK by 50%, its cohesion is decreased by 50%. In general, this might lead to an overestimation of shear fracturing in BK and an underestimation in SPK. Nevertheless, Jaeger et al. ([17]: 93) suggests for a friction coefficient of $\mu = 0.6$ and a rock strength of $\sigma_u = 80$ MPa a cohesion of 24 MPa, which is used as lowest value. For "limestone" and "dolostone", this thesis uses a $\mu = 1$ and a σ_u of 250 and 200 MPa, resulting in a cohesion of 50 and 40 MPa. Triaxial tests would be necessary for true values.

The values for the tensile strength represent the maximum of the equivalent rocks. Derived from the tensile strength, fracture energy Mode I—the energy needed to overcome the fracture toughness, calculates after Whittaker et al. [40] with a plane strain constitutive law including the static Young's modulus and Poisson's ratio. Mode II fracture energy is by default 10 times higher. Specific values for both fracture energies need comprehensive laboratory testing, not feasible within this work.

6.5.1.2 Model Setup

The model setup for scenario 1 is a single set of material parameters that considers the carbonates either with or without a DFN. "Weak limestone" and "limestone" include three DFN sets (Sect. 5.6.1.3, Table 5.15). Both represent a **structurally-controlled** rock mass, where the bedding plane JS1 evidentially (Sect. 6.2.3) has a huge impact on fracture initiation and propagation. Whereas "Dolostone," adopting the insights of Core Run 6, neglects JS1 since it characterizes an unstratified, massive, **stress-controlled** rock mass.

Scenario 2 combines the concept of a **structurally-** and **stress-controlled** rock mass into one model. In addition, it adds the concept of a **lithology-controlled** rock mass for two reasons:

1. The 5-cm overcored samples (Sect. 5.2.2, DC 2, Fig. 5.16) show lower values for elasticity and derived strength parameter and higher values than the encompassing in situ rocks (Sect. 5.2.2, DC 1 and DC 3, Fig. 5.16), indicating a weak—strong alternation within the **structurally-controlled** rock mass.
2. A numerical verification of the irregular borehole breakouts might relate to lithological features, as Teufel [36] suggests

This argumentation results in scenario 2 geometries, where the **stress-controlled** "dolostone" is the bedrock in the footwall, where DFN-2 and -3 apply in the DFN scenarios. Above, a sequence of "limestone" and "weak limestone," with thicknesses of 60 and 30 cm, respectively (Sects. 1.4.2 and 6.1.2). The bedding planes separate the units, which dip with the apparent angle of JS1 (Sect. 5.6.1.3, Table 5.15).

6.5.1.3 DFN Parameters

Due to its slickensided, planar surface but unaltered appearance, the broken set DFN-1 has a friction coefficient of $\mu = 0.3$ ($\varphi = 17°$, Barton [8]: 260). Since none of the single joint sets, JS2, JS3, JS4, and JS5, are indistinguishable in the core runs (Sect. 6.3), it is justifiable to combine them in DFN-2 and DFN-3 and assign identical properties. However, the strength properties assigned to DFN-2 and -3 are underestimated. Therefore, in the numerical model, the joints already break before excavation, while neither stress-induced fractures nor US-measurements show a weakening of these joint sets in reality (Sect. 6.2.3). More realistic DFN properties would be more rigid but absolute parameters are not available. A spacing distance of 0.5 m for all DFN sets is based on a conservative generalization by the largest occurring frequency (closest spacing) of 2/m of JS3 (Sect. 6.3.3).

6.5.1.4 In situ Stresses

While Thuro et al. ([38]: 111ff) used absolute stresses in their numerical models, this thesis uses effective stresses with a lower radial pressure, resulting in a more realistic fracturing before excavation. Also, by the stress field restrictions in Sect. 6.4, only the rotated stress values after Backers et al. [4] for 4600 m TVD apply.

6.5.2 Comparison of the Elastic (Empirical) Approaches with the Numerical Models

Linear elastic approaches (Sect. 2.3.1.1) calculate the maximum and minimum tangential stresses at a cavity wall without considering plastic deformation, i.e., yielding, and fracturing (Sect. 2.3.1.2). Therefore, they give an estimation about the stability of the borehole, but do not about the habitus of the plastic deformation as numerical models do. Still, a comparison of elastic and empirical methods with numerical models may ensure the quality and reliability of those.

6.5.2.1 Occurring Stresses around the Borehole

The same effective stresses from Sect. 5.5.2 determine the tangential stresses in Sect. 5.5.3 and serve as input for the numerical models in Sect. 5.6.1.2. Hence, the absolute stresses that develop due to the borehole's excavation in the numerical model should correspond to the magnitude of maximum and minimum tangential stresses of the linear elastic consideration after Kirsch [18].

The tangential stress in 4600 m TVD from Fig. 5.43b for GEN-1 is $\sigma_{\theta,max} =$ 366 MPa and $\sigma_{\theta,min} = 43$ MPa, and for GEN-1ST-A1 $\sigma_{\theta,max} = 265$ MPa and $\sigma_{\theta,min} =$ 76 MPa. The single lithology models from Scenario 1 without DFNs (Sect. 5.6.2.1) allow a rational verification and comparison of the occurring stresses. The FEM run of GEN-1 establishes an initial stress magnitude of 150 MPa. These stresses already cause fracturing in the "weak limestone", resulting in a serrated line, which characterizes the fracturing without any change of the ambient conditions (Fig. 6.2, light blue lines). However, "limestone" and "dolostone" (Fig. 6.2, dark blue and pink lines) can bear these stresses, and the stresses grow at the top of the borehole and decline at the borehole wall with the beginning of the core modulus reduction (compare 4.4.3.2). Stresses build up in a shallow slope and steeply increase shortly before failure, perfectly aligning with the plastic deformation description by Feder and Arwanitakis ([13]: 104, Fig. 1.2).

Up to the excavation (Fig. 1.2, green vertical line), stresses reconstitute, but continuous fracturing leads to a progressing decline both in the top and bottom as well as at the walls. After the excavation, the stresses are influenced by remaining shear resistances and pore pressure, which declines and reduces the vertical direction's stresses to zero (after the second orange line in Fig. 6.2). The stresses around the borehole in "dolostone" equalize, confirming the impression, described in Sect. 5.6.2.1, Fig. 5.48 of a uniform disintegration of the rock around the borehole. On the contrary, in the "limestone", the stresses stay differential until, to the end of the simulation, the walls fall to zero as soon as the pore pressure deteriorates (Fig. 5.47, left). Hence, none of the SC-1 models from GEN-1 reach the peak stress $\sigma_{\theta,max}$ of 366 MPa. Emerging fractures degrade the stresses around the borehole and propagate and coalesce in

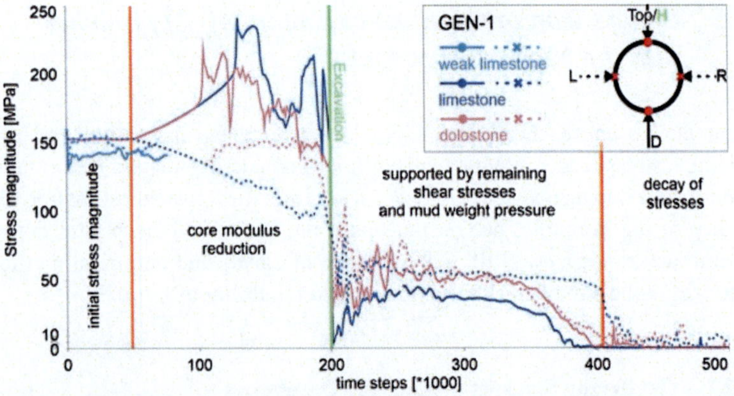

Fig. 6.2 Stresses at the borehole wall in all models of GEN-1 over all timesteps in scenario 1. Solid lines represent stresses at the top and bottom of the borehole (marked with a circle at the borehole perimeter in the legend), dashed lines signify stresses at the walls of the borehole (marked with an x). The green line marks the excavation, the orange line sections described in the text

its vicinity. Particularly none of the stress magnitudes in Fig. 6.2 from GEN-1, SC-1, (compare to Figs. 5.46 and 5.48, left) come close to these peak stresses since the rock disintegrates with a dense net of fractures. Still, single elements at the tips of the propagating fractures from SC-1 "limestone" (Fig. 5.47, left) show as peak stresses, which might be as high as 350 MPa matches $\sigma_{\theta,max}$.

The minimum tangential stress fits well with the elastic predictions. Hence, data from "limestone" and "dolostone" show initial values after excavation a little above 50 MPa, which deviates by a few MPa of $\sigma_{\theta,min}$ with 43 MPa plus the pore pressure. However, the walls' stresses also decline as soon as fracturing coalescence around the borehole and disconnect the lateral elements from the cohering rock mass.

In GEN-1ST-A1, fracturing develops similarly to GEN-1, although fracturing appears at a later time step under a lesser absolute stress magnitude. The stresses before excavation in the vertical walls to the horizontal walls differ the most in "limestone" but hardly in "weak limestone", since progressing fracturing distributes the stresses uniformly (Fig. 6.3, up to a time step of 200).

Interestingly, in the "dolostone", stresses reconstitute at the top and bottom (Fig. 6.3, pink line after the excavation), indicating jamming of the elements at the top and bottom presumably by shearing, allowing them to stay in place, while the material at the side loosens and the stresses vanish. On the contrary, in "limestone", severe fracturing also occurs at the top and bottom, but initial fracturing does not encompass the borehole. Hence, the vertical direction stresses diminish, while the sidewall stresses do not drop at all and even rise with further pore pressure reduction (Fig. 6.3, blue dashed line from time steps 200 to 480). These stresses rise to almost 75 MPa, which would equal the elastic minimum tangential stress 76 MPa, calculated

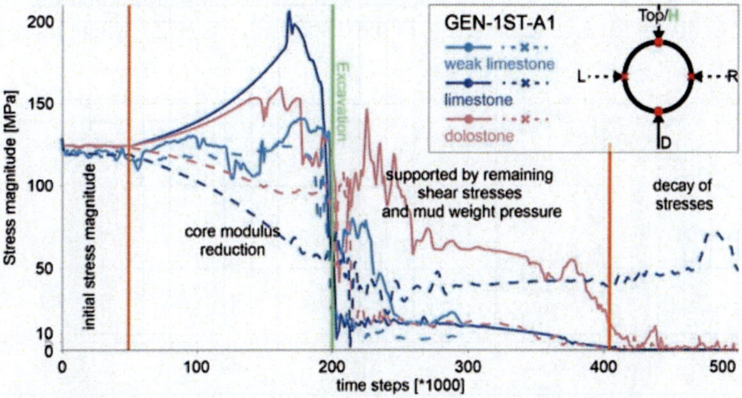

Fig. 6.3 Stresses at the borehole wall of all scenario 1 models of GEN-1ST-A1 over all timesteps. Solid lines represent stresses at the top and bottom of the borehole (marked with a circle in the legends' borehole perimeter), dashed lines signify stresses at the walls of the borehole (marked with an x). The green line marks the excavation, the orange line sections separate distinct sections, explained in the text

in Sect. 5.5.3.1, before single fractures propagate to the walls and the stresses drop (Fig. 6.3, blue dashed line from time steps 480).

A comparison with the DFN-linked models and the models from scenario 2 is only limited possible since these not only consider the transition from elastic to plastic deformation but also the stress degradation along pre-existing discontinuities (in the DFN-infused models) and the heterogeneous stress distribution due to various elastic properties (in the models of scenario 2). Hence, "limestone" and "dolostone" from scenario 1 apply for the estimation of the excavation damaged zone (EDZ). Since the simulations from "weak limestone" prematurely stopped, the model SC2-2 from Sect. 5.6.2.4 applies, where the borehole is in this lithology only.

6.5.2.2 Depth of the Excavation Damaged Zone (EDZ)

Section 5.5.4.3 estimates the depth of the excavation zone, for a rock strength of σ_u = 80, 200 and 250 MPa, which correspond to "weak limestone", "dolostone", and "limestone" in the numerical models. Determining the boundary of the EDZ is not trivial since the demarcation of stable and unstable areas is blurry. However, plotting the stresses through the borehole might compare the empirical EDZ (eEDZ) with a disturbed zone from numerical models, which comprises the fracture process zone (FPZ) and a simulated EDZ (sEDZ).

Hence, Fig. 6.4 plots the stresses from GEN-1 of a line centric intersecting the borehole in the indicated area, for the "weak limestone" in the direction of bedding and the isotropic, homogeneous SC1 models in a vertical direction, with the maximum stresses at the final stages of the models.

To the left of Fig. 6.4, the development of the stresses shows that the empirical EDZ of 40 cm fits well to the range of the disturbed zone (DZ), where the simulated

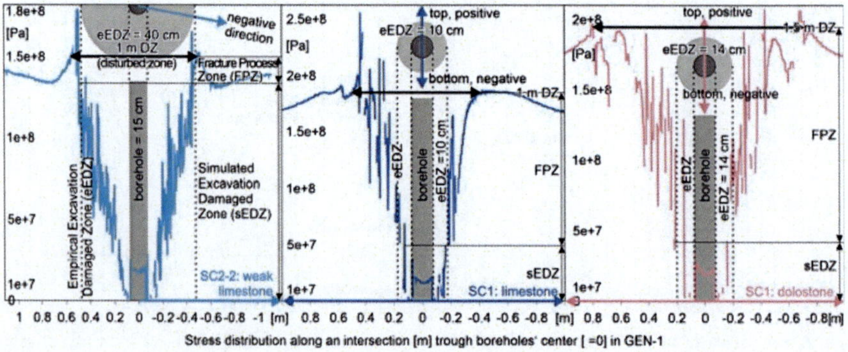

Fig. 6.4 Stress magnitudes (y-axes) GEN-1 along a line intersecting the borehole (dark grey circle intersected by an arrow, projected to a dark grey rectangle), surrounded by the empirical estimated excavation damaged zone (eEDZ) from Sect. 5.5.4.3, illustrated by a grey circle projected to dashed lines. From the left to the right: SC2-2 "weak limestone", SC1 "limestone", and SC1 "dolostone" with simulated fracture process zones (FPZ) and excavation damaged zones (sEDZ)

EDZ (sEDZ) prevails with a diameter of 1 m around the borehole. The strongly jagged, narrow-graded line indicates the high degree of disintegration. Although "limestone", from the mid of Fig. 6.4 shows an equal DZ, the less jagged, wide-ranged stress development indicates a smaller sEDZ. However, defining a limit where the EDZ begins is difficult.

One indicator may be a first long, uninterrupted vertical line, illustrating a large change in stress at a small distance around the borehole. Hence, the 10 cm of the empirical EDZ would slightly overestimate the actual EDZ. The same applies to the "dolostone", where a first uninterrupted line is within the range of 14 cm (Fig. 6.4, right). However, the FPZ reaches further in the rock mass with a diameter of 1.5 m since the rock cannot disintegrate as densely as "weak limestone" but cannot withstand fracturing as well as "limestone". Hence, fractures in "dolostone" propagate the furthest.

In GEN-1ST-A1, the DZ in the "weak limestone" is with 1.4 by 0.4 m larger in diameter than in GEN-1 (Fig. 6.5, left). A lower stress magnitude cannot transit fractures into the limestone (Fig. 5.57) and expands in the lateral direction of "weak limestone", enlarging the FPZ. Following the previous hypothesis that an EDZ is defined by the first uninterrupted vertical line, this would also correspond to the eEDZ. However, the sEDZ may be deeper since the line is again strongly jagged and narrow-graded, marked in Fig. 6.5, left with "FPZ or EDZ?". Here, the empirical EDZ agrees well with "limestone" and "dolostone" with the first vertical modeled spike right at the edge of the dashed lines in Fig. 6.5, mid and right. The width of the DZ is matches the order of rock strengths.

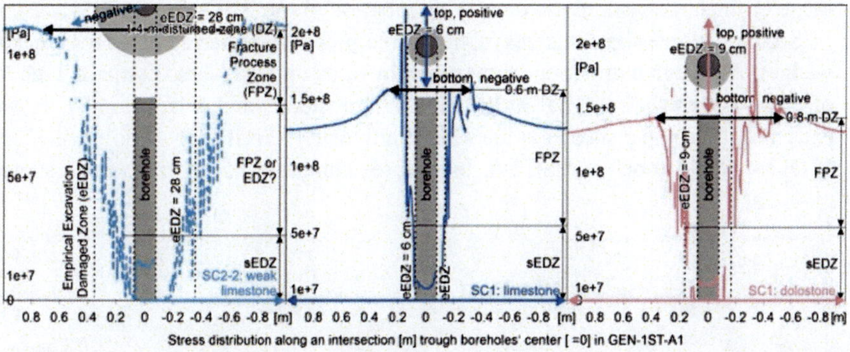

Fig. 6.5 Stress magnitudes (y-axes) GEN-1ST-A1 along a line intersecting the borehole (dark grey circle intersected by an arrow, projected to a dark grey rectangle), surrounded by the empirical estimated excavation damaged zone (eEDZ) from Sect. 5.5.4.3, illustrated by a grey circle projected to dashed lines. From the left to the right: SC2-2 "weak limestone", SC1 "limestone", and SC1 "dolostone" with simulated fracture process zones (FPZ) and excavation damaged zones (sEDZ)

6.5.3 *Key Findings from Numerical Models*

6.5.3.1 Is the Occurrence of Borehole Breakouts controlled by other Factors than Stress?

The numerical models suggest that. Firstly, great differences in alternating material properties, as it is the case between "weak limestone" and "limestone," lead to a severe and concentrated fracturing in the weak material (e.g., SC2-2, Fig. 5.57). Independently from the in situ stress gradients in SC2-2, the loosening of the rock mass (EDZ) happens uniformly around the borehole, as Fig. 6.6a schematically illustrates. The HMI log of GEN-1 shows breakout structures that encompass the whole borehole diameter, mapped as breakout widths with 180° (Fig. 5.31, the second peak of the BBOs). Unfortunately, a thinning of the "weak limestone" was not modeled. However, the comparison of SC2-2 and SC2-3 (compare Figs. 5.57 and 5.58) shows the mechanical context. Hence, during thinning of the weak layers, the EDZ extends, and fractures increasingly propagate in distinct paths in a conjugated form in the hard layers. Figure 6.6b indicates an elongated EDZ, where fracturing progresses further in the weak layers, and distinct fracture paths in the hard layers accumulate. However, borehole breakouts would still have a width of 180°. Finally, further thinning would cause the EDZ even further in the weak layer and a coequal increase of shear fractures in the hard layer (Fig. 6.6c). Figure 6.6c schematically shows the dip of the bedding in GEN-1. Compared to the rose plot in Fig. 5.28 (adopted in the borehole in Fig. 6.6c), the direction of breakouts fits perfectly to the direction of bedding. Depending on the bedding thickness, the angle of the breakout widths would be applicable for the range of the large peak in the density distribution of the BBOs in Fig. 5.31.

Secondly, large slip distances might be interpreted as preferred paths for breakout wedges. Although unmentioned up to now by other authors, some numerical models of GEN-1 with the material properties of "dolostone" and a lack of JS1 showed extremely high slip distances at the walls. Particularly, this is SC1—"dolostone," SC-1_DFN—"dolostone" and SC2-6. In all three simulations, the final stages showed

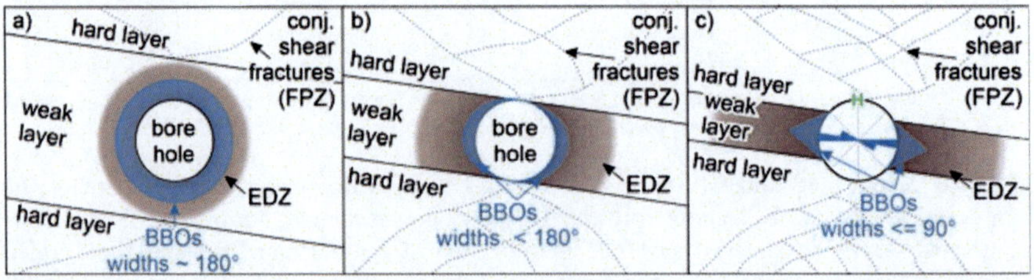

Fig. 6.6 Borehole breakouts controlled by lithological features with **a** uniform development of an EDZ around a borehole in a weak layer enclosed by a hard layer, **b** weak layer with the thickness of the borehole, where a laterally expanded EDZ leads to breakouts (BBOs) in the weak layers with widths of 180°, and **c** weak layer thinner than the borehole, resulting in an even further expanded EDZ and BBOs widths become equal or smaller than 90°

large slip distances at the wall, whereas at the top and bottom, the fragmentation was very dense, but also the slip distances much smaller. Hence, Fig. 6.7 marks areas with coalescing black slip lines (slip movement > 1 mm). The ratio of the horizontal to the vertical surfaces illustrates the extent. In SC1—"dolostone," there are two large potential breakout wedges with widths of 82° and 129°, whereas small areas crop out in the top with 37°, and at the bottom with less than 10°. SC1_DFN—"dolostone" shows slightly smaller widths of 82° and 117°, where no wedges form. The top remains at 37° and the bottom coalescent to 56°. Remarkably, the breakout wedge with a width of 117° forms at the tip of the DFN. For SC2-6, where "limestone" follows in the hanging wall, and the fracture network is denser, the horizontal breakout widths rotate dextral and decrease to 73° and 84°, while top and bottom do not change.

Although connecting the lines of large slip movements is prone to human error and interpretation, the overall direction of all breakout wedges corresponds well with the rose plot in Fig. 5.28 (illustrated in the background within the borehole of Fig. 6.7). The range of breakout widths from 73° to 129° covers perfectly with the first peak of the breakout width density distribution in Fig. 5.31. A DFN limits breakout width and, therefore, also influences the development and form of breakouts. However, this is not further discussed in this thesis, but might pose future research questions for a better understanding of the mechanics of breakouts and an enhanced understanding of borehole stability in a structurally-controlled rock mass.

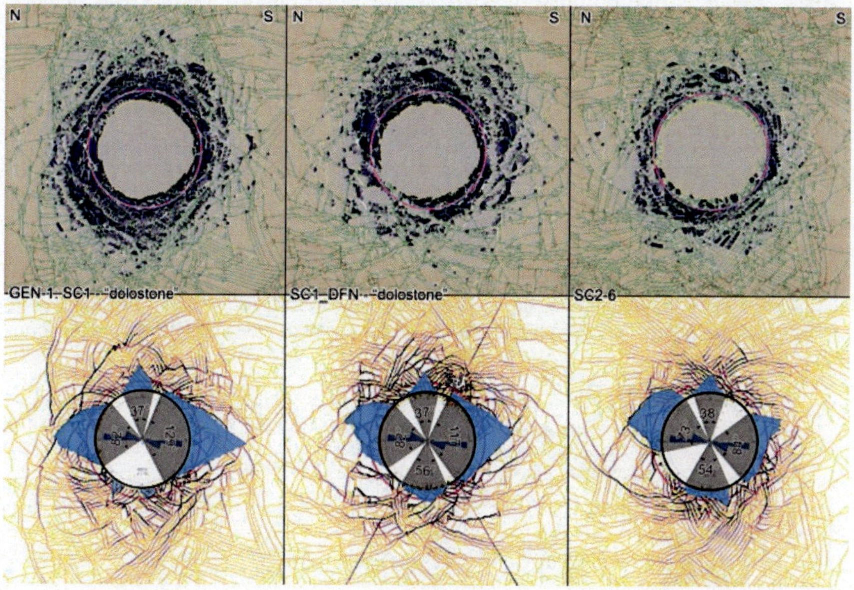

Fig. 6.7 Phenomenon of irregular and unexpected borehole breakouts (BBOs) due to increased, connected slip amounts at the horizontal walls of a borehole despite larger horizontal stresses than vertical stresses. For each model, the width of possible occurring breakout widths show within the borehole

An explanation, why the breakouts form differently than expected in the models might be due to severe initial fracturing at the top and bottom, which corresponds to their supposed location. While fracturing continues densely vertically, the mesh elements cannot break any further and interlock. Due to the high horizontal stresses, the interlocked particles shear laterally and cause large slips, which can be confirmed by the absolute stress magnitudes in Sect. 6.5.2.1 (Fig. 6.3, stress magnitude of "dolostone").

6.5.3.2 What is the Appearance of Unilateral Patterns linked with?

The mapping of single appearing features is mostly linked to vuggy structures to estimate a vugular porosity and deduce the rock's permeability. However, besides a preliminary theoretical exclusion of karstification, there is no description of it, nor do the cores indicate any cavity porosity. Hence, these Unilateral Patterns (UPs) may result from a mechanical deformation bound to any stress-, structural, or lithology-controlled influences.

Indicated by the density distributions in Figs. 5.32 and 5.33, the UPs focus on where the BBOs occur less (Sect. 5.3.2.2). According to the argumentation from the previous Sect. 6.5.3.1, borehole breakouts focus, independent from their orientation, in weak rocks (**lithology-controlled**) and unstratified, **stress-controlled** areas. Hence, after eliminating two of the three major controlling factors, a **structural-controlled** reason may cause these UPs.

A kinematic analysis shows that the five joint sets from Sect. 5.4.3 allow ten possible wedge combinations. Along GEN-1, with a borehole axis of 080/80 where most UPs appear (5600–6000 m MD, Fig. 5.32), five possible wedges form. From these five wedges, three do not match with the orientation of the UPs in the rose plots (Fig. 5.29), and two possible wedges remain (Table 6.3). According to the drill path

Table 6.3 Possibility of the occurring five joint sets of forming a wedge for the well paths of GEN-1 and GEN-1ST-A1, without highlighting: no wedge forms in both wells; highlighted in grey: wedge forms in **one** of the wells; highlighted in black: a wedge, concurrent with the UPs forms in both wells

#	Combination of joint sets forming a wedge					GEN-1 080/10	GEN-1ST-A1 110/20
	JS1	JS2	JS3	JS4	JS5		
1	×	×	×			no wedge forms	✓
2	×	×		×		no match with UPs	
3	×	×			×	no wedge forms	
4	×		×	×		no wedge forms	✓
5	×		×		×	no wedge forms	✓
6	×			×	×	no match with UPs	no wedge forms
7		×	×	×		no wedge forms	
8		×	×		×	✓ (JS2, JS3 & JS5)	
9		×		×	×	no match with UPs	no wedge forms
10			×	×	×	✓	no wedge forms

orientation in the HMI log's area of GEN-1ST-A1, five wedges form. One wedge does not fit the UPs from Fig. 5.30. Five wedges are geometrically impossible. From the four remaining wedges, #8 in Table 6.3 coincides with a potential wedge in GEN-1. Therefore, the interaction of JS2, JS3, and JS5 could cause a wedge-failure in both wells GEN-1 and GEN-1ST-A1 that agree with the UPs in both HMI logs.

However, a pure wedge failure under high compressional forces is unlikely, and a combination of stress- and structural controlled fragmentation, as suggested by the numerical models, would be more plausible. Although fracturing focuses on the top and bottom in SC1—"limestone," there is no clear correlation with the UPs (Fig. 6.8, SC1). Nevertheless, this correlation shows in the DFN-infused scenarios where SC1-DFN—"limestone" and SC2-5_DFN clearly show a correspondence with the UPs (Fig. 6.8, SC1_DFN and SC2-5_DFN). Particularly in SC1-DFN—"limestone," the DFN lines from DFN2 and DFN3 exactly hedge a small-scale fragmentation at the position of majorly occurring UPs to the top, but also minorly to the bottom (Fig. 6.8, SC1_DFN). However, also in SC-5_DF, at the transition from "limestone" to "dolostone" where the DFN lines are rambling compared to SC1_DFN, fracturing occurs in the exact position of the UPs (Fig. 6.8, SC2-5_DFN). Conclusively, even if the natural fracture network is not as fragile as the implemented DFN in the models, fracturing indicatively follows these predefined weakness paths also through the intact rock.

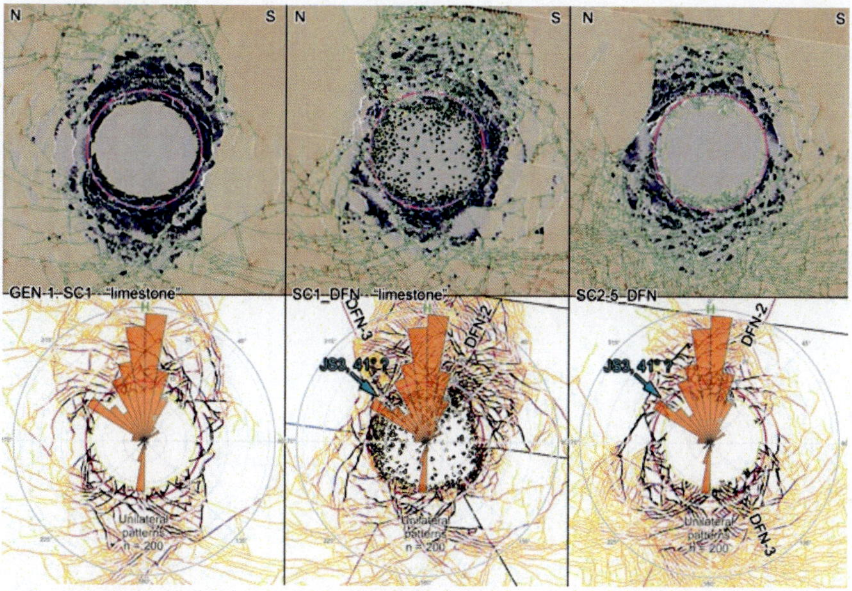

Fig. 6.8 Mapped Unilateral Patterns from HMI logs of GEN-1 (Sect. 5.3.2.1) compared to the models of GEN-1 from the left to the right: SC-1—"limestone", SC1_DFN—"limestone", and SC2-5_DFN

Apart from most UPs concentrated on the top, there is another maximum on the upper left abutment that does not match the fractures but would coincide well with JS3 solely (Fig. 6.8, SC1_DFN).

Although the numerical models do not consider gravity, its influence on the strongly disintegrated rock mass in a nearly horizontal borehole should not be neglected and might cause the loose rock mass caving into the borehole. However, the UPs in GEN-1ST-A1 show that gravity might not only be the only factor for their appearance.

As the results of UPs in GEN-1ST-A1 present (Sect. 5.3.2.2, Fig. 5.30), the sole orientation towards the top of the borehole does not exist. Instead, a much more diverse orientation concentrated inclining towards NE and dropping towards the borehole's bottom and SW. A comparison between the rose plots and the models required a horizontal flip of the enlarged sections in Fig. 6.9. SC1 does not cohere with the direction of the UPs (Fig. 6.9, SC1). However, comparable to GEN-1, the two scenarios, SC1-DFN "limestone" and SC2-5_DFN, develop fracture patterns that match the UPs. In SC1_DFN, fractures concentrate on the DFN sets' tips and to the bottom of the borehole. The fractures towards DFN-2 coincide with most of the UPs. These appear at the top left and the bottom right abutment, and also at the bottom. Towards DFN-3 are no correlating directions. Almost simultaneous is the fracture pattern in SC2-5_DFN, where further fracturing also includes the shallowly inclining UP-set, which might associate with the Bedding JS1 (Fig. 6.9, SC2-5_DFN).

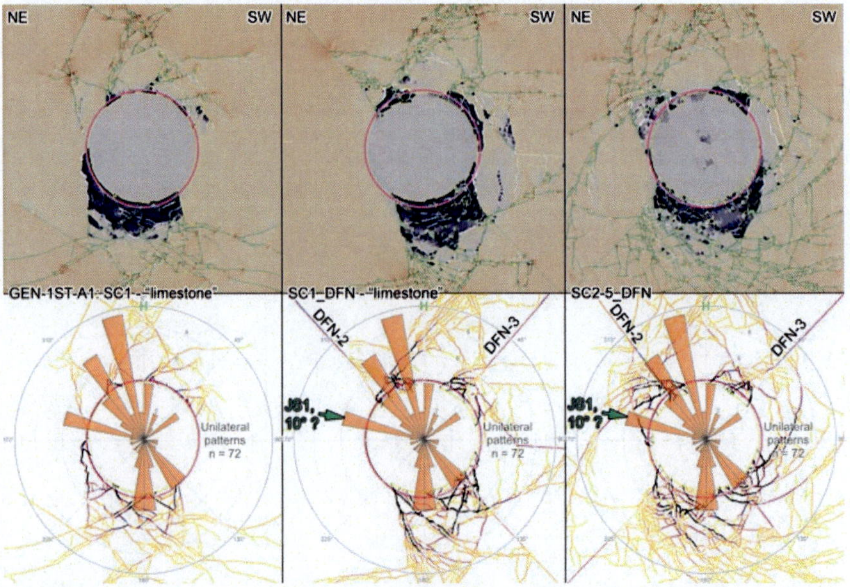

Fig. 6.9 Mapped Unilateral Patterns from HMI logs of GEN-1ST-A1 (Sect. 5.3.2.1) compared to the models of GEN-1ST-A1 from the left to the right: SC-1—"limestone", SC1_DFN—"limestone", and SC2-5_DFN

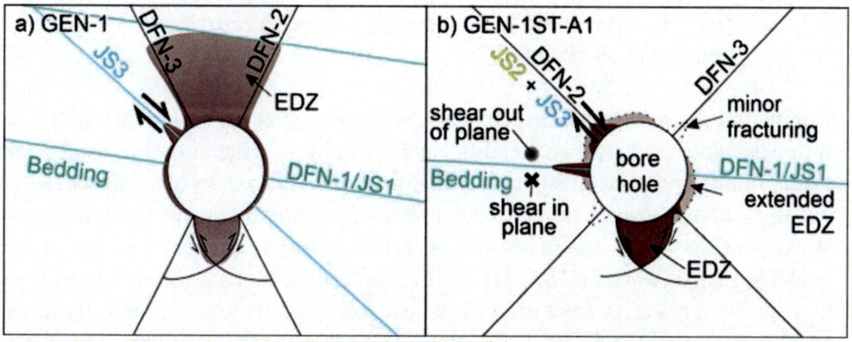

Fig. 6.10 Schematical drawing of asymmetrical breakouts so-called Unilateral Patterns (UPs), in the borehole. In **a**: UPs concentrate in the top of GEN-1, delimited by DFN-2 and -3 with further appearances at the bottom as conjugated shear fractures and along JS3 as shear-induced breakouts. In **b**: UPs in GEN-1ST-A1 mainly along DFN-2, in the bottom as conjugated shear fractures and in/out of plane along JS1

Fundamentally different and independent methods, where the joint sets come from cores and the UPs from image log from two different wells, establish the comparative results. This compliance between the UPs and the numerical models' fracture patterns is remarkable and can hardly be arbitrary. The further interpretation might allow differentiation of the joint sets, where the theories of a wedge failure and the fracture initiation and propagation along structural elements do not contradict and may rather complement each other. JS2, JS3, and JS5 were identified for potential wedge failure, while no specific DFN set could be identified for the UPs in GEN-1, although JS3 could be interpreted separately. Figure 6.10a schematically shows the fracture patterns for GEN-1 that emerge in combination with the preexisting joint sets. Hence, the Bedding (DFN-1/JS1), DFN-2, and -3 delimit fracturing in the direction of the maximum tangential stress within a steep wedge to the top. At the bottom, where JS1 is missing, a typical breakout wedge forms from two conjugated shear fractures. Around the borehole, minor spalling takes place. UPs accumulate along the apparent orientation of JS3 not captured by the numerical models. Assumingly, increased shear deformation occurs explicitly along JS3 in the real borehole.

In GEN-1ST-A1, almost all Unilateral Patterns (UPs) develop along DFN-2, consisting of JS2 and JS3 (Fig. 6.10b). Fractures that form parallel to the Bedding JS1 in GEN-1ST-A1 (Fig. 6.9, JS1, 10°?), but not in GEN-1, might confirm emerging shear stresses deteriorate the borehole stability and causing the core jamming, as discussed in Sect. 6.2.3. As in GEN-1, emerging shear stresses at the bottom show as two conjugated shear fractures. However, these slip movements cannot be seen in the models since these shear stresses move perpendicular to the models' 2D cross-section and do not apply in the simulation (Fig. 6.10b shear in and out of plane). Therefore, qualitatively, either JS3 is weaker than the other joint sets, or the shallow angle of 41° promotes shearing. Deducing on these interpretations, UPs identify JS3 and JS2, which are involved in all instability indicators.

6.5.3.3 How do alternating Lithologies influence Fracturing and Borehole Stability?

Comparing the numerical models from Scenario 2 (Sect. 5.6.2.4) shows that fracturing develops differently depending on the lithology. In hard lithology, represented by the "limestone," fracturing starts in distinct paths at the points of maximum tangential stress and propagate into the rock mass. Depending on the maximum tangential stress, fractures may sprout as one branch or multiple branches, propagate, and coalesce to a network (SC2-1, Fig. 5.56). For a weak material, proximal to a borehole in a hard material, fracturing can also occur remotely without being initiated by another fracture (SC2-5, Fig. 5.60, Time: 200). Once the fractures transit from a hard to soft material, they penetrate this weak material and increase in their lateral extent (Fig. 5.56; SC2-4, Fig. 5.59). Even if the material where the fractures transit is slightly weaker (e.g., "dolostone"), fractures extent laterally (Fig. 5.59). Reversely from a soft to hard material, the fractures propagate primarily laterally in the soft rock and concentrate at the hard rock interface. The plastic deformation relaxes the stresses until fracturing stops or shatters the rock mass completely, and remaining stresses cause fractures leaping to the harder material (Figs. 5.57 and 5.58). Although weaker, this effect can be seen between the "dolostone" and the "limestone" in Fig. 5.61.

Concludingly, the simulations show that if a hard-soft stratified alternation exists, the stresses and the thickness of the hard layers limit the lateral fracture extent. Contrarily, the soft layer boundaries act as a natural limitation for vertical fracture propagation in the rock mass since the stresses within the soft layers deplete due to lateral plastic deformation. Conveying scenario 2 in a 3D-context, the initial fracturing of the soft layers might already deplete so much stress that the hard layers may stay intact. The borehole breakouts, discussed in Fig. 6.6, would be one representative. The fluctuating BBO-density with increasing stresses in the PCs from 5070 m MD to 5370 m MD (Fig. 5.32) supports the numerical interpretation. The UPs in the PBCs from 4800 m MD to 5070 m MD might represent caving from soft layers (Fig. 5.62).

6.5.3.4 What Role do Discontinuities play in the Reservoir Rock?

As the previous Sect. 6.5.3.3 claims, lithology is a limiting factor for fracture propagation. However, the layer boundaries and the other discontinuities control not only fracture propagation but also initiation. Exemplarily, Sect. 6.5.3.2 indicates a correlation of UPs with the joint sets. The simulations with implemented DFNs show that fracturing initiates at the borehole wall, closest to DFN-2 or -3 (e.g., Figs. 5.50, 5.51 and 5.52). Primarily, stresses converge at the tips of the DFNs, and at their intersections, leading to a breakage of the rock bridges or to stress maxima that promote and attract further fracturing.

Although DFN-1 is the weakest of all DFN sets, no simulation shows initial fracturing directly towards or along with it. Under the prevailing stress regime, the bedding (DFN-1) is subparallel to the minimum tangential stress. Logically, no fractures emerge in that direction since fractures in shear spread from the max. tangential stress to DFN-2 or -3 and fractures in tension cannot establish under the given stresses (compare to Fig. 5.44). Hence, the bedding does not influence the initiation of fractures. However, as before the lithological boundary, it acts as prevention for vertical fracture propagation, although it separates two identical lithologies (Figs. 5.62 and 5.63).

Generally, the presence of a mechanically effective DFN leads to a confinement of the fracture extent. Regarding the simulations without a DFN, the models show well-developed conjugated shear planes that spread—according to the material parameters and the stresses—more or less far from the borehole (e.g., Figs. 5.47, 5.48 and 5.60). In the corresponding examples with a DFN (Figs. 5.50, 5.51 and 5.63), the DFNs restrict fracturing. Formerly distinct paths convert into larger and denser fracture networks in the borehole's vicinity through this confinement, decreasing its stability. Under high stresses, as in GEN-1, this effect specifies to hard rock (e.g., "limestone" in Fig. 5.54, or SC25 in Fig. 5.66), anew fracture cannot propagate in an intact rock. For weaker rock under high stresses, this effect cannot be seen (e.g., "dolostone" in Fig. 5.54, or SC2-3 in Fig. 5.66). In a lower stress environment, as in GEN-1ST-A1, hard rocks are still vulnerable to an increased fracturing in the vicinity of the borehole. However, a lesser rock strength becomes more prone to failure since the stresses are insufficient to overcome the material properties. Hence, in a highly jointed rock, less strong rock also becomes unstable and poses instability issues under decreasing stresses due to severe near-field disintegration.

6.5.3.5 Do the Numerical Models indicate natural Openings of Fractures?

Tensile failures in the numerical models only appear as secondary fractures in the loose rock mass around the borehole, while an opening is only achieved due to shearing. The maximum aperture of 2 cm is in the vicinity of the borehole. For the single lithology models in SC1, the opening declines in a 0.25 m radius around to borehole to less than 1 μm (Figs. 5.52 and 5.53). For the multiple-lithology models, the radius remains for GEN-1ST-A1 (Fig. 5.65) but increases for GEN-1. In GEN-1 (Fig. 5.64), fracture openings retrace up to half a meter from the borehole.

Nevertheless, numerical models in 2D are unsuitable for determining fracture openings. Since the boundary conditions inflict compressional forces from all directions, there cannot be a direction where potential openings may dilate, particularly since the 2D cross-section of GEN-1 is parallel with the potential opening face. Hence, for a true opening mode, also in shear, a third dimension is necessary.

6.5.3.6 How does the Permeability develop in the Rock Mass due to Excavation?

As the previous Sect. 6.5.3.5 examines, the fracture opening declines in the different scenarios and under different stresses from 2 cm to less than 1 μm in a radius from 0.25 to 0.5 m around the borehole. Assuming that the loosened rock mass is stable and pressure within the well is steady, the created voids may enhance permeability. A simplified approach by Wittke ([42]: 141) suggests a hydraulic conductivity K_D for apertures of a smooth set of persistent fractures, with a normalized spacing of 1 m. The opening results in Figs. 5.52, 5.53, 5.64 and 5.65 present an opening width with a maximum of 2 mm and a minimum of $1.0*10^{-6}$ mm. For these opening widths, Table 6.4 presents the open fractures' hydraulic conductivity with water at 10 °C with a viscosity of $1.3*10^{-6}$ m^2/s after Wittke ([42]: 141) and at 130 °C (after Sect. 3.3.6) with $2.3*10^{-7}$ m^2/s [19]. The results show that the hydraulic conductivity in the loosened rock mass around the borehole declines rapidly from a k_D of $3*10^{-2}$ m/s, which corresponds to gravel's conductivity, to $4*10^{-12}$ m/s, which relates to the conductivity of clay. This steep decrease happens at a distance of 25 to 50 cm in a non-linear curve, as indicated by the logarithmic legend in the opening figures. Assuming the spacing distance is the element size of 5 mm, the conductivity would increase by a factor of 200.

The assumption is based on only one persistent joint set. However, as discussed in Sect. 6.5.3.5, no distinct single joint sets identify as conductive in the numerical simulation. Hence, this applies to the disintegrated rock mass around the borehole,

Table 6.4 Hydraulic conductivity of ascending opening fractures with water at a viscosity 10 °C and 130 °C with the equivalent hydraulic conductivity in soil, modified after Wittke ([42]: 141)

Fracture opening/aperture		Hydraulic conductivity k_D (m/s)		Equivalent to the hydraulic conductivity of soil k_{soil} (m/s) for viscosity at 10 °C	
(m)	(mm)	Viscosity at 10 °C	Viscosity at 130 °C		
$1.0*10^{-6}$	0.001	$6*10^{-13}$	$4*10^{-12}$		clay
$2.5*10^{-5}$	0.025	$1*10^{-8}$	$6*10^{-8}$	silt	
$5.0*10^{-5}$	0.05	$8*10^{-8}$	$4*10^{-7}$		
$1.0*10^{-4}$	0.1	$6*10^{-7}$	$4*10^{-6}$		
$2.0*10^{-4}$	0.2	$5*10^{-6}$	$3*10^{-5}$		sand
$4.0*10^{-4}$	0.4	$4*10^{-5}$	$2*10^{-4}$		
$7.0*10^{-4}$	0.7	$2*10^{-4}$	$1*10^{-3}$	gravel	
$1.0*10^{-3}$	1.0	$6*10^{-4}$	$4*10^{-3}$		
$2.0*10^{-3}$	2.0	$5*10^{-3}$	$3*10^{-2}$		
$2.0*10^{-2}$	20	5	30		

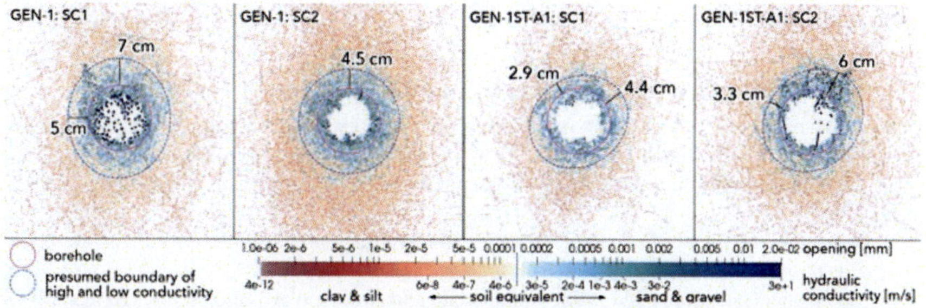

Fig. 6.11 Conductivity caused by natural fracturing around the borehole, superimposed for each scenario and each well, compared with soil equivalents. The blue dashed line marks the boundary of high and low conductivity

where a laminar flow is questionable. Therefore, the values should not be used quantitatively but qualitatively to underline the narrow range and rapid decline connecting the borehole to the host rock without artificial assistance.

6.5.3.7 How does the Borehole hydraulically connect to the Fractured Rock Mass?

As the previous chapter shows, an opening of natural fractures creates new flow paths for fluids. Consequently, the fracture openings can be categorized in a specific hydraulic conductivity (k_D), listed in Table 6.4. Figure 6.11 colors the openings with a diverging colormap, distinguishing conductivities lower than 4e-6 m/s in red (equivalent to clay andsilt) and greater than 3e-5 m/s in blue (equivalent to sand and gravel) for water at a viscosity at 130 °C. In order to simplify the illustration, each scenario for each well consists of all sub-scenarios superimposed. Hence, GEN-1, SC1 has a conductivity above 3e-5 m/s to the walls with 5 cm and vertically with 7 cm. The highly differential stresses seem to form larger fluid flows towards the least principal stress (Fig. 6.11, GEN-1: SC1). For GEN-1 in SC2, the distribution of highly conductive fractures is uniform around the borehole with 4.5 cm (Fig. 6.11, GEN-1: SC2). Hence, multiple lithologies decrease the irregularity of hydraulic effective fracturing.

In GEN-1ST-A1, the highly conductive area is smaller than in GEN-1, with a maximum of 4.4 cm and a minimum of 2.9 cm. The acting shear might cause the conductivity's elliptic distribution (Fig. 6.11, GEN-1ST-A1: SC1). The same observation conveys SC2 in GEN-1ST-A1. However, a minorly tilted elliptic distribution of k_D and an overall larger range contradicts the small uniform distribution of GEN-1. Hence, GEN-1ST-A1 shows a minimum of 3.3 cm and a maximum of 6 cm around the borehole. The occurring shear stresses tilt the distribution, yet multiple lithologies diminish this effect.

References

1. Alber M, Backers T, Bartmann K, Brenne S, Dinter S, Dresen G, Gipper P, Grühser C, Heft S, Meier T, Molenda M, Röske F, Stöckhert F. (2015) Erforschung der Mechanismen und Simulation hydraulisch induzierter Risse in geklüfteten Gesteinen für die Optimierung des Aufschlusses geothermischer Lagerstätten. Ruhr Universität Bochum, geomecon GmbH, p 136
2. Bachmann GH, Dohr G, Müller M (1982) Exploration in a classic thrust belt and its foreland: Bavarian Alps Germany. AAPG Bull 66(12):2529–2542
3. Bachmann GH, Müller M (1992) Sedimentary and structural evolution of the German Molasse Basin. Eclogae Geologicae Helvetiae 85:519–530
4. Backers T, Meier T, Gipper P, Munsch P, Bücken D, Nokar K (2017) Abschlussbericht zum Teilprojekt B: Struktur- und Spannungsfeld im Verbundprojekt MAFA: Parametrisierung von Fazies, Diagenese, Struktur- und Spannungsfeld sowie Optimierung der Testabläufe im Malm zur Verringerung des Erfolgsrisikos. Berlin, geomecon GmbH, p 44
5. Baecher G, Einstein H (1981) Size effect in rock testing. RGeophys Res Lett 8:671–674
6. Barthel KW (1978) Solnhofen: ein Blick in die Erdgeschichte. p 393
7. Barton N (1976) The shear strength of rock and rock joints. Int J Rock Mech Min Sci Geomech Abstracts 13(9):255–279
8. Barton N (2013) Shear strength criteria for rock, rock joints, rockfill and rock masses: Problems and some solutions. J Rock Mech Geotech Eng 5(4):249–261
9. Boersma Q, Hardebol N, Barnhoorn A, Bertotti G (2018) Mechanical factors controlling the development of orthogonal and nested fracture network geometries. Rock Mech Rock Eng 51(11):3455–3469
10. Budach I, Moeck I, Lüschen E, Wolfgramm M (2018) Temporal evolution of fault systems in the upper jurassic of the central german molasse basin: case study unterhaching. Int J Earth Sci 107(2):635–653
11. Byerlee J (1978) Friction of Rocks. Pure Appl Geophys 116:615–626
12. Cuisiat FD, Haimson BC (1992) Scale effects in rock mass stress measurements. Int J Rock Mech Min Sci Geomech Abstracts 29(2):99–117
13. Feder G, Arwanitakis M (1976) Zur Gebirgsmechanik ausbruchsnaher Bereiche tiefliegender Hohlraumbauten. Berg- u. Huettenmaennische Monatshefte 121(4):103–117
14. GeoService GmbH (2017) Geothermiebohrung GEN-1ST-A1 Abschlussbericht. unpublished report, p 20
15. Gischig V, Bethmann F, Hertrich M, Wiemer S, Mignan A, Broccardo M, Villiger L, Obermann A, Diehl T (2019) Induced seismic hazard and risk analysis of hydraulic stimulation experiments at the Bedretto Underground Laboratory for Geoenergies (BULG). ETH Zurich, Swiss Competence Center for Energy Research – Supply of Electricity (SCCER-SoE), p 40
16. Griffith AA (1921) The phenomena of rupture and flow in solids. Philosophical Trans Royal London. Series A, Containing Papers Math Phys Character (1896–1934) 221(582):163–198
17. Jaeger JC, Cook NG, Zimmerman R (2007) *Fundamentals of rock mechanics,* 4th ed. Blackwell Publishing, p 475
18. Kirsch (1898) Die Theorie der Elastizität und die Bedurfnisse der Festigkeitslehre. Zeitschrift des Vereines Deutscher Ingenieure 42:797–807
19. Korson L, Drost-Hansen W, Millero FJ (1969) Viscosity of water at various temperatures. J Phys Chem 73(1):34–39
20. Kraemer C (2009): Identifikation horizontaler Blattverschiebungen im Untergrund des Molassetrogs - neue Interpretationsergebnisse auf Grundlage von 2D-Seismik. Der Geothermiekongress 2009, Bochum, Deutschland (Bundesverband Geothermie e.V.)
21. Lemcke K (1988) Das Bayerische Alpenvorland vor der Eiszeit: Erdgeschichte. Geologie von Bayern I. Stuttgart, Schweizerbart, Bau, Bodenschätze, p 175
22. Lim SS, Martin CD (2010) Core disking and its relationship with stress magnitude for Lac du Bonnet granite. Int J Rock Mech Min Sci 47(2):254–264

23. Maurer WC (1966) Shear failure of rock under axial and hydrostatic pressure. In: Proceeding 1st Congress International Socity Rock Mech., Lissbon, pp 337–341
24. Megies T, Wassermann J (2014) Microseismicity observed at a non-pressure-stimulated geothermal power plant. Geothermics 52:36–49
25. Moeck IS, Dussel M, Buness H, Wawerzinek B (2019) Verbundprojekt Dolomitkluft - Erschließung, Test und Analyse des ersten kluftdominierten Dolomitaquifers im tiefen Malm des Molassebeckens; Teilprojekt E: Spannungsfeldanalyse und Charakterisierung der Störungs- und Kluftzonen. Hannover,Leibniz-Institut für Angewandte Geophysik, p 61
26. Mogi K (1966) Pressure dependence of rock strength and transition from brittle fracture to ductile flow. Bull Earthquake Res Instit 44:215–232
27. Mraz E, Bohnsack D, Stockinger G, Käsling H, Zosseder K, Thuro K (2018) Die Bedeutung von Analogaufschlüssen des Oberjura für die Interpretation der Lithologie der geothermalen Tiefbohrung Geretsried. Jahresberichte und Mitteilungen des Oberrheinischen Geologischen Vereins 100:517–548
28. Ramsey JM, Chester FM (2004) Hybrid fracture and the transition from extension fracture to shear fracture. Nature 428(6978):63–66
29. Ratschbacher L, Frisch W, Linzer H-G, Merle O (1991) Lateral extrusion in the eastern Alps, Part 2: Structural analysis. Tectonics 10(2):257–271
30. Santarelli FJ, Dusseault MB (1991) Core quality control in petroleum engineering. In: Roegiers JC (ed) Rock mechanics as a multidisciplinary science, proceedings of the 32nd U.S. symposium. Rotterdam, A.A. Balkema, pp 111–120
31. Schubert A (2013) Geothermieprojekt Geretsried-Nord - Kurzbericht Reservoircharakterisierung Geretsried GEN 1. unpublished report, p 34
32. Seithel R, Steiner U, Müller B, Hecht C, Kohl T (2015) Local stress anomaly in the Bavarian Molasse Basin. Geothermal Energy 3(1):22
33. Shipilin V, Tanner DC, von Hartmann H, Moeck I (2019) Temporal Evolution of faults in the southern German Molasse Basin: a case study of Wolfratshausen, Germany. In: 81st EAGE conference and exhibition, pp 1–5
34. Steiger T, Uhlig S (2018) Bio- und Lithostratigraphie der Geothermie-Bohrung Geretsried GEN- 1 mit Sidetrack GEN-1ST-A1. (Anlage 1). In: Thuro K, Zosseder K, Bohnsack D, Heine F, Konrad F, Mraz E, Stockinger G (eds) Dolomitkluft - Erschließung, Test und Analyse des ersten kluftdominierten Dolomitaquifers im tiefen Malm des Molassebeckens zur Erhöhung der Erfolgsaussichten. Teilprojekt B: Geomechanische und hydro-geologische Parametrisierung und Modellierung, München (Technische Universität München - Ingenieurfakultät Bau Geo Umwelt - Lehrstuhl für Ingenieurgeologie.)
35. StMWIVT (2010) Bayerischer Geothermieatlas - Hydrothermale Energiegewinnung. München, Bayerisches Staatsministerium für Wirtschaft, Infrastruktur, Verkehr und Technologie, p 94
36. Teufel LW (1989) Influence of lithology and geologic structure on in situ stress: Examples of stress heterogeneity in reservoirs. In: 2. international reservoir characterization technical conference. Dallas, TX, pp 565–576
37. Thuro K, Plinninger R, Zah S, Schutz S (2001) Scale effects in rock strength properties. Part 1: Unconfined compressive test and Brazilian test. In: Särkkä P, Eloranta P (eds) Proceedings of the ISRM regional Symposium Eurock 2001. Espoo, Finland, A.A. Balkema, Lisse, pp 169–174
38. Thuro K, Zosseder K, Bohnsack D, Heine F, Konrad F, Mraz E, Stockinger G (2019) Dolomitkluft - Erschließung, Test und Analyse des ersten kluftdominierten Dolomitaquifers im tiefen Malm des Molassebeckens zur Erhöhung der Erfolgsaussichten. Teilprojekt B: Geomechanische und hydro-geologische Parametrisierung und Modellierung. München, Technische Universität München - Ingenieurfakultät Bau Geo Umwelt - Lehrstuhl für Ingenieurgeologie, p 413
39. Unger HJ (1999) Die tektonischen Strukturen der bayerischen Ostmolasse. Documenta naturae 124:1–16

40. Whittaker BN, Singh RN, Sun G (1992) Rock fracture mechanics: principles, design, and applications. Amsterdam, Elsevier BV, p 591
41. Wieser C (2016) Quantifying the effect of stress changes on the deformation and cracking behavior of solid rock using acoustic emission techniques. Dissertation, Chair of Engineering Geology, Technical University of Munich, Munich, p 166
42. Wittke W (2014) Rock mechanics based on an anisotropic jointed rock model (AJRM). Berlin, Ernst, p 875
43. Wolfgramm M, Thiem S, Zimmermann J, Budach I, Buse C, Kabus F (2018) Forschungsbericht GTN: Erschließung, Test und Analyse des ersten kluftdominierten Dolomitaquifers im tiefen Malm des Molassebeckens. Berlin, Geothermie Neubrandenburg GmbH, p 93

Standards

44. ISRM (1978) Suggested methods for the quantitative description of discontinuities in rock masses. Int J Rock Mech Min Sci Geomech Abstracts 15:319–368

Chapter 7
Conclusions, Implementation and Outlook

7.1 Analog Rocks versus in situ Rocks

Although some of the analog rocks, as discussed in Sect. 6.1, are lithologically familiar, e.g., DD and PFD, QK and SPK, DK and BO, or BK and OKL, they still differ in decisive parameters that trace back to external, natural, and artificial influences. Artificial factors, such as sampling and mining, need special focus and explicit documentation, consideration, and discussion when substituting in situ rocks. Natural factors need distinct evaluation. While granites or generally polymineralic rocks are prone to weathering and a reduction in rock strength [1], weathering effects in carbonates mainly affect the rock mass (consisting of porosity, karstification, and discontinuities). By contrast, the rock itself in a weathered matrix stays intact, and its strength stays unaffected. Hence, comparing in situ carbonates with analog rocks should not be limited by weathering if the structural component also matches [1]. Pyrite oxidation, carbonate dissolution or reprecipitation happen along structural features and alter the rock's structure (e.g., pyrite-swelling). Still, if these external influences are known for the analog rocks and the in situ rocks, a limitation and discussion of an equivalent rock with the corresponding parameters is possible. An improved comparability matrix from Table 6.1 may accommodate supplemental information like accessory minerals, tectonic deformation, or paleo/present-environmental conditions. For instance, QK fits the overall lithology and the accessory minerals, but its tectonic history prematurely disqualified its static parameters. Although the parameter range may experience some expansion, e.g., complementing and comparing shear parameters of the in situ specimen and the analog rocks, the initial approach to work with a minimum and maximum, together with an intermediate strength, creates input parameters for numerical models that fit well with in situ observations, e.g., in the HMI logs. Therefore, the parameters elucidate the basic mechanics of fracturing and failure in the rock mass around the borehole.

G. M. Stockinger, *Fracturing in Deep Boreholes*, Springer Theses,
https://doi.org/10.1007/978-3-030-94569-5_7

7.2 Borehole Stability and Weakening of the Rock Mass

All results confirm that borehole stability decreases and the degree of fracturing increases with low rock strength and vice versa. Yet the diagrams in 6.5.2 show on the one hand that fracturing occurs in both geothermal wells and the rock mass weaken due to fracturing (Figs. 6.2 and 6.3) and on the other hand that the depth of the loosened rock mass is not necessarily dependent on rock strength (Figs. 6.4 and 6.5).

Typical stress-related phenomena such as BBOs only form lateral, symmetrical breakouts when high rock strength prevents fractures coalescing around the borehole but exceeds occurring tangential stresses, or for a weak layer between two hard layers (Fig. 6.6). A full circumferential breakout happens only when the whole rock (mass) is weak and stresses around the borehole remain equal due to fracturing and loosening of the rock, which it is worth noting applies in a highly differential stress field. This is a unique phenomenon in sedimentary rocks, where a quick alteration of lithology takes place in contrast to granitoide rocks. Hence, borehole instabilities also differentiate stratigraphy using breakout orientation, width, density, and drill path orientation. Due to the thin-bedded weak layers, a maximum resolution of the HMI is needed. A resolution of 5 mm, as in GEN-1ST-A1, is sufficient and in keeping with modern applications. The resolution in GEN-1 is too coarse to assume the stratigraphy. Furthermore, a complementary caliber log would be helpful to synchronize HMI logs and cores.

The extent of the disturbed zone (DZ, Figs. 6.4 and 6.5) depends on the absolute stresses as is the excavated damaged zone (EDZ). Hence, the DZ of "limestone" and "dolostone" is 1 m and 1.5 m in GEN-1 and 0.6 m and 0.8 m in GEN-1ST-A1 and the EDZ is slightly beyond the empirical prediction. Depending on the well, the EDZ is slightly less than 10–14 cm in GEN-1 and 6–9 cm in GEN-1ST-A1.

Interestingly, the DZ in the "weak limestone" is larger in GEN-1ST-A1 than in GEN-1. The lower stresses in GEN-1ST-A1 prevent the transition of fractures in harder layers. Hence, the overall extent of the DZ becomes larger and the transition from the sensitive EDZ to the fracture process zone (FPZ) becomes indistinct. However, in GEN-1, the EDZ occupies the overall weakened rock mass, which is almost equal to the extent of the EDZ in GEN-1ST-A1.

Notably, the "weak limestone" and "limestone" share the extent of the DZ at 1 m. While severe fracturing disintegrates the "weak limestone" completely, fractures cannot propagate further than maximum 0.5 m from the borehole since occurring stresses fail to overcome the high failure requirements. Whereas in "dolostone" of GEN-1, fractures propagate at a maximum from the borehole while still reducing the EDZ to a minimum (Fig. 6.4, right). Hence, an intermediate rock strength keeps boreholes stable and simultaneously encourages fractures to propagate into the rock (compare Figures 5.48, 5.60 and 5.63). However, a weak preexisting fracture network diminishes this effect (Figs. 5.51 and 5.63).

Although residual stresses generally might prevent a collapse of the loosened material, a change in pore pressure (e.g., by a drawdown as the result of a pump test or the pressurization of the borehole), temperature changes (explicitly not considered in this thesis, Sect. 4.5.5), or gravitational effects may influence the borehole stability negatively.

7.3 Core Drilling

Whether core jamming emerges from stress releases of weak discontinuities within the core (as discussed in Sect. 6.2.3) or from external shear motions (Sect. 6.5.3.2) cannot be finally answered—possible it is both equally. Nevertheless, both result in the same inevitable consequences, thus core jamming will persistently occur if:

- shear stresses emerge due to a wellbore oblique to the principal stresses **and**
- either a weak discontinuity plane cannot withstand theses shear stresses, **or**
- the shear strength of the rock cannot bear these stresses.

Despite unveiling the mechanical behavior of rock cores in a highly differential in situ stress field, a satisfying recommendation for future success of core drilling cannot be made. On the one hand, shear-induced discontinuities, which allow deformation and stress degradation, positively prevent the core disking, and ensure the cores' maximum integrity. Thus, specimens allow the measurement of dynamic properties from the cores. At the same time, core jams lead to a deterioration of the core within the barrel, milling of the rock at the coring's front, and finally to a compulsory abort of the coring operation. On the other hand, a lack of effective discontinuities prevents stress degradation and core jamming but promotes core disking. Thus, the cores disintegrate in disks of a few cm and generally become disqualified for dynamic and static rock properties testing.

A coring direction on a drill path, oblique to any of the principal stresses is certainly bad since emerging shear stresses may trigger certain discontinuities, which causes core jamming. Hence, coring parallel to any of the principal stresses would be recommended. From 3D FEM and 2D DEM simulation results, Bahrani et al. [2] concluded that cores retrieved parallel to σ_1 experience the highest tensile stress and parallel to σ_3 the highest differential stress. Micro-cracks usually form parallel to σ_1 and perpendicular to σ_3. Bahrani et al. [2] argue that cores parallel to σ_3 are more damaged than parallel to σ_1. Hence, both avoiding shear stresses and obtaining the best core quality a NS horizontal wellbore trajectory in the Molasse Basin, independent of NF or SS stress regime, would be the best option. However, core disking would still occur in a NS striking horizontal well. The largest perpendicular stress would be σ_v, with a value of 74 MPa, resulting in larger discs than anticipated in GEN-1ST-A1 and a saddle-shaped form, shifted by 90°. So, the rock cores would not be much better, and questions remain as to whether the discontinuities still become active without the missing shear stresses.

7.4 Productivity of the Wells and Permeability of the Rock Mass

Several authors suggest that faults are either hydraulically active under shear stresses, particularly when close to slipping [3–6], or under extension [7, 8]. Based on an adapted Coulomb criterion (Sect. 2.1.2.1), Morris et al. [9] calculate the potential for slip on a surface—the so-called slip tendency T_s. The slip tendency T_s, valid for planar discontinuities (faults, fractures, bedding), is defined by the ratio of shear stress to normal stress and cohesion is usually neglected [9]:

$$T_s = \frac{\tau}{\sigma_n}$$

And for a normalized slip tendency:

$$T_{s,norm} = \frac{\tau}{\sigma_n} / T_{s,max}$$

Hence, after Byerlee [10], faults with a $T_s \geq 0.6$ (or above) are likely to slip. The slip tendency requires only the stress tensor and the orientation of the discontinuity. The dilatation potential of a discontinuity requires the magnitudes of the maximum and minimum principal stress, together with the effective normal stress acting on a plane. The resulting dilatation tendency T_d is the ratio of the maximum principal stress minus the effective normal stress, normalized by the differential stress [11]:

$$T_d = \frac{(\sigma_1 - \sigma_n)}{(\sigma_1 - \sigma_3)}$$

The larger T_d is, the higher the potential for dilatation. Stephens [12], who also provide the FracTend MATLAB code to create the stereographic plots in Fig. 7.3, note that faults with both a high T_s (0.8–1.0) and T_d (0.8–1.0) suggest a zone of potential extensional shear, and thus, an enhanced hydraulic activity. Figure 7.3 shows the normalized slip potential (Fig. 7.3a) and dilatation potential (Fig. 7.3b) for a strike-slip (SS) stress regime after the principal stress magnitudes from Backers et al. [13]. Therefore, JS2 and JS4 have both high T_s and T_d and would correspond with an extensional shear regime, while the normal faults JS3 and JS5 show low slip and dilatation tendencies. Logically, the shallow-dipping bedding, JS1, has a zero slip tendency but tends to a higher dilatation potential than the normal faults.

T_d in Fig. 7.3 coincides with the high flow rate in the wells of Sauerlach, where the NS-striking, vertical faults dominate [14]. Besides these purely vertical faults, shear faults with a NNE-SSW orientation, such as JS2 and JS4, also dilate, yet not as intensely as purely N-S striking faults. Logically, the least compressive stress, here in an EW direction, has its lowest magnitude perpendicular to the open flow pathways. Hence, the more the joints rotate away from a NS direction, the bigger the effective compressional stresses become and the dilation and, thus, the fluid flow decreases. Principally, the manner in which stresses act on a fault corresponds to the stress rotation described in Sect. 4.3.

Misleadingly, the HMI logs only show antithetic EW-striking faults (JS5) with mainly NS striking, sub-vertical faults (JS2 and JS4). Figure 7.1 shows that in the area of the HMI logs of GEN-1ST-A1, the drill path runs parallel to the Gartenberg-Branch Nord, which corresponds to the joint set JS3. Consequently, JS3 may be invisible in the HMI log since it runs sub-parallel (orange and purple sections in Fig. 7.1) and, thus, does not appear in the stereographic projections from Sect. 5.4.2. As can be seen in the lower left of Fig. 7.1, JS3 appears with the highest frequency of all tectonic-related joint sets, marked in the coring section with the stereographic net from the drill cores described in Sect. 5.4.3.

Apparently, the pre-existing extensional faults caused by an ancient NF stress regime are under compression in a modern SS stress regime with an extremely low dilatation tendency below 0.1 for JS3 and 0.2 for JS5 (Fig. 7.3b). In favor of the compressional interpretation of these stresses are, on the one hand, the markers on the drill cores (Fig. 5.37) and, on the other hand, the indicated shear failure along JS3 (Fig. 6.8). Along with JS3, JS2 is also identified as mechanically active, which therefore entails a dilatational movement in shear with a tendency between 0.65 and 0.75 in Fig. 7.3b. Although JS4 has an even higher dilatation tendency of 0.9 (Fig. 7.3b), any proof of being mechanically active is missing. Therefore, either these fractures are sealed by mineral precipitation, or any sign may be undetected due to their perpendicular strike and a steep incline to the borehole path as well as the parallel strike to the 2D modeled cross-sections.

Recommended drill direction for wells on the latitude of Geretsried, Icking & Weilheim in the NAFB

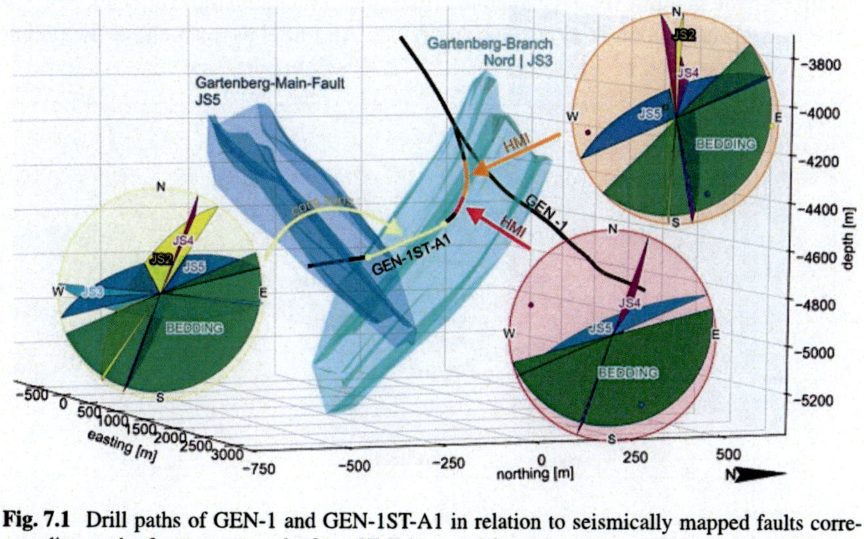

Fig. 7.1 Drill paths of GEN-1 and GEN-1ST-A1 in relation to seismically mapped faults corresponding to the fracture networks from HMI logs and from the core runs. 3D data from Shipilin et al. [15], LIAG

In conclusion, the NS striking faults dilate and act as flow paths as long as **no** EW-striking antithetic or synthetic faults are present. For instance, the well Th1 in Sauerlach shows very satisfactory flow rates (Fig. 6.1c, [14]). From a mechanical point of view, derived from the near-field numerical models with DFNs in Sect. 5.6.2, the presence of pre-existing conjugated shear fractures, such as JS3, degrades the stresses, and existing shear fractures cannot dilate although they are present and mechanically weaker than the surrounding rock. Interestingly, according to the Geothermie Atlas [16], an almost EW-striking synthetic normal fault is also mapped close to Weilheim's and Icking's geothermal wells, which were abandoned due to a lack of fluid flow [17, 18].

Hence, the results show that in order to be economically prosperous, drilling operations in the Upper Jurassic carbonates of the North Alpine Foreland Basin should aim for an ESE-WNW direction with a final azimuth between 111° to 124° and 291° and 304° in the open hole section, facing perpendicular to the strike-slip faults JS2 and JS4 (Fig. 7.2b), which have the highest dilatational tendency (Fig77...7.2b). Hence, the mechanical results from modelling indicate that although JS2 and J4 are continuously present, no dilatation applies when a lager fault, corresponding to JS3, appears. Therefore, an additional condition for the drill path trajectory requires the absence of JS3. Neither HMI logs, nor numerical models, however, indicate any negative influence of JS5.

For Geretsried, the drill trajectory should cross the normal faults as quickly as possible, preferably in a NS direction (Fig. 7.3a, red, dashed lines). After reaching a certain distance, remote from the normal faults, the recommended drill trajectory should be adopted, depending on the course and continuity of the synthetic normal faults (Fig. 7.3a, red area, enclosed by red arrows). The local effect of the normal faults on the stress field requires far-field simulations, which would, on the one hand, help with the planning of a reasonable drill trajectory and, on the other hand, evaluate the true dilatation potential of the shear faults. An initial assessment of the planned

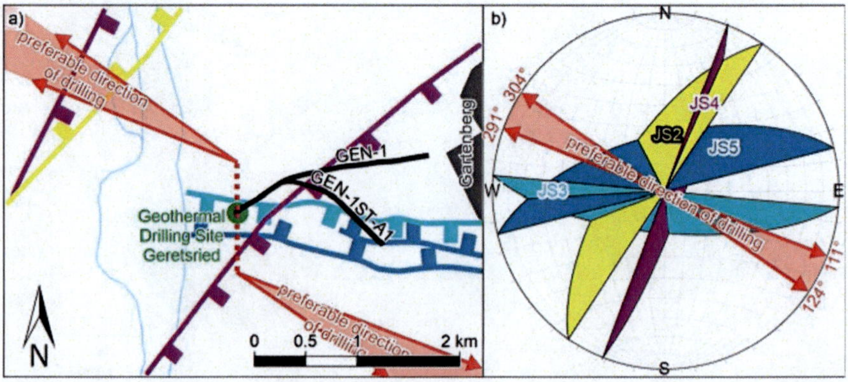

Fig. 7.2 Recommended direction of drill paths, aiming for a preferable connection on potentially dilatating joint set JS2 and JS4, under the slightest influence of formally extensional faults JS3 and JS5

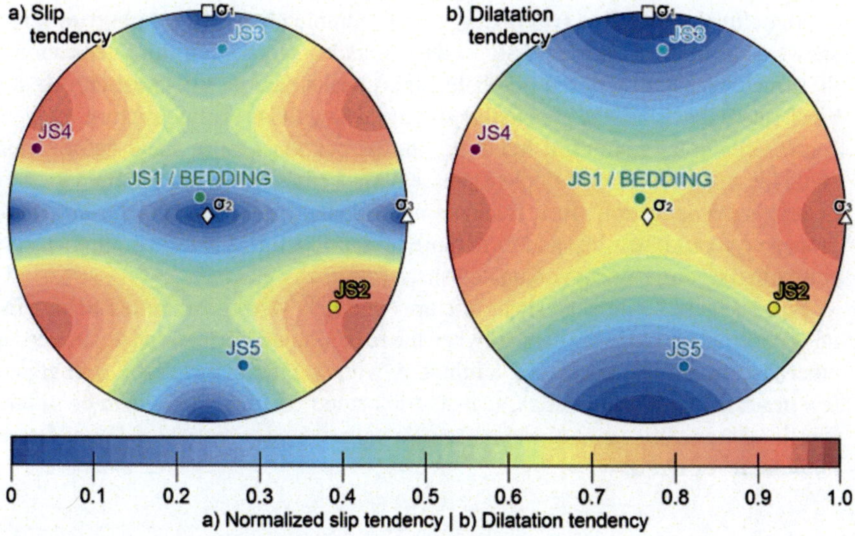

Fig. 7.3 Slip tendency (**a**, left) and dilatation tendency (**b**, right) of the occurring joint sets JS1 to JS5 for a friction coefficient of 0.6. Created after Stephens et al. [12]

borehole's stability can be obtained by the stress transformation/rotation method from Sect. 4.3. Still coring may likely fail.

Extreme caution is required before realizing these preferred drilling directions. Extensive investigations on the slip tendency of all faults should be conducted to mitigate the risk of artificially induced seismic events. A simple normalized slip tendency shows that JS2 and JS4 are prone to slip under a strike-slip regime (Fig. 7.3a). Therefore, extensive triaxial testing on the remaining drill core specimens from GEN-1ST-A1 should foster shear parameters, serving as plausible parameters for calculating the mechanical slip potential and estimating possible risks.

7.5 Characteristic Types of Fracturing around the Borehole

Section 6.4.2 concludes that three main factors control fracturing around the borehole, namely: **stress, structural elements,** and **lithology**. Fracturing in the reservoir rock, however, does not occur in isolation from any one of these factors but mostly from various combinations.

Fracturing in a combined **stress- and lithology-controlled** environment results in a heavily isolated disintegration of weak layers. The severe fracturing in isolated weak layers leads to a relief of stresses that prevents fracturing propagation in more competent, stronger rocks. Thus, depending on the orientation of the drill path, occurring borehole breakout might lead to the misinterpretation of the in situ stress field.

Structural-controlled fracturing, be it in combination **with lithology or stress**, shows as asymmetrical breakouts in the borehole. Along with the intersection of the structural plane with the borehole, stress peaks emerge that inevitably lead to breakouts at the tips of the preexisting structural flaws. Generally, structural elements decrease the total amount of fracturing in the rock mass since shear deformations along these linear elements deplete stresses. Also, structural elements mold the form of fracture propagation since fractures align along pre-existing joints, initiate, or propagate at or to their tips and form more wedge-like forms, whereas purely **stress-controlled** fractures form conjugated shear fractures.

Besides these conjugated shear fractures in purely **stress-controlled** fields, fractures penetrate the rock the deepest when the rock is homogeneous and not pierced by other discontinuities. Therefore, a lithology with an intermediate rock strength and few pre-existing discontinuities, such as "dolostone" in Fig. 5.48, would be suitable for stimulation. Still, some questions remain unanswered, which should be addressed in future research.

7.6 Recommendations for Future Research

Although the fundamental causes for fracturing have been outlined in this thesis, some remaining questions go beyond the scope of this work. Although 2D FEMDEM models solve satisfactorily the plastic deformation and the initiation and propagation fracture around a borehole, the processes that happen in-plane stay suppressed. Full descriptions, however, regarding how the rock behaves in a tensile motion, how faults dilate, and their tendency to shear in a highly differential stress field, require 3D simulations. Therefore, future research should implement the rock-mechanical parameters obtained in this thesis to 3D models to refine and verify the discoveries reported here.

Thermal stresses, not eligible for consideration in the applied version of the simulation code, are now incorporated in the newest versions and might apply accordingly.

The investigations of the initial questions, posed in Sect. 1.1, has led to a thorough understanding of the mechanical behavior of the fractured carbonate rock mass around the borehole. Closing the circle, future research should deploy this data and these findings to simulate the feasibility of EGSs. 3D finite-discrete element models might deploy tactics to adopt the orientation of the drill path to the in situ stress field and the discontinuity sets to produce the maximum of clean geothermal energy as efficiently as possible and simultaneously keep both geological and economic risks to a minimum.

References

1. Heitfeld K-H (1985) *Ingenieurgeologische Probleme im Grenzbereich zwischen Locker–und Festgesteinen.* Berlin, Springer, pp 696
2. Bahrani N, Valley B, Maloney S, Kaiser PK (2012) Numerical investigation of the influence of borehole orientation on drilling-induced core damage. Eurock, 13, Stockholm, Sweden
3. Barton CA, Zoback MD, Moos DB (1995) Fluid flow along potentially active faults in crystalline rock. Geology 23:683–686
4. Ito T, Zoback MD (2000) Fracture permeability and in situ stress to 7 km depth in the KTB scientific drillhole. Geophys Res Lett 27(7):1045–1048
5. Zhang X, Sanderson DJ, Barker AJ (2002) Numerical study of fluid flow of deforming fractured rocks using dual permeability model. Geophys J Int 151(2):452–468
6. Zhang X, Koutsabeloulis N, Heffer K (2007) Hydromechanical modeling of critically stressed and faulted reservoirs. AAPG Bulletin 91:31–50
7. Gudmundsson A (2000) Active fault zones and groundwater flow. Geophys Res Lett 27(18):2993–2996
8. Gudmundsson A, Fjeldskaar I, Brenner SL (2002) Propagation pathways and fluid transport of hydrofractures in jointed and layered rocks in geothermal fields. J Volcanol Geothermal Res 116(3–4):257–278
9. Morris AP, Ferrill DA, Henderson DB (1996) Slip-tendency analysis and fault reactivation. Geology 24:275–278
10. Byerlee J (1978) Friction of rocks. Pure and Appl Geophys 116:615–626
11. Ferrill DA, Winterle J, Wittmeyer G, Sims D, Colton S, Armstrong A, Morris AP (1999) Stressed rock strains groundwater at Yucca Mountain, Nevada. GSA Today 9:1–8
12. Stephens TL, Walker RJ, Healy D, Bubeck A, England RW (2018) Mechanical models to estimate the paleostress state from igneous intrusions. Solid Earth 23
13. Backers T, Meier T, Gipper P, Munsch P, Bücken D, Nokar K (2017) Abschlussbericht zum Teilprojekt B: Struktur- und Spannungsfeld im Verbundprojekt MAFA: Parametrisierung von Fazies, Diagenese, Struktur- und Spannungsfeld sowie Optimierung der Testabläufe im Malm zur Verringerung des Erfolgsrisikos. Berlin (geomecon GmbH), pp 44
14. Seithel R, Steiner U, Müller B, Hecht C, Kohl T (2015) Local stress anomaly in the Bavarian Molasse Basin. Geothermal Energy 3(1):22
15. Shipilin V, Tanner DC, von Hartmann H, Moeck I (2019) Temporal evolution of faults in the Southern German Molasse Basin: a case study of Wolfratshausen, Germany. In: 81st EAGE conference and exhibition 2019, pp 1–5
16. StMWIVT (2010): Bayerischer geothermieatlas—hydrothermale energiegewinnung. München (Bayerisches Staatsministerium für Wirtschaft, Infrastruktur, Verkehr und Technologie), pp 94
17. www-03: https://www.tiefegeothermie.de/news/projekt-in-weilheim-gestoppt Accessed on: 09.12.2020
18. www-04: https://www.tiefegeothermie.de/news/bohrplaetze-in-icking-werden-zurueckgebaut Accessed on: 09.12.2020

Appendices

Appendix A: Fully Applicable Python Code for Stress Rotation

Section 4.3 explains the methodology and the mathematical background of the stress rotation. The following code, divided by its different units and mathematical operations, is explained in Sect. 4.3.4. The full code runs with this python script, called “calcTrafoMat”:

```
import numpy as np
import sys
def generateTransMatrix(phi, theta, psi, chi):
#-------------------------------------------------
# in-situ stress field
Sigmaxx = 0.0276 #MPa/m
Sigmayy = 0.0154 #MPa/m
Sigmazz = 0.023 #MPa/m
Sigmapp = Sigmazz*0.4 #MPa/m
Sigma = np.matrix([
[Sigmaxx-Sigmapp,0,0], #Sigmax-xx; x-Achse
[0,Sigmayy-Sigmapp,0], #Sigma-yy; y-Achse
[0,0,Sigmazz-Sigmapp] #Sigmax-zz; z-Achse
])
### Strike of the horizontal principal stresses, for NS Strike = 0, a max. angle of +-45° is possible
### for a clockwise rotation, use positive values; for counter clockwise rotation, use negative values
```


G. M. Stockinger, *Fracturing in Deep Boreholes*, Springer Theses,
https://doi.org/10.1007/978-3-030-94569-5

```
### angle chi
if chi >= -45.0 and chi <= 45.0:
chi = chi
elif chi < -45.0 or chi > 45.0:
print("The stress field cannot be rotated by more than 45 degree, please per form the changes in the in situ stress matrix")
sys.exit(1)
#print(chi)
### angle phi, theta und psi as input for the 3D-rotation
#-----------------------------------------------
# phi: azimuth of the borehole 0 (N) - 360 (N) degree
if theta <= 3 and theta >= -3:
phi = 0
elif theta > 3 or theta < -3:
if phi <= 45:
phi = -1*phi+chi #for rotation when azimuthal direction is in the 1st quadrand
elif phi > 45 and phi <= 135:
phi = 90-phi+chi #for rotation when azimuth is in the 4th quadrand
elif phi > 135 and phi <= 225:
phi = 180-phi+chi #for rotation when azimuthal direction is in the 3rd quadrand
elif phi > 225 and phi <= 315:
phi = 270-phi+chi #for rotation when azimuthal direction is in the 2nd quadrand
elif phi > 315 and phi <= 360:
phi = 360-phi+chi
#print(phi) #-----------------------------------------------
# theta: inclination of the drill path 0 = vertical, 90 = horizontal, limited to 90°
if theta <=45:
theta = theta*(-1)
elif theta > 45 and theta < 90:
theta = theta-90
elif theta >= 90:
print("The Inclination is limited to a range of 0 to not more than 90 degree and excludes 90!")
sys.exit(1)
#print(theta)
#print(theta)
#-----------------------------------------------
# psi: no rotation around this axis!
psi = 0
```

```
60. #-----------------------------------------------
61. # Rotation matrices r1, r2 and r3
62. r1 = np.matrix([
63. [np.cos(np.radians(phi)), np.sin(np.radians(phi)), 0],
64. [-np.sin(np.radians(phi)), np.cos(np.radians(phi)), 0],
65. [0, 0 ,1]
66. ])
67. r2 = np.matrix([
68. [1, 0, 0],
69. [0, np.cos(np.radians(theta)), np.sin(np.radians(theta))],
70. [0, -np.sin(np.radians(theta)), np.cos(np.radians(theta))]
71. ])
72. r3 = np.matrix([
73. [np.cos(np.radians(psi)), np.sin(np.radians(psi)), 0],
74. [-np.sin(np.radians(psi)), np.cos(np.radians(psi)), 0],
75. [0, 0 ,1]
76. ])
77. #-----------------------------------------------
78. # Rotational matrices combined and transposed
79. R = np.dot(r1,r2,r3)
80. RT = np.transpose(R)
81.
82. #-----------------------------------------------
83. # newly calculated stress matrix at a specific point for a specific drill path azimuth, inclination and depth
84.
85. Sigma_Prime = R*Sigma*RT
86. print(Sigma_Prime)
87. return Sigma_Prime
```

This code above is intended to work as a package, imported in an extended python script that calculates the stresses along a drill path for each line of azimuth, inclination, and depth in an Excel-file. Nevertheless, for single and simple calculations, **line 4** can be replaced by defining absolute values for chi (correction angle), phi (azimuth of the drill path), and theta (inclination of the drill path). Other values than zero for psi, which is not applicable for boreholes, must be entered manually by replacing 0 with the desired value (**line 58**).

The extend python script requires an Excel-file with rows named accordingly and with integers or floats in the cells. The sequence of rows is arbitrary. The labeling of the first cell of each row, however, is important:

- **depth, tvd (m)**depth of the drill trajectory in meter, total vertical depth (TVD)
- **phi** (°)azimuthal direction of the drill trajectory in degree, N = 0, E = 90, S = 180…
- **theta** (°)inclination of the drill trajectory in degree, 0 = vertical, 90 = horizontal
- **psi** (°)is always 0

- **chi** (°)deviation of the max. horizontal stress from any main cardinal direction

depth, tvd (m)	phi (°)	theta (°)	psi (°)	chi (°)
…	…	…	…	…
4558.53	112.8	69.8	0.0	0.0
4563.36	114.6	69.9	0.0	0.0
4568.20	116.2	69.7	0.0	0.0
4573.04	117.5	69.9	0.0	0.0
4578.21	119.7	69.8	0.0	0.0
4583.06	121.2	69.7	0.0	0.0
…	…	…	…	…

The extended python script below should be in the same folder as the previous script. The script imports the previous script with an **import** command in **line 4** (here: calcTrafoMat). As a second step, the file path of the Excel-file with the drill trajectory orientations must be added in **line 130** and the output name for the csv-file and the pdf-file must be assigned in **line 132** and **line 134**. Two options are available in **lines 44** and **46** for normalized or absolute stresses.

```
1. from xlrd import open_workbook
2. #________________________
3. #imports the mathematical Transformation Matrix
4. import calcTrafoMat
5. #________________________
6. import matplotlib.pyplot as plt
7. import locale
8. locale.setlocale(locale.LC_NUMERIC, "de_DE")
9. from matplotlib import rcParams
10. rcParams["font.family"]="sans-serif"
11. rcParams["font.sans-serif"]=["Arial"]
12. rcParams["font.size"]=12
13. rcParams["pdf.fonttype"]=42
14.
15. class drillDataSet :
16.
17. def __init__(self, filename):
18.
19. self.filename = filename
20. self.data = []
21.
22. self.parseAnglesAndDepthFromCSV()
23. self.calcTransformationMatrix()
24. self.normTransformationMatrix()
25.
```

```
def calcTransformationMatrix(self):
for pos, dt in enumerate(self.data):
Sigma_Prime = calcTrafoMat.generateTransMatrix(self.data[pos]['phi'],
self.data[pos]['theta'],
self.data[pos]['psi'],
self.data[pos]['chi'])
self.data[pos].update({"Sigma_Prime": Sigma_Prime})
def normTransformationMatrix(self):
for pos, dt in enumerate(self.data):
currSigma = self.data[pos]["Sigma_Prime"]
currDepth = self.data[pos]["depth"]
#either for absolute stresses
self.data[pos].update({ "Sigma_Normed": currSigma * currDepth })
#or for normalized stresses
#self.data[pos].update({ "Sigma_Normed": currSigma}) #---nur normiert
def parseAnglesAndDepthFromCSV(self):
dataIdentifiers = []
try:
wb = open_workbook(self.filename)
except FileNotFoundError:
print(self.filename + " could not be found")
return []
sheet = wb.sheet_by_name("Tabelle1")
number_of_rows = sheet.nrows
number_of_columns = sheet.ncols
# Get order of angles
for pos in range(number_of_columns):
dataIdentifiers.append(sheet.cell(0, pos).value.replace(" [°]", "").repl
ace(", tvd [m]", ""))
for row in range(number_of_rows - 1):
self.data.append({dataIdentifiers[0]: "", dataIdentifiers[1]: "",
dataIdentifiers[2]: "", dataIdentifiers[3]: "",dataIdentifiers[4]: ""})
```

```
for col, data in enumerate(dataIdentifiers):
self.data[row][data] = sheet.cell(row + 1, col).value
def saveNormedMatrix(self, outfile):
for pos, currData in enumerate(self.data):
List1 = str(list(currData["Sigma_Normed"].flat))
with open(outfile, "a", newline="") as outF:
outF.write(List1+"\n")
def plotData(self, outfile, showPlot = False):
s11 = []
s22 = []
s33 = []
s12 = []
s13 = []
s23 = []
depths = []
for pos, currData in enumerate(self.data):
s11.append(currData["Sigma_Normed"].item(0, 0))
s22.append(currData["Sigma_Normed"].item(1, 1))
s33.append(currData["Sigma_Normed"].item(2, 2))
s12.append(currData["Sigma_Normed"].item(0, 1))
s13.append(currData["Sigma_Normed"].item(0, 2))
s23.append(currData["Sigma_Normed"].item(1, 2))
depths.append(currData["depth"])
plt.plot([abs(i) for i in s11], depths, label="sigma xx")
plt.plot([abs(i) for i in s22], depths, label="sigma yy")
plt.plot([abs(i) for i in s33], depths, label="sigma zz")
plt.plot([abs(i) for i in s12], depths, label="tau xy/yx")
plt.plot([abs(i) for i in s13], depths, label="tau xz/zx")
plt.plot([abs(i) for i in s23], depths, label="tau yz/zy")
plt.ylim(plt.ylim()[::-1])
```

```
plt.ylabel("depth [m]")
plt.xlabel("stress [MPa/m]")
plt.title("Stress transformation")
plt.grid(True)

plt.legend()

if showPlot :
plt.show()
plt.savefig(outfile)



if __name__ == "__main__" :
# Excel-file with drill path azimuth and inclination;
myData = drillDataSet("folderpath_to_an_excel-file.xls")
# .csv-file of the the stress matrix with the 6 stresses (S11, S22, S33, S12, S13, S23), delimited with commas;
myData.saveNormedMatrix("drill_path_matrix_a-specific-stress-regime_a-specific-drill-path.csv")
# .pdf-file with a line-plot of the stresses (absolute or normalized values) plotted vs. depth
myData.plotData("SigmaVsDepths_a-specific-stress-regime_a-specific-drill-path.pdf")
print("finished")
```

Appendix B: Rock Mechanical Properties of the Analog Rocks

Bankkalk (BK)

Parameters	n	Min	Mean	Median	Max	stdv
Uniaxial Compressive Strength σ_u (MPa)	8	67.0	84.8	87.4	98.0	11.2
Average Young's modulus V (GPa)	8	20.0	25.1	23.7	32.2	4.5
Static Young's modulus E_{stat} (GPa)	8	31.1	33.8	33.4	36.8	2.3
Static Poisson's ratio ν_{stat} (−)	8	0.10	0.47	0.28	1.82	0.57
Indirect tensile strength σ_t (MPa)	4	4.9	5.6	5.5	6.4	0.7
Dynamic Young's modulus E_{dyn} (GPa)	5	38.5	40.4	40.3	42.6	2.0
Dynamic Poisson's ratio ν_{dyn} (−)	5	0.25	0.28	0.28	0.32	0.03
P-wave velocity v_p (m/s)	5	4300.0	4568.2	4587.0	4711.0	166.3
S-wave velocity v_s (m/s)	5	2438.0	2527.0	2519.0	2594.0	65.5
Bulk density (g/cm^3)	2.49					0.0
Mineral density (g/cm^3)	2.71					
Porosity (%)	5	8.0	8.2	8.2	8.5	0.2
Anisotropy	Isotropic					

Basisoolith (BO)

Parameters	n	Min	Mean	Median	Max	stdv
Uniaxial Compressive Strength σ_u (MPa)	11	132.0	151.6	149.9	173.9	15.4
Average Young's modulus V (GPa)	11	43.8	46.5	45.2	52.9	3.3
Static Young's modulus E_{stat} (GPa)	11	49.5	51.9	52.3	56.7	2.4
Static Poisson's ratio ν_{stat} (−)	13	0.01	0.20	0.25	0.33	0.13
Indirect tensile strength σ_t (MPa)	11	5.1	8.4	8.7	11.0	1.7
Dynamic Young's modulus E_{dyn} (GPa)	15	61.0	66.6	65.6	70.9	3.3
Dynamic Poisson's ratio ν_{dyn} (−)	15	0.32	0.33	0.33	0.35	0.01
P-wave velocity v_p (m/s)	15	6001.0	6194.9	6198.0	6396.0	153.5
S-wave velocity v_s (m/s)	15	2993.0	3127.9	3125.0	3204.0	65.7
Bulk density (g/cm^3)	2.56					0.0
Mineral density (g/cm^3)	2.72					
Porosity (%)	15	4.4	5.4	5.7	6.2	0.7
Anisotropy	Isotropic					

Dietfurter Dolomit (DD)

Parameters	n	Min	Mean	Median	Max	stdv
Uniaxial Compressive Strength σ_u (MPa)	18	84.5	130.0	131.0	213.6	33.6
Average Young's modulus V (GPa)	18	36.3	48.3	49.0	59.3	6.1
Static Young's modulus E_{stat} (GPa)	8	45.1	57.4	56.7	65.5	6.4
Static Poisson's ratio ν_{stat} (−)	13	0.00	0.12	0.06	0.49	0.15
Indirect tensile strength σ_t (MPa)	10	8.2	10.9	10.2	15.5	2.4
Dynamic Young's modulus E_{dyn} (GPa)	16	56.6	69.0	69.4	78.1	5.7
Dynamic Poisson's ratio ν_{dyn} (−)	16	0.30	0.33	0.33	0.35	0.02
P-wave velocity v_p (m/s)	16	5627.0	6174.1	6206.0	6662.0	269.8
S-wave velocity v_s (m/s)	16	2824.0	3103.0	3129.5	3265.0	119.2
Bulk density (g/cm^3)	2.67					0.0
Mineral density (g/cm^3)	2.81					
Porosity (%)	10	3.3	4.5	4.6	5.8	0.7
Anisotropy	Isotropic					

Dietfurter Kalkstein (DK)

Parameters	n	Min	Mean	Median	Max	stdv
Uniaxial Compressive Strength σ_u (MPa)	12	109.1	153.2	155.1	227.8	34.9
Average Young's modulus V (GPa)	12	36.1	39.6	39.5	44.0	2.2
Static Young's modulus E_{stat} (GPa)	2	47.5	49.0	49.0	50.4	2.1
Static Poisson's ratio ν_{stat} (−)	12	0.01	0.09	0.10	0.15	0.04
Indirect tensile strength σ_t (MPa)	9	7.7	10.5	9.7	13.5	2.0
Dynamic Young's modulus E_{dyn} (GPa)	12	57.5	61.2	62.0	63.8	2.4
Dynamic Poisson's ratio ν_{dyn} (−)	12	0.32	0.34	0.35	0.36	0.01
P-wave velocity v_p (m/s)	12	5859.0	6105.9	6039.5	6353.0	175.2
S-wave velocity v_s (m/s)	12	2896.0	2982.5	2997.0	3054.0	55.8
Bulk density (g/cm^3)	2.56					0.0
Mineral density (g/cm^3)	2.71					
Porosity (%)	11	4.6	5.4	5.2	6.9	0.7
Anisotropy	Isotropic					

Quintner Limestone (QK)

Parameters	n	Min	Mean	Median	Max	stdv
Uniaxial Compressive Strength σ_u (MPa)	5	95.9	129.0	114.5	170.5	36.4
Average Young's modulus V (GPa)	5	35.3	43.5	42.6	51.9	6.1

(continued)

(continued)

Parameters	n	Min	Mean	Median	Max	stdv
Static Young's modulus E_{stat} (GPa)	5	39.9	51.6	53.5	56.5	6.8
Static Poisson's ratio ν_{stat} (−)	5	0.03	0.19	0.16	0.41	0.15
Indirect tensile strength σ_t (MPa)	8	2.6	4.6	3.8	8.8	2.0
Dynamic Young's modulus E_{dyn} (GPa)	5	60.8	62.6	61.7	68.0	3.1
Dynamic Poisson's ratio ν_{dyn} (−)	5	0.37	0.38	0.38	0.39	0.01
P-wave velocity v_p (m/s)	5	6370.0	6604.8	6696.0	6856.0	208.8
S-wave velocity v_s (m/s)	5	2875.0	2908.8	2883.0	3026.0	65.7
Bulk density (g/cm^3)	2.68					0.0
Mineral density (g/cm^3)	2.70					
Porosity (%)	5	0.4	0.8	0.8	1.3	0.3
Anisotropy	Isotropic					

Kelheimer Auerkalk (KAK)

Parameters	n	Min	Mean	Median	Max	stdv
Uniaxial Compressive Strength σ_u (MPa)	14	59.2	79.4	80.7	98.6	13.2
Average Young's modulus V (GPa)	14	34.2	39.3	39.2	43.6	2.6
Static Young's modulus E_{stat} (GPa)	9	41.7	46.2	47.5	49.7	2.9
Static Poisson's ratio ν_{stat} (−)	14	0.10	0.37	0.30	1.32	0.32
Indirect tensile strength σ_t (MPa)	26	3.6	6.5	6.6	10.0	1.5
Dynamic Young's modulus E_{dyn} (GPa)	14	55.1	63.3	62.7	78.4	6.0
Dynamic Poisson's ratio ν_{dyn} (−)	14	0.25	0.32	0.33	0.36	0.03
P-wave velocity v_p (m/s)	14	5715.0	6042.3	6086.0	6283.0	169.1
S-wave velocity v_s (m/s)	14	2916.0	3085.1	3050.5	3507.0	171.0
Bulk density (g/cm^3)	2.53					0.0
Mineral density (g/cm^3)	2.71					
Porosity (%)	14	4.1	7.0	6.9	9.7	1.3
Anisotropy	Isotropic					

Solnhofener Plattenkalk (SPK)

Parameters	n	Min	Mean	Median	Max	stdv
Uniaxial Compressive Strength σ_u (MPa)	14	191.6	228.8	232.7	244.2	15.0
Average Young's modulus V (GPa)	14	44.5	46.0	45.8	48.8	1.1
Static Young's modulus E_{stat} (GPa)	14	47.1	50.0	49.0	63.4	4.0
Static Poisson's ratio ν_{stat} (−)	14	0.00	0.09	0.07	0.18	0.07

(continued)

(continued)

Parameters	n	Min	Mean	Median	Max	stdv
Indirect tensile strength σ_t (MPa)	30	6.6	13.7	14.4	18.3	2.6
Dynamic Young's modulus E_{dyn} (GPa)	14	60.9	63.1	62.7	67.2	1.9
Dynamic Poisson's ratio ν_{dyn} (−)	14	0.30	0.31	0.31	0.36	0.02
P-wave velocity v_p (m/s)	14	5775.0	5899.8	5802.5	6560.0	252.7
S-wave velocity v_s (m/s)	14	3021.0	3064.9	3063.0	3129.0	31.7
Bulk density (g/cm^3)	2.60					0.0
Mineral density (g/cm^3)	2.70					
Porosity (%)	14	4.0	5.3	5.5	5.7	0.5
Anisotropy	Isotropic					

Pfraundorfer Dolomit (PFD)

Parameters	n	Min	Mean	Median	Max	stdv
Uniaxial Compressive Strength σ_u (MPa)	19	137.2	180.4	181.0	229.4	26.8
Average Young's modulus V (GPa)	19	45.2	52.5	52.3	59.7	4.7
Static Young's modulus E_{stat} (GPa)	19	54.1	60.0	59.7	65.2	3.4
Static Poisson's ratio ν_{stat} (−)	23	0.00	0.17	0.08	0.78	0.20
Indirect tensile strength σ_t (MPa)	26	7.4	11.0	10.9	14.8	1.8
Dynamic Young's modulus E_{dyn} (GPa)	24	50.0	66.2	67.6	77.6	7.5
Dynamic Poisson's ratio ν_{dyn} (−)	24	0.21	0.27	0.27	0.32	0.02
P-wave velocity v_p (m/s)	24	4861.1	5501.0	5543.5	6086.0	323.3
S-wave velocity v_s (m/s)	24	2680.0	3096.8	3142.0	3354.5	175.0
Bulk density (g/cm^3)	2.71					0.0
Mineral density (g/cm^3)	2.82					
Porosity (%)	13	2.3	3.9	4.3	5.2	0.9
Anisotropy	Isotropic					

Wachenzeller Dolomit (WAD)

Parameters	n	Min	Mean	Median	Max	stdv
Uniaxial Compressive Strength σ_u (MPa)	11	73.8	116.5	117.6	137.5	19.4
Average Young's modulus V (GPa)	11	41.4	44.4	44.1	52.2	3.0
Static Young's modulus E_{stat} (GPa)	9	50.4	52.6	52.1	56.4	2.0
Static Poisson's ratio ν_{stat} (−)	11	0.07	0.11	0.09	0.26	0.06
Indirect tensile strength σ_t (MPa)	10	4.8	7.3	7.3	10.1	2.0
Dynamic Young's modulus E_{dyn} (GPa)	11	45.3	50.8	52.0	54.8	3.2

(continued)

(continued)

Parameters	n	Min	Mean	Median	Max	stdv
Dynamic Poisson's ratio ν_{dyn} (−)	11	0.22	0.28	0.28	0.33	0.03
P-wave velocity v_p (m/s)	11	4745.0	4956.6	4982.0	5087.0	107.6
S-wave velocity v_s (m/s)	11	2579.0	2742.7	2788.0	2845.0	97.6
Bulk density (g/cm^3)	2.62					0.0
Mineral density (g/cm^3)	2.83					
Porosity (%)	11	5.6	6.8	6.5	8.6	1.1
Anisotropy	(an-)-isotropic					

Wasserzeller Dolomit (WD)

Parameters	n	Min	Mean	Median	Max	stdv
Uniaxial Compressive Strength σ_u (MPa)	9	67.3	102.1	105.4	122.4	16.1
Average Young's modulus V (GPa)	9	34.3	38.8	38.5	43.9	2.9
Static Young's modulus E_{stat} (GPa)	8	43.3	46.3	46.2	51.8	3.0
Static Poisson's ratio ν_{stat} (−)	9	0.09	0.24	0.21	0.62	0.19
Indirect tensile strength σ_t (MPa)	8	4.5	6.0	5.8	7.7	1.0
Dynamic Young's modulus E_{dyn} (GPa)	9	33.6	48.6	47.9	56.5	7.1
Dynamic Poisson's ratio ν_{dyn} (−)	9	0.29	0.32	0.32	0.37	0.03
P-wave velocity v_p (m/s)	9	4738.0	5194.4	5268.0	5561.0	252.6
S-wave velocity v_s (m/s)	9	2152.0	2673.1	2698.0	2903.0	223.2
Bulk density (g/cm^3)	2.60					0.0
Mineral density (g/cm^3)	2.81					
Porosity (%)	9	4.6	8.0	7.2	11.2	2.3
Anisotropy	Isotropic					

Obere Krumme Lage (OKL)

Parameters	n	Min	Mean	Median	Max	stdv
Uniaxial Compressive Strength σ_u (MPa)	4	77.0	95.1	98.1	107.2	14.5
Average Young's modulus V (GPa)	4	38.4	43.4	43.8	47.4	3.9
Static Young's modulus E_{stat} (GPa)	4	47.1	52.0	52.8	55.1	3.8
Static Poisson's ratio ν_{stat} (−)	4	0.26	0.33	0.31	0.44	0.07
Indirect tensile strength σ_t (MPa)	6	6.6	8.7	8.5	12.0	2.1
Dynamic Young's modulus E_{dyn} (GPa)	6	37.7	42.7	43.4	45.9	2.9
Dynamic Poisson's ratio ν_{dyn} (−)	6	0.29	0.31	0.31	0.32	0.01
P-wave velocity v_p (m/s)	6	4824.0	5017.8	5058.5	5152.0	131.0

(continued)

(continued)

Parameters	n	Min	Mean	Median	Max	stdv
S-wave velocity v_s (m/s)	6	2507.0	2649.0	2675.0	2724.0	76.8
Bulk density (g/cm^3)	2.35					0.0
Mineral density (g/cm^3)	2.71					
Porosity (%)	6	12.5	13.5	13.4	14.8	0.8
Anisotropy	Anisotropic					

References

Standards

1. ASTM D4543-19 (2019) Standard practices for preparing rock core as cylindrical test specimens and verifying conformance to dimensional and shape tolerances. ASTM International, West Conshohocken, PA. www.astm.org
2. DIN 51220 (2003–08) Werkstoffprüfmaschinen - Allgemeines zu Anforderungen an Werkstoffprüfmaschinen und zu deren Prüfung und Kalibrierung. Beuth Verlag GmbH, Berlin, 2003–08
3. DIN EN 14146 (2004–06) Prüfverfahren für Naturstein - Bestimmung des dynamischen Elastizitätsmoduls (durch Messung der Resonanzfrequenz der Grundschwingung) (Deutsche Fassung EN 14146:2004). Beuth Verlag GmbH, Berlin, 2004–06
4. DIN EN 14579 (2005–01) Prüfverfahren für Naturstein - Bestimmung der Geschwindigkeit der Schallausbreitung; Deutsche Fassung EN 14579:2004. Beuth Verlag GmbH, Berlin, 2005–01
5. VDI 4640 (2010) Thermal Use of the Underground. VDI-Gesellschaft Energie und Umwelt (GEU), Berlin

Web Sources

6. www-01: https://www.geotis.de/geotisapp/templates/heatsumstatistic.php, Accessed on: 03.03.2020.
7. www-02: https://www.geotis.de/geotisapp/templates/powersumstatistic.php, Accessed on: 03.03.2020.
8. www-03: https://www.tiefegeothermie.de/news/projekt-in-weilheim-gestoppt, Accessed on: 09.12.2020.
9. www-04: https://www.tiefegeothermie.de/news/bohrplaetze-in-icking-werden-zurueckgebaut, Accessed on: 09.12.2020
10. www-05: https://www.tiefegeothermie.de/projekte/geretsried, Accessed on: 03.03.2020.

G. M. Stockinger, *Fracturing in Deep Boreholes*, Springer Theses,
https://doi.org/10.1007/978-3-030-94569-5

11. www-06: https://www.worldoil.com/news/2012/12/10/corpro-s-core-barrel-technology-debuts-in-thegulf-of-mexico, Accessed on: 02.02.2021.
12. www-07: https://www.weatherford.com/en/documents/technical-specification-sheet/products-andservices/formation-evaluation/wireline-services/compact-microimager/, Accessed on: 11.11.2020
13. www-08: https://www.geotron.de/software-ultraschall-messtechnik.html#umpc, Accessed on: 05.04.2020.
14. www-09: https://www.geotron.de/software-ultraschall-messtechnik.html#lhdw, Accessed on: 05.04.2020
15. www-10: https://www.geotron.de/schallgeschwindigkeitsmessung-p-und-s-wellen.html, Accessed on: 05.04.2020
16. www-11: https://www.glossary.oilfield.slb.com/en/Terms/m/multifinger_caliper.aspx, Accessed on: 12.12.2020.